AF522689

MEASURE THEORY

MEASURE
THEORY

By

A.K. Sharma

DISCOVERY PUBLISHING HOUSE
NEW DELHI-110002

Published by:

DISCOVERY PUBLISHING HOUSE
4383/4B, Ansari Road, Darya Ganj
New Delhi-110 002 (India)
Phone : +91-11-23279245; 23253475; 43596065
E-mail : discoverybooksindia@gmail.com
discoverypublishinghouse@gmail.com
namitwasan9@gmail.com
web : www.discoverypublishinggroup.com

***First Published:* 2007**
***Reprinted:* 2023**

ISBN: 978-81-8356-221-8

Measure Theory

Printed at:
Infinity Imaging Systems
Delhi

Preface

This book "Measure Theory" has been written for the students of B.A./B.Sc., students of all Indian university. It is the experience of author that the awareness students need the treatment of theory in a way that should be easily comprehensive to him. Therefore an effort has been made in this book to put the matter in a very lucid and simple way to that even a beginner has no difficulty in grasping the subject.

The author will feel amply rewarded if the book serve. The purpose for which it is mean suggested for the importance of this book are always welcome.

Author

Preface

CONTENTS

1

SEPARATION AXIOM AND CONVERGENCE

INTRODUCTION

A topological space may be very sparsely endowed with open sets. As we know, some spaces have only two, the empty set and the full space. In a discrete space, on the other hand, every set is open. Most of the familiar spaces of geometry and analysis fall somewhere in between these two artificial extremes. The so-called *separation properties* enable us to state with precision that a given topological space has a rich enough supply of open sets to serve whatever purpose we may have in mind.

The separation properties are of concern to us because the supply of open sets possessed by a topological space is intimately linked to its supply of continuous functions; and since continuous functions are of central importance in topology, we naturally wish to guarantee that enough of them are present to make our discussions fruitful. If, for instance, the only open sets in a topological space are the empty set and the full space, then the only continuous functions present are the constants, and very little of interest can be said about these. In general terms, the more open sets there are, the more continuous functions a space has. Discrete spaces have continuous functions in the greatest possible abundance, for all functions are continuous. However, few really important spaces are discrete, so this goes a bit too far. The separation properties make it possible for us to be sure that our spaces have enough continuous functions without committing ourselves to the excesses of discrete spaces.

The T_i space nomenclature for i = 0, 1, 2, 3, 4, 5 was introduced by Alexandroff and Hopf. The word 'T' refers to the German word 'ternnugs axiom' which means 'Separation axiom'. In this chapter, we treat the point x and the singleton set {x} in the same sense.

Every sub-space of T_1-space is a T_1-space. Every sub-space of a T_2-space is a T_2-space.

T_1-SPACES AND HAUSDORFF SPACES

One of the most natural things to require of a topological space is that each of its points be a closed set. The separation property which relates to this is the following. A T_1-space is a topological space in which, given any pair of distinct points, each has a neighbourhood which does not contain the other. It is obvious that any subspace of a T_1-space is also a T_1-space. Our first theorem shows that T_1-spaces are precisely those topological spaces in which points are closed.

Theorem 1:

Every compact subspace of a Hausdorff space is closed.

Proof:

Let C be a compact subspace of a Hausdorff space X. We prove that C is closed by showing that its complement C' is open. C' is open if it is empty, so we may assume that it is non-empty. Let x be any point in C'. By x has a neighbourhood G such that $x \varepsilon G \subseteq C'$. This shows that C' is a union of open sets and is therefore open itself. One of the most useful consequences of this result is

Theorem 2(a):

The product of any non-empty class of Hausdorff spaces is a Hausdorff space.

Proof:

Let $X = P_i X_i$ be the product of a non-empty class of Hausdorff spaces X_i. If $x = \{x_i\}$ and $y = \{y_i\}$ are two distinct points in X, then we must have $x_{i0} \neq y_{i0}$ for at least one index in.

Since X_{i0} is a Hausdorff space, x_{i0} and y_{i0} can be separated by open sets in X_{i0}. These two disjoint open subsets of X_{i0} give rise to two disjoint sets in the defining open subbase for X, each of which contains one of the points x and y.

Most of the important facts about Hausdorff spaces depend on the following theorem.

Theorem 2(b):

A one-to-one continuous mapping of a compact space onto a Hausdorff space is a homeomorphism.

Proof:

Let $f : X \to Y$ be a one-to-one continuous mapping of a compact space X onto a Hausdorff space Y. We must show that f(G) is open in V whenever

G is open in X, and for this it suffices to show that f(F) is closed in Y whenever F is closed in X. If F is empty, f(F) is also empty and therefore closed, so we may assume that F is non-empty. F is a compact subspace of X' or f(F) is a compact subspace of Y; and we complete the proof by using the preceding theorem to conclude that f(F) is a closed subspace of Y.

Compact Hausdorff spaces are among the most important of all topological spaces, and in the following sections and chapters we shall become thoroughly acquainted with their major properties.

Theorem 3:

In a Hausdorff space, any point and disjoint compact subspace can be separated by open sets, in the sense that they have disjoint neighbourhoods.

Proof:

Let X be a Hausdorff space, x a point in X, and C a compact subspace of X which does not contain x. We construct a disjoint pair of open sets G and H such that $x \varepsilon G$ and $C \subseteq H$. Let y be a point in C. Since X is a Hausdorff space, x and y have disjoint neighbourhoods G_x and H_y. If we allow y to vary over C, we obtain a class of H_y's whose union contains C; and since C is compact, some finite subclass, which we denote by $(H_1, H_2, ..., H_n\}$, is such that $C \subseteq \bigcup_{i=1}^{n} H_i$. If $G_1, G_2, ..., G_n$ are the neighbourhoods of a: which correspond to the H_i's, we put

$$G = \bigcap_{i=1}^{n} G_i$$

and $H = \bigcap_{i=1}^{n} H_i$ and observe that these two sets have the required properties.

We proved that every closed subspace of a compact space is compact, and in Problem 21-4 we saw that a compact subspace of a compact space need not be closed. We now use the preceding theorem to show that compact subspaces of Hausdorff spaces are always closed.

Theorem 4:

A topological space is a T_1-space $\Leftrightarrow$ each point is a closed set.

Proof:

If X is a topological space, then an arbitrary point x in X is closed $\Leftrightarrow$ its complement is open $\Leftrightarrow$ each point y different from x has a neighbourhood which does not contain x $\Leftrightarrow$ X is a T_1-space.

Our next separation property is slightly stronger. A *Hausdorff space* is a topological space in which each pair of distinct points can be separated by open sets, in the sense that they have disjoint neighbourhoods. Every

Hausdorff space is clearly a T_1-space, and every subspace of a Hausdorff space is also a Hausdorff space.

REGULAR SPACES AND NORMAL SPACES

Let X be an arbitrary topological space, and consider the set $\mathcal{C}(X, R)$ of all bounded continuous real functions defined on X. If for each pair of distinct points x and y in X there exists a function f in $\mathcal{C}(X, R)$ such that $f(x) \neq f(y)$, we say that $\mathcal{C}(X, R)$ separates points. It is easy to see that if $\mathcal{C}(X, R)$ does separate points, then X is necessarily a Hausdorff space; for assuming that $f(x) < f(y)$ and that r is a real number such that $f(x) < r < f(y)$, then the sets $[z : f(z) < r\}$ and $\{z : f\{z) > r)$ are disjoint neighbourhoods of x and y.

It is convenient to strengthen this separation property slightly by allowing one of the points to be an arbitrary closed subspace of X . A *completely regular space* is a T_1-space X with the property that if x is any point in X and F any closed subspace of X which does not contain x, then there exists a function f in $\mathcal{C}\{X, R)$, all of whose values lie in the closed unit interval [0,1], such that $f(x) = 0$ and $f(F) = 1$.

It is worth noticing that since constants are continuous, we could just as well have required here that f be 1 at x and 0 on F, for the function $g = 1 - f$ has these properties. We may think of completely regular spaces as T_1-spaces in which continuous functions separate points and disjoint closed subspaces.

Since points are closed in a completely regular space, it is permissible to take the closed subspace F to be a point, and it is clear by the above paragraph that every completely regular space is a Hausdorff space. It is also easy to see that every subspace of a completely regular space is completely regular. We give an explicit characterization of all completely regular spaces in terms of product spaces.

Our next (and last) separation property is similar to that which defines a Hausdorff space, except that it applies to disjoint closed sets instead of merely to distinct points. A *normal space* is a T_1-space in which each pair of disjoint closed sets can be separated by open sets, in the sense that they have disjoint neighbourhoods. We shall see in the next section (as a consequence of Urysohn's lemma) that every normal space is completely regular.

Intended to illustrate and clarify the relations among our various separation properties: a topological space which possesses any one property, in the order of their definition, also possesses all properties which precede it; in other words, they have been defined in order of increasing strength.

Theorem:

Every Compact Hausdorff space is normal.

Proof:

Let X be a compact Hausdorff space, and A and B disjoint closed subsets of X. We must produce a disjoint pair of open sets G and H such that A $\subseteq$ G and B $\subseteq$ H. If either closed set is empty, we can take the empty set as a neighbourhood of it and the full space as a neighbourhood of the other. We may therefore assume that both A and B are non-empty. Since X is compact, A and B are disjoint compact subspaces of X. Let a; be a point of A.

By hypothesis that X is Hausdorff, x and B have disjoint neighbourhoods G, and H_B. If we allow x to vary over A, we obtain a class of G_x's whose union contains A; and since A is compact, some finite subclass, which we denote by $\{G_1, G_2, ..., G_n\}$, is such that A $\subseteq \bigcap_{i=1}^{n} G_i$. If $H_1, H_2, ..., H_n$ are the neighbourhoods of 5 which correspond to the G_i's, it is clear that G = $\bigcap_{i=1}^{n} G_i$ G_i and H = $\bigcap_{i=1}^{n} H_i$ are disjoint neighbourhoods of A and B.

We investigate the manner in which normal spaces, compact Hausdorff spaces, and metric spaces are related to one another.

URYSOHN'S LEMMA AND THE TIETZE EXTENSION THEOREM

As we suggested in the introduction to this chapter, one of the main purposes served by assuming that a topological space is rich in open sets is to guarantee that it is also rich in continuous functions. The following is the fundamental theorem in this direction.

Theorem 1:

Let X be a normal space, and let A and B be disjoint closed subspaces of X. If [a,b] is any closed interval on the real line, then there exists a continuous real function f defined on X, all of whose values lie in [a,b], such that f {A) = a and f(B) = b.

Proof:

If a = b, we have only to define f by f(x) = a for every x, so we may assume that a < b. If g is a function with the properties stated in Urysohn's lemma, then f = (b – a)g + a has the properties required by our theorem.

If there is given a continuous function defined on a subspace of a topological space, Urysohn's lemma has an important bearing on the interesting question of whether this function can be extended continuously to the full space. The following is the classic theorem along these lines.

Theorem 2(a):

(The Tietze Extension Theorem) : Let X be a normal space, F a closed subspace, and f a continuous real function defined on F whose values lie in the closed interval [a,b]. Then f has a continuous extension f defined on all of X whose values also lic in [a,b].

Proof:

If a = b, the conclusion of our theorem is obvious, so we may assume that a < b. We may clearly assume that [a, b] is the smallest closed interval which contains the range of f. Furthermore, the device used in the proof of Theorem B enables us to assume that a = –1 and b = 1. We begin by defining f_0 to be f. The domain of f_0 is our closed subspace F, and we define two subsets A_0 and B_0 of F by

$$A_0 = \{x : f_0(x) < -1/3\}$$

and $B_0 = \{x : f_0(x) > 1/3\}$. A_o and B_o are disjoint, non-empty, and closed in F; and since F is closed, they are closed in X. A_0 and B_0 are thus a disjoint pair of closed subspaces of X, and by Theorem B there exists a continuous function $g_0 : X \to \left[-\frac{1}{3}, \frac{1}{3}\right]$ such that $g_0(A_0) -\frac{1}{3}$ and $g_0(B_0) = 1/3$. We next define f_1 on F by $f_1 = f_0 - g_0$, and we observe that $|f_1(x)| \leq 2/3$. if $A_1 = [x:f_1(x) \leq \left(-\frac{1}{3}\right)\left(\frac{2}{3}\right)\}$ and

$$B_1 = \{x:f_1(x) \geq \left(\frac{1}{3}\right)\left(\frac{2}{3}\right)\},$$

then in the same way as above there exists a continuous function

$$g_1:x \to \left[\left(-\frac{1}{3}\right),\left(\frac{2}{3}\right),\left(\frac{1}{3}\right),\left(\frac{2}{3}\right)\right]$$

such that $\quad g_1(A_1) = \left(-\frac{1}{3}\right),\left(\frac{2}{3}\right)$

and $\quad g_1(B_1) = \left(\frac{1}{3}\right)\left(\frac{2}{3}\right).$

We next define f_2 on F by $f_2 = f_1 - g_1 = f_0 - (g_0 + g_1)$, and we observe that $|f_2(x)| < (2/3)^2$. By continuing in this manner, we get a sequence $\{f_0, f_1, f_2, \ldots\}$ defined on F such that $|f_n(x)| < (2/3)^n$, and a sequence $\{g_0, g_1, g_2 \ldots\}$

defined on X such that $|g_n(x)| \le (1/3)(2/3)^n$, with the property that on F we have $f_n = f_0 - (g_0 + g_1 + \ldots + g_{n-1})$. We now define s_n by $s_n = g_0 + g_1 + \ldots + s_{n-1}$, and we regard the s_n's as the partial sums of an infinite series of functions in $\mathcal{C}(X, R)$. We know that $\mathcal{C}(X, R)$ is complete, so by $|g_n(x)| \le \left(\frac{1}{3}\right)\left(\frac{2}{3}\right)^n$ and the fact that $\sum_{n=0}^{\infty}\left(\frac{1}{3}\right)\left(\frac{2}{3}\right)^n = 1$ we see that s_n converges uniformly on X to a bounded continuous real function f' such that $|f'(x)| \le 1$. We conclude our proof by noting that since $|f_n(x)| < (2/3)^n$, Sn converges uniformly on F to $f_0 = f$, and that therefore f' equals f on F and is a continuous extension of f to the full space X which has the desired property.

It is of some interest to observe that this theorem becomes false if we omit the assumption that the subspace F is closed. This is easily seen by means of the following example. Let X be the closed unit interval [0,1], F the subspace (0,1], and f the function defined on F by f(x) = sin (1/x). X is clearly normal, F is not closed as a subspace of X, and f cannot be extended continuously to X in any manner whatsoever.

Theorem 2(b):

(Urysohn's Lemma) : Let X be a normal space, and let A and B be disjoint closed subspaces of X. Then there exists a continuous real function f defined on X, all of whose values lie in the closed unit interval [0,1], such that f(A) = 0 and f(B) = 1.

Proof:

B' is a neighbourhood of the closed set A, so by the normality of X and A has a neighbourhood $U_{1/2}$, such that

$$A \subseteq U_{1/2} \subseteq \overline{U}_{1/2} \subseteq B'.$$

$U_{1/2}$ and B' are neighbourhoods of the closed sets A and $\overline{U}_{1/2}$, so in the same way there exist open sets $U_{1/4}$ and $U_{3/4}$ such that

$$A \subseteq U_{1/4} \subseteq \overline{U}_{1/4} \subseteq U_{1/2} \subseteq \overline{U}_{1/2} \subseteq U_{3/4} \subseteq \overline{U}_{3/4} \subseteq B'.$$

If we continue this process, for each rational number of the form $t = m/2^n$ (where n = 1, 2, 3, . . . and m = 1, 2, . . . , 2^{n-1}) we have an open set of the form U_t, such that

$$t_1 < t_1 \Rightarrow A \subseteq U_{t_1} \subseteq \overline{U}_{t_1} \subseteq U_{t_2} \subseteq \overline{U}_{t_2} \subseteq B'.$$

We now define our function f by f(x) = 0 if x is in every U_t and

$$f(x) = \sup \{t : x \notin U_t\}$$

otherwise. It is clear that the values of f lie in [0, 1], and that f(A) = 0 and f(B) = 1. All that remains to be proved is that f is continuous. All intervals of the form [0, a) and (a, I], where $0 < a < 1$, constitute an open subbase for [0,1]. It therefore suffices to show that $f^{-1}([0, a))$ and $f^{-1}((a,1])$ are open. It is easy to see that $f(x) < a \Leftrightarrow x$ is in some U_t for $t < a$; and from this it follows that $f^{-1}([0, a)) = \{x : f(x) < a\} = \cup_{t<a} U_t$, which is an open set. Similarly, $f(x) > a \Leftrightarrow x$ is outside of $\overline{U}_i$ for some $t > a$; and therefore $f^{-1}((a, 1]) = \{x : f(x) > a\} = \cup_{t>a} \overline{U}_i'$, which is an open set.

It is clear from this theorem that every normal space is completely regular: all that is necessary is to take the closed subspace A to be a single point and to observe that the function f is exactly what is required in the definition of complete regularity.

The following slightly more flexible form of Urysohn's lemma will be useful in applications.

THE URYSOHN IMBEDDING THEOREM

We generalized metric spaces to topological spaces. We now reverse this procedure and seek out simple conditions which guarantee that a topological space is essentially a metric space, that is, which imply that it is metrizable. Shows that we must look for properties of a topological space X which enable us to construct a homeomorphism f to X onto a subspace of some metric space; for the metric on this subspace can then be carried back by f to X, and we can infer that X is metrizable. The simplest property of this kind is discreteness; for if X is a discrete space, then its underlying set of points, equipped with the metric defined is a homeomorphic image of X under the identity mapping. We can lift our discussion to a more meaningful level by observing that since every metric space is normal, normality must be among the properties assumed of X, or it must be implied by them.

As motivation for our main theorem, we note that since the metric space R^∞ is second countable, every subspace of it is also second countable. It turns out that second countability, in addition to normality, suffices to guarantee that a topological space is homeomorphic to a subspace of R^∞. In effect, we imbed such a space homeomorphically in R^∞.

Theorem 1:

(The Urysohn Imbedding Theorem) : If X is a second countable normal space, then there exists a homeomorphism f of X onto a subspace of R^∞, and X is therefore metrizable.

Proof:

We may assume that X has infinitely many points, for otherwise it would be finite and discrete, and clearly homeomorphic to any subspace of R^∞ with the same number of points. Since X is second countable, it has a countably infinite open base $\{G_1, G_2, G_3, ...\}$ each of whose sets is different from the empty set and the full space. If G_j and $x \varepsilon G_2$ are given, then by normality there exists a G_i such that $x \varepsilon G_i \subseteq \overline{G}_i \subseteq G_j$. The set of all ordered pairs (G_i, G_j) such that $\overline{G}_i \subseteq G_j$ is countably infinite, and we can arrange them in a sequence $P_1, P_2, ..., P_n ....$ By Urysohn's lemma, for each ordered pair $P_n = (G_i, G_j)$ there exists a continuous real f unction $f_n: X \to [0,1]$ such that $f_n(\overline{G}_i) = 0$ and $f_n(G_j') = 1$. For each x in X we define f(x) to be the sequence given by $f(x) = (f_1(x), f_2(x)/2, ..., f_n(x)/n, ...\}$. If we recall that the infinite series $\sum_{n=1}^{\infty}\frac{1}{n^2}$ converges, it is easy to see that f is a one-to-one mapping of X into R^{∞}. It remains to be proved that f and f^{-1} are continuous.

To prove that f is continuous, it suffices to show that given x_0 in X and $\in > 0$, there exists a neighbourhood H of x_0 such that $y \varepsilon H \Rightarrow \|f(y) - f(x_0)\| < \in$. Since an infinite series of functions converges uniformly if its terms are bounded by the terms of a convergent infinite series of constants, it is easy to see that there exists a positive integer n_0 such that for every y in X we have

$$\|f(y) - f(x_0)\|^2 = \sum_{n=1}^{\infty}|[f_n(y) - f_n(x_0)]/n|^2$$

$$< \sum_{n=1}^{n_0}|[f_n(y) - f_n(x_0)]/n|^2 + \frac{\in^2}{2}.$$

By the continuity of the f_n's, for each $n = 1, 2, . . . , n_0$ there exists a neighbourhood H_n of x_0 such that $y - H_n \Rightarrow |[f_n(y) - f_n(x_0)]/n|^2 < \frac{\in^2}{2n_0}$. If we define H by $H = \bigcap_{n=1}^{n_0} H_n$, it is clear that H is a neighbourhood of x_0 such that $y \varepsilon H \Rightarrow \|f(y) - f(x_0)\|^2 < \in^2 \Rightarrow \|f(y) - f(x_0)\| < \in$.

We conclude our proof by showing that f^{-1} is continuous as a mapping of f(X) onto X. It suffices to show that given x_0 in X and a basic neighbourhood G_j of x_0, there exists $\in > 0$ such that $\|f(y) - f(x_0)\| < \in \Rightarrow y \varepsilon G_j$. G_j is the second member of some ordered pair $P_{n0} = (G_i, G_j)$ such that $x_0 \varepsilon G_i; \subseteq \overline{G}_i \subseteq G_j$. If we choose $\in < 1/2n_0$, then we see that $\|f(y) - f(x_0)\| < \in \Rightarrow \sum_{n=1}^{\infty}|[f_n(y) - f_n(x_0)]/n|^2 << (1/n_0)^2 \Rightarrow |f_{n0}(y) - f_{n0}(x_0)| < 1/2$. Since x_0 is in G_i, $f_{n0}(x_0) = 0$, and therefore $|f_{n0}(y)| < 1/2$. Since $f_{n0}(G_j') = 1$, we see that y is in G_j.

THE STONE-CECH COMPACTIFICATION

In the preceding section we showed that if a normal space is second countable, then it can be imbedded as a subspace in the familiar metric space

R^∞. We now develop a similar imbedding theorem for completely regular spaces.

In order to motivate this theorem properly, we remark that if X is a topological space which occurs as a subspace of a compact Hausdorff space y, then since Y is completely regular, X is also completely regular, and is a dense subspace of the compact Hausdorff space $\overline{X}$. We see in this way that many completely regular spaces are dense subspaces of compact Hausdorff spaces.

Our purpose in this section is to show that any completely regular space X can be imbedded as a dense subspace in a special compact Hausdorff space denoted by β(X), and that β(X) has the remarkable property that every bounded continuous real .function defined on X has a unique extension to a bounded continuous real function defined on β(X).

How truly remarkable this extension property is can be seen considering the example. Here the completely regular space X is the interval (0,1]. This space is clearly a dense subspace of the compact Hausdorff space [0,1]. The function f defined on (0,1] by f(x) = sin (1/x) is a bounded continuous real function defined on X, but it cannot be extended continuously to [0,1].

The space [0,1], though it is a compact Hausdorff space which contains (0,1] as a dense subspace, is evidently not the space β(X). The latter is much too complicated for any simple description of it to be possible.

Before we start our discussion, we recall two items from the previous sections:

(1) if X is a completely regular space, then the weak topology generated by $\mathscr{C}(X, R)$ equals the given topology;

(2) the relative topology on a subspace of a product space equals the weak topology generated by the restrictions of the projections so that subspace.

These facts are the basic principles on which the following analysis rests.

We begin with an arbitrary topological space X and the set $\mathscr{C}(X, R)$ of all bounded continuous real functions defined on X. Let the functions in $\mathscr{C}(X, R)$ be indexed by a set of indices so that $\mathscr{C}(X, R) = \{f_i\}$. For each index i, let I_i be the smallest closed interval which contains the range of the function f_i. Each I_i is a compact Hausdorff space, so their product $P = P_i I_i$ is also a compact Hausdorff space, and every subspace of P is completely regular. We next define a mapping f of X into this product space by means of f(x) = $\{f_i(x)\}$, that is, in such a way that f(x) is that point in the product space P whose zth coordinate is the real number p_i (x) the fact that $p_i f = f_i$ for each projection p_i, it is clear that f is a continuous mapping of X into P.

We now assume that $\mathscr{C}(X,R)$ separates the points of X. This is a weaker assumption than complete regularity and is exactly the requirement that f be a one-to-one mapping. At this stage, we use f to replace f(X) as a set by X; that is, *we imbed* X in P as a set. X is thus a subset of P which has two topologies: its own, and the relative topology which it has as a subspace of P. We observe two features of this situation. First, since f is continuous, the given topology on X is stronger than its relative topology. Second, $\mathscr{C}(X, R)$ is precisely the set of all restrictions to X of the projections p_i of P onto its coordinate spaces I_i. It is now clear that if X is completely regular, so that by statement (1) its given topology equals the weak topology generated by $\mathscr{C}(X, R)$, then by statement (2) its given topology equals its relative topology, and X can be regarded as a subspace of P.

In accordance with these ideas, we now assume that X is completely regular, and we fully identify it, both as a set and as a topological space, with the subspace f(X) of P. It is easy to see that the closure $\overline{X}$ of X in P is a compact Hausdorff space in which X is imbedded as a dense subspace. Furthermore, each f_i in $\mathscr{C}(X, R)$—that is, each projection p_i restricted to X—has an extension to a bounded continuous real function defined on $\overline{X}$; this extension is p_i restricted only to $\overline{X}$, and it is unique by Problem 26-5. The space $\overline{X}$ is commonly denoted by β(X).

We summarize these results in the following theorem.

Theorem:

Let X be an arbitrary completely regular space. Then there exists a compact Hausdorff space β(X) with the following properties: (1) X is a dense subspace of β(X); (2) every bounded continuous real function defined on X has a unique extension to a bounded continuous real function defined on β(X).

We shall prove that the space β(X) is essentially unique, in the sense that any other compact Hausdorff space with properties (1) and (2) is homeomorphic to β(X). It is called the *Stone-Cech compactification* of the given completely regular space.

Even before our work in this section, it was clear that every sub- space of a product of closed intervals is completely regular. It is worthy of special emphasis that the above discussion shows, conversely, that *every* completely regular space is homeomorphic to a subspace of such a product.

KOLOMOGOROV SPACE

A topological space (X, T) is said to satisfy the T_0-axiom of separation if given a pair of distinct points, x, y ∈ X, either

$\exists\, G \in T$ s.t. $x \in G,\, y \notin G$;

or $\exists\, H \in T$ s.t. $y \in H,\, x \notin H$.

In this case the space (X, T) is called a T_0-space.

T_1-*axiom or Frechet axiom of separation.*

A topological space (X, T) is said to satisfy the T_1-axiom of separation if given a pair of distinct points $x, y \in X$.

$\exists\, G, H \in T$ s.t. $x \in G,\, y \notin G;\, y \in H,\, x \notin H$.

In this case the space (X, T) is the T_1-*space or Frechet space.*

T_2-*axiom of separation.*

A topological space (X, T) is said to satisfy the T_2-axiom or separation if given a pair of distinct points $x, y \in X$.

$\exists\, G, H \in T$ s.t. $x \in G,\, y \in H,\, G \cap H = \varnothing$.

In this case the space (X, T) is called a T_2-*space or Hausdorff space* or separated space.

Definition : *Normal space.*

A topological space (X, T) is said to be a normal space if ; given a pair of disjoint closed sets $C_1, C_2 \subset X$.

$\exists$ disjoint open sets $G_1, G_2 \subset X$ s.t.

$$C_1 \subset G_2 \subset C_1 \subset C_2.$$

Definition : *Regular space.*

A topological space (X, T) is said to be a regular space if: given an element $x \in X$ and closed set $F \subset X$ s.t. $x \notin F$,

$\exists$ disjoint open sets $G_1, G_2 \subset X$ s.t. $x \in G_1,\, F \subset G_2$.

A regular T_1-space is called a T_3-space.

A normal T_1-space is called a T_4-space.

Completely regular space :

A topological space (X, T) is called completely regular space if: given a closed set $F \subset X$ and a point $x \in X$ s.t. $x \notin F$,

$\exists$ a continuous map $f : X \to [0, 1]$ with the property,

$$f(x) = 0,\ f(F) = \{1\}.$$

Tychonoff-space. A completely regular T_1-space is called a Tychonoff space. Tychonoff space is also called $T_{3\frac{1}{2}}$-space.

Def. Two sets, A, B are separated sets if

$$A \neq \phi, B \neq \phi, \overline{A} \cap B, \overline{A} \cap \overline{B} = \phi$$

Completely normal : Let A, B be separated sets in a topological space (X, T). Then X is said to be *Completely normal* iff given a pair of separated sets, there are disjiont open sets U and V s.t. $A \subset U$, $B \subset V$.

It can be shown that *Completely normal* $\Rightarrow$ normal

T_5-space : A topological space (X, T) is a T_5-space iff it is a T_1-space and completely normal.

T_0-space which is not a T_1-space.

Example 1:

If $X = \{a, b\}$ *and* $T = \{\phi, \{a\}, X\}$

Is the space (X, T) a T_0*-space or a* T_1*-space.*

Solution:

(i) (X, T) is T_0-space.

For given $a, b \in X$, $\exists \{a\} \in T$ s.t. $a \in \{a\}$, $b \notin \{a\}$.

(ii) (X, T) is not T_1-space. For given

$a, b \in X$, $\exists \{a\}$ and $X \in T$ s.t.

$a \in \{a\}$, $b \notin \{a\}$ and $b \in X$, $a \in X$.

Example 2:

Let $X = \{x \in R: 0 \leq x < 1\}$. *0 x k y 1*

Write $G_k = \{x \in R: 0 \leq x < k, 0 < k \leq 1\}$, $T = \{G_k : 0 < k \leq 1\}$.

Solution:

Then T is a topological on X.

Consider a pair of elements $x, y \in X$ s.t. $x < y$.

We can find out an element $k \in X$ s.t. $x < k < y$.

Then $x \in [0, k)$, $y \notin [0, k)$.

Given a pair of distinct points $x, y \in X$.

$\exists$ open set [0, k) X s.t. $x \in [0, k)$, $y \notin [0, k)$.

This proves that (X, T) is a T_0-space.

We have seen that $x < k < y$, $k \in X$.

Simultaneously we can not find another $k_1 \in X$

s.t. $y \in [0, k_1)$, $x \notin [0, k_1)$, for $x < y$.

This proves that (X, T) is not a T_1-space.

Example 3:

Define topology T on N s.t.

(i) ϕ, $N \in T$

(ii) $A_n \in T\ \forall n \in N$, where $A_n = \{1, 2, 3, ..., n\}$,

Consider 2, 4 $\in N$.

$A_2 \in T$, $2 \in A_2$ and $4 \notin A_2$.

Solution:

Given two distinct numbers 2, 4, $\exists$ open set A_2 s.t. $2 \in A_2$ but $4 \notin A_2$.

$\therefore$ (N, T) is T_0-space.

$A_4 \in T$, $2 \in A_4$, $4 \notin A_4$.

Hence $\exists$ no open set which contains 4 but not 2.

Thus (N, T) in not T_1-space.

Example T_2-space

(i) The real line R with usual topology is a T_2-space.

(ii) Every discrete space is a T_2-space.

(iii) Every indiscrete space containing more than one point is a T_2-space.

Example of Normal and Regular Spaces

Let a, b, c be arbitrary distinct point of the set R.

Let $X = \{a, b, c\}$ and $T = \{\varnothing, X, \{a\}, \{b, c\}\}$.

Then (X, T) is a topological spcae.

Evidently $\{a\}$, $\{b, c\}$ are T-open sets.

$\therefore X - \{a\}$, $X - \{b, c\}$ are T-closed sets.

i.e., $\{b, c\}$, $\{a\}$ T-closed sets.

Thus $\{a\}$ and $\{b, c\}$ are T-open sets as well as T-closed sets.

Consider a pair of distinct elements b, c $\in X$. Then the only open sets containing either of the elements b, c are X, $\{b, c\}$ s.t.

$b \in X$, $b \in \{b, c\}$ $c \in X$,

$c \in \{b, c\}$.

This proves that (X, T) is not a T_1-space.

Given a pair of disjoint closed sets {a}, {b, c} $\subset X$, we can find a pair of disjoint open sets {a}, {b, c} $\subset X$ s.t.

closed set {a} $\subset$ open set {a}.

closed set {b, c} $\subset$ open set {b, c}.

This proves that (X, T) is a normal space.

Repeating the same arguments we can show that (X, T) is a regular space.

From this example it is clear that a regular space need not be a T_1-space as well as a normal space need not be a T_1-space.

Ex. Show that (R, U) is a T_3-space.

Example of a Hausdorff space : Let a, b $\in$ R be arbitrary s.t.

$$|a - b| = \varepsilon > 0.$$

Write $G = \left(a - \frac{\varepsilon}{3}, a + \frac{\varepsilon}{3}\right)$, $H = \left(b - \frac{\varepsilon}{3}, b + \frac{\varepsilon}{3}\right)$

Let T be the usual topology on R.

Then G, H $\in$ T s.t. a $\in$ G, b $\in$ H, $G \cap H = \varnothing$.

By def. this implies that (R, T) is a Hausdorff space.

Theorem 1:

A topological space (X, T) is completely normal iff every subspace is normal.

Proof:

Let (Y, U) be a subspace of completely normal space (X, T).

To prove that Y is normal.

Let A, B be disjoint U-closed subsets of Y, then

$\overline{A}^U$ = U-closure of A = A, $\overline{B}^U = B$, $A \cap B = \phi$.

But $\overline{A}^U = \overline{A}^T \cap Y$

$$A \cap \overline{B}^T = \overline{A}^U \cap \overline{B}^T = (\overline{A}^T \cap Y) \cap \overline{B}^T = (\overline{A}^T \cap Y) \cap (Y \cap \overline{B}^T)$$

$$= \overline{A}^U \cap \overline{B}^U = A \cap B = \phi.$$

Thus $A \cap \overline{B}^T = \phi$. Similarly $\overline{A}^T \cap B = \phi$.

This $\Rightarrow$ A and B are T-separated subsets of X.

By complete normality of X, $\exists$ disjoint open sets G, H $\in$ T s.t. $A \subset G$, $B \subset H$.

Write $G_1 = G \cap Y$, $H_1 = H \cap Y$. Then

$G_1 \cap H_1 = (G \cap Y) \cap (H \cap Y) = (G \cap H) \cap Y = \phi \cap Y = \phi$.

$$A \subset G \Rightarrow A \cap Y \subset G \cap Y$$

$\Rightarrow \quad A \cap G_1$ as $A \cap Y = A$.

Similarly $B \subset H_1$. Hence given a pair of disjoint U-closed sets A, B $\exists$ disjoint U-open sets G_1, H_1 s.t. $A \subset G_1$, $B \subset H_1$. This means that (Y, U) is normal.

Conversely suppose that every subspace (Y, U) of (X, T) is normal. to prove that X is completely normal. Let A, B be separated subsets of X. Then

$\overline{A} \cap B = \phi$, $A \cap \overline{B} = \phi$. Write $Y = (\overline{A} \cap \overline{B})'$.

Then Y is T-open as Y is complement of a T-closed set. By assumption (Y, U) is normal s.t. $\overline{A} \cap Y$, $\overline{B} \cap Y$ are disjoint U-closed sets.

By normality of Y, $\exists$-disjoint open sets $G, H \in U$ s.t. $\overline{A} \cap Y \subset G$, $\overline{B} \cap Y \subset H$.

$H, G \in U$ and Y is T-open $\Rightarrow$ G, H are T-open.

$$A \cap Y = A \cap (\overline{A} \cap \overline{B})' = A \cap [(\overline{A}) \cup (\overline{B})']$$

$$= [A \cap (\overline{A})'] \cup [A \cap (\overline{B})'] = \phi \cup [A \cap (\overline{B})'] = \phi \cup A = A$$

Thus $A \cap Y = A$, similarly $B \cap Y = B$.

$A = A \cap Y = A \subset \overline{A} \cap Y \subset G$ or $A \subset G$. Similarly $B \subset H$.

Finally, given separated sets A, B, $\subset X$ $\exists$ disjoint T-open sets G, H s.t. $A \subset G$, $B \subset H$.

This $\Rightarrow$ X is completely normal.

[For $A \cap (\overline{B})' = A \cap (X - \overline{B}) = A \cap \overline{B} = A - \phi = A$]

Theorem 2(a):

The property of a space being a T_0-space is a topological property. *i.e.*, the homeomorphic image of a T_0-space is a T_0-space.

Proof:

Let $f : (X, T) \to (Y, U)$ be a homeomorphism. Let (X, T) be a T_0-space.

Let $y_1, y_2 \in Y$ be arbitrary s.t. $y_1 \neq y_2$.

f is one-one $\Rightarrow \exists\ x_1, x_2 \in X$ s.t. $f(x_1) = y_1$, $f(x_2) = y_2$

X is T_0-space $\Rightarrow$ there exists $G \in T$ s.t. $x_1 \in G$, $x_2 \notin G$.

$$\Rightarrow f(x_1) \in f(G),\ f(x_2) \notin f(G)$$

$\Rightarrow y_1 \in f(G),\ y_2 \notin f(G),\ f(G) \in U.$

[For f is homeomorphism $\Rightarrow$ f is open map]

$\Rightarrow$ (Y, U) is T_0-space.

Theorem 2(b):

A topological space (X, T) is a T_1-space iff {x} closed for each $x \in X$. In a topological space, show that

T_1-space $\Leftrightarrow$ each point is a closed set.

Proof:

(i) Let (X, T) be a topological space s.t. {x} is closed $\forall\ x \in X$

To prove that X is T1-space.

Consider x, y, $\in$ X s.t. $x \neq y$.

Then, by hypothesis, {x} and {y} are disjoint closed sets. This means that X – {x} and X – {y} are T-open sets.

Write G = X – {y}, H = X – {x}.

Then G, H $\in$ T s.t. $x \in G,\ y \notin G,\ y \in H,\ x \notin H$.

This proves that (X, T) is a T_1-space.

(ii) Conversely suppose that (X, T) is a T_1-space.

To prove that {x} is closed $\forall\ x \in X$.

Since X is a T_1-space.

$\therefore$ given a pair of distinct points $x, y \in X$; $\exists$ G, H $\in$ T.

s.t. $x \in G,\ y \notin G$ and $y \in H,\ x \notin H$.

Evidently, $G \subset X - \{y\},\ H \subset X - \{x\}$.

Given any $x \in X - \{y\} \Rightarrow \exists\ G \in T$ s.t. $x \in G \subset X - \{y\}$.

This proves that every point x of X – {y} is an interior point of X – {y}, meaning thereby X – {y} is open, *i.e.*, {y} is closed.

Furthermore,

given any $y \in X - \{x\} \Rightarrow \exists\ H \in T$ s.t. $y \in H \subset X - \{x\}$.

This implies that every point y of X–{x} is an interior point of X–{x}. Hence X – {x} is open, *i.e.*, {x} is closed.

Finally {x}, {y} are closed sets in X.

Generalising this result,

{x} is closed $\forall\ x \in X$.

Theorem 3:

The property of a space being a T_1-space is a topological property, *i.e.*, the homeomorphic image of T_1-space is a T_1-space.

Proof:

Let (X, T) be a T_1-space and let

$$f : (X, T) \rightarrow (Y, U)$$

be a homeomorphism.

To prove that the T_1-space is a topological property, it is ;enough to prove that (f(X) = Y, U) is a T_1-space.

$\because$ X is a T_1-space.

Given a pair of distinct elements $p, q \in X$;

$$\exists\ G, H \in T \text{ s.t. } p \in G, q \notin G;$$

$$q \in H, p \notin H.$$

Therefore given $f(p), f(q) \in Y$ s.t. $f(p) \neq f(q)$;

$$\exists\ f(G), f(H) \in U \text{ s.t. } f(p) \in f(G), g(q) \notin g(G)$$

$$f(q) \in f(H), f(p) \notin f(H).$$

This proves that (Y, U) is a T_1-space.

Theorem 4(a):

The property of a space being a T_2-space is a topological property.

Proof:

Let (X, T) be a T_2-space and let

$$f : (X, T) \rightarrow (Y, U)$$

be a homeomorphism so that any $G \in T \Rightarrow f(G) \in U$. ...(1)

To prove that T_2-space is a topological property, it suffices to prove that (f(X) = Y, U) is a T_2-space.

$\because$ X is a T_2-space.

$\because$ Given a pair of points

$$p, q \in X \text{ s.t. } p \neq q, \exists\ G, H \in T$$

s.t. $p \in G, q \in H, G \cap H = \varnothing$.

Therefore

$G, H \in T \Rightarrow f(G), f(H), \in U$.

We claim $f(G) \cap f(H) = \varnothing$.

Suppose not.

Then $\exists\ p_1 \in f(G) \cap f(H)$, so that $p_1 \in f(G)$, $p_1 \in f(H)$,

$p_1 \in f(G)$, f is one-one $\Rightarrow \exists\ p_2 \in G$ s.t. $p_1 = f(p_2)$,

$p_1 \in f(H)$, f is one-one $\Rightarrow \exists\ p_3 \in H$ s.t. $p_1 = f(p_3)$.

$p_1 = f(p_3), p_1 = f(p_2) \Rightarrow f(p_2) = f(p_3)$

$\Rightarrow p_2 = p_3$ $\quad\because$ f is one-one

$\Rightarrow p_1 \in G, p_2 \in H$.

$\Rightarrow p_1 \in G \cap H$

A contradiction. For $G \cap H = \varnothing$.

$\therefore f(G) \cap f(H) = \varnothing$.

Finally $p \neq q \Rightarrow f(p) \neq f(q)$.

Given a pair of points $f(p), f(q) \in Y$ s.t. $f(p) \neq f(q)$; we are able to find out disjoint sets $f(G), f(H) \in U$

s.t. $\quad f(p) \in f(G), f(q) \in f(H)$.

This proves that (Y, U) is a T_2-space.

Theorem 4(b):

Regularity is a toplogical property.

Proof:

Let : $(X, T) \to (Y, U)$ be a homeomorphism and let (X, T) be a regular space.

If we show that (Y, U) is a regular space, the result will follow. Let $y \in Y$ and F be U-closed subset of Y s.t. $y \notin Y$, f; $X \to Y$ is one-one onto.

$\Rightarrow \exists\ x \in X$ s.t. $f(x) = y$

$\Rightarrow x = f^{-1}(y)$

f is a homeomorphism $\Rightarrow$ f and f^{-1} both are continuous maps.

$F \subset Y$ is U-closed, f is continuous

$\Rightarrow f^{-1}(F) \subset X$ is T-closed.

$y \notin F \Rightarrow f^{-1}(y) \notin f^{-1}(F) \Rightarrow x \notin f^{-1}(F)$

Now $x \in X$ s.t. $x \notin f^{-1}(F)$, $f^{-1}(F)$ is T-closed. By def. of regularity, $\exists$ G, H $\in$ T s.t.

$x \in G, {}^{-1}(F) \subset H, G \cap H = \varnothing$

This $\Rightarrow f(x) \in f(G), f[f^{-1}(F)] \subset f(H), f(G \cap H) = f(\varnothing)$,

$\Rightarrow y \in f(G), F \subset f(H), f(G) \cap (H) = f(\varnothing)$,

$f(G), f(H) \subset U$

[For f is homeomorphism $\Rightarrow$ f is an open map]

$\Rightarrow$ (Y. U) is regular.

Theorem 5:

To show that normality is a topological property.

Proof:

Let f (X, T) $\Rightarrow$ (Y, U) be a homeomorphism and let (X, T) be a normal space.

Let A, B be U-disjoint closed sets. f is continuous $\Rightarrow f^{-1}(A), f^{-1}(B)$ are T-closed sets.

$f^{-1}(A) \cap f^{-1}(B) = f^{-1}(A \cap B) = f^{-1}(\varnothing) = \varnothing$

Now $f^{-1}(A), f^{-1}(B)$ are T-disjoint closed subsets of X. Also X is normal

Hence $\exists$ G, H $\in$ T s.t.

$$f^{-1}(A) \subset G,\ f^{-1}(B) \subset H,\ G \cap H = \varnothing$$

From which $f[f^{-1}(A)] \subset f(G), f[f^{-1}(B)] \subset f(H)$,

$$f(G) \cap f(H) = f(\phi) = \phi,$$

Or, $\quad A \subset f(G), B \subset f(H), f(G) \cap f(H) = \phi$,

where $\quad$ f(G), f(H) are U-open sets.

[For f is hoomeomorphism $\Rightarrow$ f is open map]

This proves that (Y, U) is normal.

Theorem 6(a):

Show that every Hausdorff space is a T_1-space. Is the converse true? Illustrate your answer by an example.

Or, To prove that

X is a T_2-space $\Rightarrow$ X is a T_1-space

but not conversely, i.e., if (X, T) is a T_1-space then need not be a T_2-space.

Proof:

First part. Suppose (X, T) is a T_2-space.

To prove that (X, T) is a T_1-space.

By def. of T_2-space, given $x, y \in X$ s.t. $x \neq y$, we can find disjoint sets $G, H \in T$ s.t. $x \in G$, $y \in H$.

$G \cap H = \varnothing$, $x \in G$, $y \in H \Rightarrow x \in G$, $y \notin G$

$y \in H$, $x \notin H$.

Hence given $x, y \in X$ s.t. $x \neq y$, $\exists$ $G, H \in T$

s.t. $x \in G$, $y \notin G$; $y \in H$, $x \notin H$.

This prove that (X, T) is a T_1-space.

To prove that (X, T) is not a T_2-space.

We need only given an example of a space which does satisfy the T_1-axiom of separation but not the T_3-axiom of separation.

Now write the solution.

Theorem 6(b):

A normal space is a regular space.

That is to say, X is a T_4-space $\Rightarrow$ X is a T_3-space.

Proof:

Let (X, T) be a T_4-space so that

(i) X is a T_1-space, (ii) X is a normal space

To prove that X is a T_3-space.

For this we must show that

(iii) X is a T_1-space, (iv) X is a regular space.

Evidently (i) $\Rightarrow$ (iii).

If we show that (ii) $\Rightarrow$ (iv) the result will follow.

X is a T_1-space $\Rightarrow$ $\{x\}$ is closed in X.

(Later on we shall prove this).

By normality, given a pair of disjoint closed sets $\{x\}$ and F in X, $\exists$ disjoint open sets G, H in X, s.t. $\{x\} \subset G$, $F \subset H$,

i.e., given a closed set $F \subset X$ and $x \in X$ s.t. $x \notin F$:

$\exists$ disjoint open sets G, H in X, s.t. $\{x\} \subset G$, $F \subset H$.

This proves that (X, T) is a regular space.

Theorem 6(c):

A topological space (X, T) is a T_0-space iff for any distinct arbitrary points x, y of X, the closures of $\{x\}$ and $\{y\}$ are distinct.

Proof:

Let (X, T) be a T_0-space. Let x, y $\in$ X be arbitrary s.t. $x \neq y$.

X – G is closed set.

$\overline{\{y\}} = \cap \{F : F \text{ is a closed set s.t. } y \in F\}$

Also X – G is a closed set containing y. $\therefore\ \overline{\{y\}} \subset X - G$.

Moreover $x \notin X - G$ so that $x \notin \overline{\{y\}}$. ...(1)

Obviously $\{x\} \subset \overline{\{X\}}$. For $A \subset \overline{A}$. This $\Rightarrow x \in \overline{\{X\}}$. ...(2)

From (1) and (2), it follows that $\overline{\{X\}} \neq \overline{\{y\}}$.

Conversely let x, y be any two distinct points of the topological space (X, T).

Also let $\overline{\{X\}} \neq \overline{\{y\}}$. ...(3)

To prove that (X, T) is T_0-space.

(3) $\Rightarrow \exists\ p \in X$ s.t. $p \in \overline{\{X\}}$ and $p \notin \overline{\{y\}}$. ...(4)

We claim $x \notin \overline{\{y\}}$.

Suppose not. Then $x \in \overline{\{y\}}$.

$x \in \overline{\{y\}} \Rightarrow \{x\} \subset \overline{\{y\}} \Rightarrow \overline{\{X\}} \subset \overline{\overline{\{y\}}} = \overline{\{y\}}$

$\Rightarrow \overline{\{X\}} \subset \overline{\{y\}}$. Also $p \in \overline{\{X\}}$.

$\Rightarrow p \in \overline{\{y\}}$. A contradiction. For $p \notin \overline{\{y\}}$.

Hence (4) holds.

(4) $\Rightarrow x \in X - \overline{\{y\}}$. $y \in \overline{\{y\}} \Rightarrow y \notin X - \overline{\{y\}}$, $X - \overline{\{y\}}$, is open.

For $\overline{\{y\}}$ is closed.

This $X - \overline{\{y\}}$ is an open set s.t. $x \in X - \overline{\{y\}}$ and $y \notin X - \overline{\{y\}}$.

This proves that (X, T) is T_1-space.

Theorem 7:

If (X, T) is a T_0-space and T_1 is finer than T, then (X, T_1) is also T_0--space.

Proof:

Let (X, T) be a T_0-space and $T \subset T_1$.

To prove that (X, T_1) is T_0-space.

(X, T) is T_0-space $\Rightarrow$ given distinct points x, y $\in$ X.

there is T-open set G s.t. $x \in G$, $y \notin G$.

Since any $A \in T \Rightarrow A \in T_1$

Hence every T-open set is T_1 open set.

Then G is T_1-open set with $x \in G$, $y \notin G$.

Theorem 8(a):

A finite subset of a T_1-space has no cluster point.

Proof:

Let (X, T) be a T_1-space and let

$$A = \{a_1, a_2, \ldots\ldots, a_n\}$$

be a finite subset of X.

To prove that A has no cluster point.

X is T_1-space $\Rightarrow$ A is closed $\Rightarrow D(A) \subset A$.

$\Rightarrow$ every limit point of A belongs of A.

$\Rightarrow$ A contains all its limit points. ...(1)

Let ai $\in$ A be arbitrary and

Write Gi = $\{a_1, a_2, \ldots, a_{i+1}, \ldots, a_n\}$

G_i is a finite subset of X $\Rightarrow G_i$ is closed.

$\Rightarrow G'_i = X - G_i$ is open.

$a_i \notin G_i \Rightarrow a_i \in G'_i$

Thus G'i is an open set with $a_i \notin G_i$.

Also $\{G'_i - \{a_i\} \cap A = \phi$.

By def. this $\Rightarrow a_i$ is not a limit point of A.

But a_i is an arbitrary point of A.

Hence every point of A is not a limit point of A. Now (1) declares that A has no limit point.

Theorem 8(b):

Show that X is T_3-space so that

(i) X is a T_2-space,

(ii) X is a regular space.

To prove that (X, T) is a T_2-space.

(ii) Shows that

given a closed set $F \subset X$ and $x \in X$ s.t. $x \notin F$. we can find disjoint sets $G, H \in T$ s.t. $x \in G$, $F \subset H$.

Let $y \in F$ be arbitrary, then $y \in H$. For $F \subset H$.

Now $y \in F$, $x \notin F \Rightarrow x \neq y$.

$\therefore$ Given a pair of distinct points $x, y \in X$, $\exists$ $G, H \in T$ s.t. $x \in G$, $y \in H$, $G \cap H = \varnothing$.

This proves that X is a T_2-space.

Theorem 9:

To prove that

(X, T) is a T_4-space $\Rightarrow$ (X, T) is a T_2-space.

Let (X, T) be a T_4-space so that

(i) X is a T_1-space.

(ii) X is a normal space.

To prove that (X, T) is a T_2-space.

Let $x, y \in X$, be arbitrary s.t. $x \neq y$.

Then (i) $\Rightarrow$ $\{x\}$, $\{y\}$ are disjoint closed sets in X.

Using normality, we can find disjoint open sets $G, H \in T$

s.t. $\{x\} \subset G$, $\{y\} \subset H$,

i.e., $x \in G$, $y \in H$.

Given $x, y \in X$, s.t. $x \neq y$, we are able to discover disjoint open sets $G, H \in T$ s.t. $x \in G$, $y \in H$. This means that X is a T_2-space.

Def. dyadic fraction. A real number of the form $\frac{m}{2^n}$ (where $m, n \in N$ s.t. $m \leq 2^n$) is called a dyadic fraction.

For instance

$$\frac{1}{4} \qquad 1$$

$$\frac{1}{2^2}, \frac{2}{4}, \frac{3}{4} \qquad 1$$

$$\frac{1}{8}, \frac{1}{4}, \frac{3}{8}, \frac{1}{2}, \frac{5}{8}, \frac{6}{8}, \frac{7}{8}, \quad 1 \text{ etc}$$

Theorem 10:

A completely regular space is regular.

Proof:

Let (X, T) be a completely regular space; then given any closed set $F \subset X$ and $p \in X$ s.t. $p \notin F$; $\exists$ continuous map $f: X \to [0, 1]$ with the property that

$$f(p) = 0, \; f(F) = \{1\}.$$

To prove that (X, T) is a regular space.

Consider the set [0, 1] with usual topology. It is easy to verify that [0, 1] is a T_2-space, then we can find out disjoint open sets G, H in [0, 1] s.t. $0 \in G$, $1 \in H$.

By hypothesis, f is continuous, hence $f^{-1}(G)$, $f^{-1}(H)$ are open in X.

$f^{-1}(G) \cap f^{-1}(H) = f^{-1}(H \cap G) = f^{-1}(\varnothing) = \phi$.

$$f^{-1}(G) = \{x \in X : f(x) \in G\}$$

Furthermore

$$f(p) = 0 \in G \Rightarrow f(p) \in G$$
$$\Rightarrow p \in f^{-1}(G)$$
$$f(F) = 1 \in H \Rightarrow f(F) = \{1\} \subset H$$
$$\Rightarrow f(F) \subset H$$
$$\Rightarrow F \subset f^{-1}(H)$$

Given any closed set $F \subset X$ and $p \in X$ s.t. $p \notin F$; $ disjoint open sets $f^{-1}(G)$, $f^{-1}(H)$ in X s.t. $p \in f^{-1}(G)$, $F \subset f^{-1}(H)$, showing thereby X is regular.

Theorem 11:

A Tychonoff space is a T_3-space.

Or. Complete regular space $\Rightarrow$ regular space.

Proof:

Let (X, T) be a Tychonoff space then

(i) X is a T_1-space.

(ii) X is a completely regular space.

To prove that (X, T) is a T_3-space, it suffices to show that

(iii) X is a T_1-space.

(iv) X is a regular space.

Evidently (i) $\Rightarrow$ (iii)

Prove as in Theorem (9) that (ii) $\Rightarrow$ (iv).

HAUSDORFF SPACE PROBLEMS

Theorem 1:

In a T_2-space, a convergent sequence has a unique limit.

If a sequence in a Hausdorff space has limit, prove that the limit must be unique.

Proof:

Let (X, T) be a Hausdorff space. Let (a_n) be a convergent sequence in X.

To prove that the sequence $\langle a_n \rangle$ has a unique limit.

Suppose the contrary.

Then $\langle a_n \rangle$ cannot have a unique limit.

Let $\langle a_n \rangle$ converge to two distinct point, say a_0, $b_0 \in X$. Then $a_0 \neq b_0$.

By definition of convergence,

$a_0 \in G \in T \Rightarrow \exists\, n_0 \in N$ s.t. $\forall\, n \geq n_0 \Rightarrow a_n \in G$

and $b_0 \in H \in T \Rightarrow \exists\, k_0 \in N$ s.t. $\forall\, n \geq k_0 \Rightarrow a_n \in H$

Set $m_0 = \max.\ (n_0, k_0)$

Then $\forall\, n \geq m_0 \Rightarrow a_n \in G,\ a_n \in H \Rightarrow a_n \in G \cap H \Rightarrow G \cap H \neq \emptyset$.

Given a pair of distinct elements a_0, b_0, $\in X$, there are open sets,

G, H $\subset$ X s.t. $a_0 \in G$, $b_0 \in H$, $G \cap H \neq \phi$.

This proves that (X, T) is not a Hausdorff space, contrary to the hypothesis that X is a Hausdorff space.

Hence $\langle a_n \rangle$ has a unique limit.

Theorem 2:

Let (X, T) be a first countable space. Then (X, T) is a Hausdorff space iff every convergent sequence in X has a unique limit.

Proof:

(i) Let (X, T) be a first countable Hausdorff space. Let $\langle a_n \rangle$ be a convergent sequence in it. To prove sequence has unique limit. Prove this as in Theorem 17.

(ii) Conversely suppose that (X, T) is first countable space. Let every convergent sequence in X converge to a unique limit.

To prove that (X, T) is a Hausdorff space.

Suppose the contrary.

Then (X, T) is not a Hausdorff space.

$\therefore$ Given a pair of distinct points p, q $\in$ X, $\exists$ G, H $\in$ T with P $\in$ G. q $\in$ H s.t. G $\cap$ H $\neq \emptyset$.

i.e., intersection of every open set containing P with every open set containing q is non-empty.

By the hypothesis (X, T) is first countable; hence $\exists$ countable local bases B_p and B_q at P and q respectively for the topology T on X.

We are justified in writing

$$B_p = \{G_n \in T : p \in G_n \ \forall\ n \in N\},$$

$$B_q = \{H_n \in T : q \in H_n \ \forall\ n \in N\}.$$

$\therefore G_n \cap H_n \neq \emptyset \ \forall\ n \in N$ in accordance with (*)

Then \$ an element $a_n \in G_n \cap H_n \ \forall\ n \in N$

$\therefore\ a_n \in G_n, a_n \in H_n \ \forall\ n \in N$.

Consider the sequence $< a_n : n \in N >$.

Now $p \in G_n \ \forall\ n \in N \Rightarrow\ a_n \in G_n \ \forall\ n \in N$.

This $\Rightarrow a_n \rightarrow p$.

Again $q \in H_n \ \forall\ n \in N \Rightarrow\ a_n \in H_n \ \forall\ n \in N$.

This $\Rightarrow a_n \rightarrow q$.

Thus the sequence $\langle a_n \rangle$ converges to distinct points p, q $\in$ X, contrary to the assumption that every convergent sequence in X has unique limit.

Hence (X, T) is a Hausdorff space.

Theorem 3(a):

Let (X, T) be any topological space and (Y, U) a Hausdorff space. Let f and g be continuous mapping of X into Y. Then the set { x $\in$ X ; f(x) = g(x)} is closed.

Proof:

Let (X, T) be any topological space and (Y, U) be Hausdorff space.

Let : X $\rightarrow$ Y and g : X $\rightarrow$ Y be continuous maps. Let

$$A = \{x \in X : f(x) = g(x)\}.$$

To prove that A is a closed set.

Evidently $X - A = \{x \in X : f(x) \neq g(x)\}$ ...(1)

Let $x \in X - A$ be arbitrary.

Take $f(x) = y_1$, $g(x) = y_2$.

According to (1), $f(x) \neq g(x)$, *i.e.*, $y_1 \neq y_2$.

Furthermore $y_1, y_2 \in Y$ and (Y, U) is a Hausdorff space, hence

$$\exists\, G, H \in T \text{ s.t. } y_1 \in G,\ y_2 \in H,\ G \cap H = \varnothing.$$

Since f and g are given to be continuous maps. Hence $f^{-1}(G)$, $g^{-1}(H)$ are open in X.

Write $W = f^{-1}(G) \cap g^{-1}(H)$.

Being a finite intersection of open sets in X, W is open in X.

$y_1 \in G \Rightarrow f^{-1}(y_1) \in f^{-1}(G) \Rightarrow x \in f^{-1}(G)$

$y_2 \in H \Rightarrow g^{-1}(y_2) \in g^{-1}(H) \Rightarrow x \in g^{-1}(H)$

$x \in f^{-1}(G),\ x \in g^{-1}(H) \Rightarrow x \in f^{-1}(G) \cap g^{-1}(H) = W$

$\Rightarrow x \in W.$

$W = f^{-1}(G) \cap g^{-1}(H) \Rightarrow W \subset f^{-1}(G),\ W \subset g^{-1}(H)$

$\Rightarrow f(W) \subset G,\ g(W) \subset H$

$\Rightarrow f(W) \subset g(W) \subset G \cap H$

$\Rightarrow f(W) \cap g(W) \subset \varnothing.$ ...(2)

But $\varnothing \subset f(W) \cap g(W).$ ...(3)

Combining (2) and (3),

$$\varnothing = f(W) \cap g(W).$$

This $\Rightarrow f(y) \neq \forall\, y \in W$.

Thus, any $y \in W \Rightarrow f(y) \neq g(y)$,

$\Rightarrow \quad y \in X - A.$

$\therefore \quad W \subset X - A.$

Finally, any $x \in X - A \Rightarrow \exists\, W \in T$ with $x \in W$ s.t. $W \subset X - A$.

This proves that every point x of X – A is ;an interior point of X + A, meaning thereby X – A is open, *i.e.*, A is closed.

Hence the theorem.

Theorem 3(b):

If f and g are continuous functions on a topological space X with values in Hausdorff space Y, then prove that the set of all points x in X such that f(x) = g(x) is closed. Deduce that if f and g agree on a dense subset on X, then f = g on the whole X.

Proof:

(i) Write the complete proof of the above Theorem 19.

(ii) Suppose f and g agree on a dense subset P of X so that

$$f(x) = g(x)\ \forall\ x \in P,\ \overline{P} = X$$

$f(x) = g(x)\ \forall\ x \in P \Rightarrow P$ is closed, be case (i)

$\Rightarrow \overline{P} = P$. Also $\overline{P} = X$

$\Rightarrow P = X$. Also $f(x) = g(x)\ \forall\ x \in P$

$\Rightarrow f(x) = g(x)\ \forall\ x \in X$

$\Rightarrow f = g$ on the whole set X.

Theorem 3(c):

A topological space (X, T) is a normal space iff each open neighbourhood G of a closed set F contains as a subset the closure of another open neighbourhood G_1 of F.

An alternate statement. A topological space (X, T) is normal iff for any closed set F and an open set G containing F, there exist an open set V s.t.

$F \subset V$ *and* $\overline{V} \subset G$.

Proof:

Let (X, T) is a normal space and let $G \subset X$ be an open neighbour hood of a closed set $F \subset X$, then $F \subset G$.

To prove that $\exists$ another open neighbourhood G_1 of F s.t.

$$F \subset G_1 \subset \overline{G}_1 \subset G.$$

F is a closed set $\Rightarrow X - F$ is open.

$G \subset X$ is open $\Rightarrow X - G$ is closed.

$F \cap (X - G) = F \cap X - F \cap G = F - F = \phi \because \quad F \subset G.$

$\therefore$ F and (X – G) are disjoint closed sets in X.

Using normality, we can find a pair of disjoint open sets

$H_1, H_2 \subset X$ s.t. $F \subset H_1,\ X - G \subset H_2$. ...(1)

$H_1 \cap H_2 = \phi \Rightarrow H_1 \subset X - H_2$

$\Rightarrow \overline{H}_1 \subset \overline{X - H_2} = X - H_2 \qquad \therefore X - H_2$ is closed.

$\Rightarrow \overline{H}_1 \subset X - H_2$

$\Rightarrow \overline{H}_1 \subset G$, from (1).

Thus the set H_1 has the following properties:

(i) H_1 is open

(ii) $F \subset H_1$,

(iii) $\overline{H}_1 \subset G$.

(i) and (ii) $\Rightarrow H_1$ is an open neighbourhood of F.

Evidently the set H_1 has the properties as the set G_1 referred to in the theorem.

Taking $G_1 = H_1$, we have the required result as follow :

G_1 is an open neighbourhood F s.t. $F \subset G_1 \subset \overline{G}_1 \subset G$.

Conversely suppose that (X, T) is a topological space s.t. given a closed set F and an open neighbourhood G of implies that $\exists$ an open neighbourhood

G_1 of F s.t. $F \subset G_1 \subset \overline{G}_1 \subset G$.

To prove that (X, T) is a normal space.

Let C_1 and C_2 be a pair of disjoint closed sets in X.

$C_1 \cap C_2 = \varnothing \Rightarrow C_1 \subset X - C_2$

$\Rightarrow X - C_2$ is a neighbourhood of the closed set C_1.

BY hypothesis, $\exists$ another open neighbourhood G_1 of C_1 s.t.

$$C_1 \subset G_1 \subset \overline{G}_1 \subset X - C_2 \qquad \text{...(2)}$$

Write $X - \overline{G}_1 = G_2$.

$\overline{G}_1$ is a closed set $\Rightarrow X - \overline{G}_1$ is an open subset of X

$\Rightarrow G_2 \subset X$ is open.

From (2), $C_2 \subset X - \overline{G}_1 = G_2$,

i.e., $C_2 \subset G_2$.

$G_1 \cap G_2 = G_1 \cap (X - \overline{G}_1) = G_1 \cap X - G_1 \cap \overline{G}_1 = G_1 - H_1 = \varnothing$

Given a pair of disjoint closed sets $C_1, C_2 \subset X$ there is a pair of disjoint open set $G_1, G_2 \subset X$ s.t. $C_1 \subset G_1, C_1 \subset G_2$.

This proves that (X, T) is a normal space.

Theorem 4:

Show that for any space X, the following conditions are equivalent.

(i) X is T_2-space.

(ii) For each pair x, y, $\in X$, $\exists$ a nbd. N_y of y s.t. x is not in $\overline{N}_y$.

(iii) For each $x \in X$, $\{x\} = \cap \overline{N}_y$, where the intersection is taken over all the nbds of x.

Proof:

Let (X, T) be a topological space.

To prove (i) $\Rightarrow$ (ii).

Suppose (X, T) is T_2-space.

Let $x, y \in X$ be arbitrary s.t. $x \neq y$.

To prove that $\exists$ nbd N_y of y s.t. $x \notin \overline{N}_y$. By def. to T_2-space.

$\exists$ G, H $\in$ T s.t. $x \in G$, $y \in H$, $G \cap H = \emptyset$.

$y \neq G$, $x \in G \in T \Rightarrow y \in G'$, $x \notin G'$, G' is closed.

$\Rightarrow N_y$ is a closed nbd of y s.t. $x \notin N_y$.

But $N_y = \overline{N}_y$. For N_y is closed.

$\Rightarrow N_y$ is a closed nbd of y s.t. $x \notin \overline{N}_y$.

To prove (ii) $\Rightarrow$ (iii).

Suppose any $x, y \in X \Rightarrow \exists$ nbd N_y of y s.t $x \notin \overline{N}_y$.

To prove that $\cap \overline{N}_x = \{x\}$ where the intersection is taken over all nbds of x. By assumption, there also exists nbd N_x of x of s.t. $y \notin \overline{N}_y$.

Take B = $\{\overline{N}_y : N_x$ is a nbd of x s.t. $y \notin \overline{N}_y\}$.

Then $\cap \overline{N}_x = \cap \{B : B \in B\} = \{x\}$.

To prove (iii) $\Rightarrow$ (i).

Suppose $\cap \overline{N}_x \Rightarrow \{x\}$.

To prove that X is a T_2-space.

Let x, y, $\in$ be arbitrary s.t. $x \neq y$.

$\{x\} \cap \overline{N}_x \Rightarrow y \notin \overline{N}_x$. Also $\overline{N}_x$ is a closed nbd of x.

$\Rightarrow \exists$ a closed nbd N of x which does not contain y

$\Rightarrow x \in N$, $y \notin N$.

By def. of nbd, $\exists$ G $\in$ T s.t. $x \in G \subset N$, $y \notin N$.

Write H = X – N.

$y \notin N \Rightarrow y \in N' = H \Rightarrow y \in H$

$G \subset N \Rightarrow G \cap (X - N) = \emptyset \Rightarrow G \cap H = \emptyset$.

N is a closed set $\Rightarrow$ N' = H is open.

∴ Given x, y ∈ X ∃ G, H ∈ T s.t.

$$x \in G,\ y \in H,\ G \cap H = \varnothing.$$

This ⇒ X is a T_2-space.

Theorem 5:

Let f : (X, T) → (Y, U) be a one-one continuous map. Also suppose that (Y, U) is a Hausdorff space. Show that (X, T) is Hausdorff space.

Proof:

Let : (X, T) → (Y, U) be a one-one continuous map. Also suppose that (Y, U) is a Hausdorff space.

To show that (X, T) is a Hausdorff space.

Consider an arbitrary pair of elements $x_1, x_2 \in X$ s.t. $x_1 \neq x_2$.

Since f is one-one and hence $f(x_1) \neq f(x_2)$.

Take $\quad y_1 = f(x_1),\ y_2 = f(x_2),$

Then $\quad y_1, y_2 \in Y$ s.t. $y_1 \neq y_2$.

By the hausdorff-space property, ∃ a pair of disjoint open sets

$$G_1, G_2 \subset Y \text{ s.t. } y_1 \in G_1,\ y_2 \in G_2.$$

Continuity of f implies that $f^{-1}(G_1)$ and $f^{-1}(G_2)$ are open in X.

$f^{-1}(G_1) \cap f^{-1}(G_2) = f^{-1}(G_1 \cap G_2) = f^{-1}(\varnothing) = \varnothing.$

$y_1 \in G_1 \Rightarrow f^{-1}(y_1) \in f^{-1}(G_1) \Rightarrow x_1 \in f^{-1}(G_1),$

$y_2 \in G_2 \Rightarrow f^{-1}(y_2) \in f^{-1}(G_2) \Rightarrow x_2 \in f^{-1}(G_2).$

Finally we have shown that

given a pair of elements $x_1, x_2 \in X$ s.t. $x_1 \neq x_2 \Rightarrow$ ∃ disjoint open sets $f^{-1}(G_1), f^{-1}(G_2)$ in X s.t.

$$x_1 \in f^{-1}(G_1),\ x_2 \in f^{-1}(G_2)$$

This leads to the conclusion that (X, T) is ;a Hausdorff space.

Theorem 6:

A topological space (X, T) is regular iff for every point x of X and every nbd N of x, ∃ a nbd M of x s.t. $\overline{M} \subset N$.

In other words, a topological space is regular iff the collection of all T-closed nbds, form a local base at x.

Proof:

Let (X, T) be a regular space.

To prove that given a nbd N of x, ∃ a nbd M of x s.t. $\overline{M} \subset N$.

By the property of a regular-space,

given a closed set F and an element $x \in X$ s.t. $x \notin F$.

∃ disjoint open sets $G_1, G_2 \subset X$ s.t.

$$x \in G_1, F \subset G_2.$$

$x \in G_1 \Rightarrow G_1$ is a nbd. of x

$G_1 \cap G_2 = \varnothing \Rightarrow G_1 \subset X - G_2$

$$\Rightarrow \overline{G_1} \subset \overline{X - G_2} = X - G_2 \qquad \therefore X - G_2 \text{ is closed.}$$

$$\Rightarrow \overline{G_1} \subset X - G_2 \subset X - F$$

For $F \subset G_2 \Rightarrow X - G_2 \subset X - F$

$$\Rightarrow \overline{G_1} \subset X - F = H \text{ (say)}$$

$$\Rightarrow \overline{G_1} \subset H$$

∴ F is closed ⇒ X − F is open ⇒ H is open

$x \notin F \Rightarrow x \in X - F \Rightarrow x \in H$.

We have the final results as :

given a nbd. H of x, ∃ a nbd. G_1 of x s.t. $x \in G_1 \subset \overline{G_1} \subset H$.

Hence the result.

Conversely suppose that (X, T) is a topological space having the property listed below:

Given a nbd. G of an element $x \in X$, ∃ a nbd. H of x s.t.

$$x \in H \subset \overline{H_1} \subset G.$$

To prove that (X, T) is a regular space.

Let $F \subset X$ be a closed set and $x \in X$ s.t. $x \notin F$.

$x \notin F$, F is closed $\Rightarrow x \in X - F$, X − F is open.

$$\Rightarrow X - F \text{ is nbd. of x.}$$

∴ By assumption, ∃ a nbd. G of x s.t.

$$x \in G \subset \overline{G_1} \subset X - F. \qquad ...(1)$$

Write $G = G_1$, $X - \overline{G_1} = G_2$.

Then $G_1 \cap G_2 = G \cap (X - \overline{G_1}) = G \cap X - G \cap \overline{G_1}$

$$= G - G = \varnothing$$

$x \in G \Rightarrow x \in G_1$.

From (1) $\overline{G}_1 \subset X - F$

or, $F \subset X - \overline{G}_1$

or, $F \subset G_2$.

G is closed $\Rightarrow X - \overline{G}_1$ is open.

Thus we have shown that

given a closed set $F \subset X$ and $x \in X$ s.t. $x \notin F$,

$\exists$ disjoint open sets $G_1, G_2 \subset X$ with $x \in G_1$, $F \subset G_2$.

This proves that (X, T) is regular.

Corollary. Every regular Frechet space is Hausdorff space.

Proof:

Let a topological space (X, T) be a regular Crechet space

To prove that (X, T) is a Hausdorgff space.

Let $x, y \in X$ be arbitrary s.t. $x \neq y$.

X is a Frechet space $\Rightarrow$ X is a T_1-space.

$\Rightarrow \{x\}, \{y\}$ are closed in X.

$\Rightarrow X - \{x\}, X - \{y\}$ are open in X.

Moreover $y \in X - \{x\}$. For $x \neq y$.

$y \in X - \{x\}$, $X - \{x\}$ is open in X $\Rightarrow \exists$ $X - \{x\}$ is a neighbourhood of y.

$y \in \overline{G} \subset X - \{x\}$.

Write $U = G^o$, $V = X - \overline{G}$.

Evidently U, V are open in X.

$U \cap V = G^o \cap (X - \overline{G}) = G^o \cap X - G^o \cap \overline{G} = G^o - G^o = \varnothing$

$U = G^o$, G is a neighbourhood of y $\Rightarrow$ U

$V = X - \overline{G}$, $\overline{G} \subset X - \{x\} \Rightarrow \{x\} \subset X - \overline{G} = V$

$\Rightarrow \{x\} \subset V$

$\Rightarrow x \in V$.

This given a pair of distinct elements $x, y \in X$, $\exists$ disjoint open sets $U, V \subset X$ s.t. $y \in U$, $x \in V$

This $\Rightarrow$ (X, T) is a Hausdorff space.

Problems related to Urysohn's lemma

Theorem 7:

State and prove Urysohn's lemma.

Urysohn's lemma. Consider the set R with usual topology, where $R = \{x \in R : 0 \leq x \leq 1\}$.

A topological space (X, T) is normal iff given a pair of disjoint closed sets A, B Ì X, there is a continuous function

$$f : X \to R \text{ s.t. } f(A) = \{0\} \text{ and } f(B) = \{1\}.$$

Proof:

(1) Let R denote the set of all real numbers lying in the closed interval [0, 1] with usual topology. Let (X, T) be a topological space and let given a pair of disjoint closed sets A, B $\subset$ X; $\exists$ a continuous map f : X – R s.t.

$$f(A) = \{0\},\ f(B) = \{1\}.$$

To prove that (X, T) is a normal space.

Let a, b $\in$ R be arbitrary s.t. $a \leq b$

Write G = [0, a], H = [b, 1].

Then G and H are disjoint open sets in R.

Continuity of implies that $f^{-1}(G)$ and $f^{-1}(H)$ are open in X.

Let A, B $\subset$ X be disjoint closed sets.

Then our assumption says that

$$f(A) = \{0\},\ f(B) = \{1\}$$

$$f(A) = \{0\} \Rightarrow f^{-1}(\{0\}) = f^{-1}(F(A)) \supset A$$

$$\Rightarrow f^{-1}(\{0\}) \supset A \Rightarrow A \subset f^{-1}(\{0\})$$

Similarly $B \subset f^{-1}\{1\}$.

Evidently

$$\{0\} \subset [0, a] \Rightarrow f^{-1}(\{0\}) \subset f^{-1}([0, a))$$

$$\Rightarrow A \subset f^{-1}(\{0\}) \subset f^{-1}([0, a))$$

$$\Rightarrow A \subset f^{-1}([0, a)) \Rightarrow A \subset f^{-1}(G)$$

$$\{1\} \subset (b, 1] \Rightarrow B \subset f^{-1}(\{1\}) \subset f^{-1}(b, 1])$$

$$\Rightarrow B \subset f^{-1}(b, 1]) \Rightarrow B \subset f^{-1}(H)$$

$$f^{-1}(G) \cap f^{-1}(H) = f^{-1}(G \cap H) = f^{-1}(\varnothing) = \varnothing .$$

Given a pair of disjoint closed sets A, B $\subset$ X.

we are able to discover a pair of disjoint open sets.

$f^{-1}(G), f^{-1}(H) \subset X$ s.t. $A \subset f^{-1}(G), B \subset f^{-1}(H)$.

This proves that (X, T) is a normal space.

(2) *Conversely* suppose that R is real numbers lying in the interval [0, 1] with usual topology. Also suppose that A, B are disjoint closed subsets of a normal space (X, T).

To prove that $\exists$ a continuous map.

$f : (x, T) \rightarrow R$ s.t. $f(A) = \{0\}, f(B) = \{1\}$.

Step 1:

(i) Firstly we shall prove that $\exists$ a map

$f : (x, T) \rightarrow R$ s.t. $f(A) = \{0\}, f(B) = \{1\}$.

Write $T = \left\{t : t = \frac{m}{2^n}, \text{ where } m, n \in N \text{ s.t. } m \leq 2^n\right\}$

Throughout the discussion we treat $t \in T$.

Making use to the fact that m takes 2^n values for a given value of n, we have

$$\sup(T) = \sup(t) = \sup\left(\frac{m}{2^n}\right) = \frac{1}{2^n}\sup(m) = \frac{2^n}{2^n} = 1 \sup(T) = 1$$

$$t = \frac{1}{2}, 1, \qquad \text{for } n = 1$$

$$t = \frac{1}{2^2}, \frac{2}{2^2}, \frac{3}{2^2}, \frac{4}{2^2}, \qquad \text{for } n = 2$$

$$A \cap B = \varnothing \Rightarrow A \subset X - B.$$

$\because X - B$ is an open set containing a closed set A. Using the normality, we can find an open set $G \subset X$ s.t.

$$A \subset G \subset \overline{G} \subset X - B \qquad \text{...(1)}$$

Writing $G = H_{1/2}$, $X - B = H_1$, we get

$A \subset H_{1/2} \subset \overline{H}_{1/2} \subset H_1$.

This is the first stage of our on instruction.

Consider the pairs of sets

$(A, H_{1/2}), (\overline{H}_{1/2}, H_1)$

Using normality, we obtain open sets, $H_{1/4}, H_{3/4} \subset X$ s.t.

$A \subset H_{1/4} \subset \overline{H}_{1/4} \subset H_{1/2}$.

$\subset \overline{H}_{1/2} \subset H_{3/4} \subset \overline{H}_{3/4} \subset H_1$

Combining the last two relations, we have

$A \subset H_{1/4} \subset \overline{H}_{1/4} \subset H_{1/2} \subset \overline{H}_{1/2} \subset H_{3/4} \subset \overline{H}_{3/4} \subset H_1$.

This is the second stage of our construction.

If we continue this process for each dyadic rational m of the function t = m/2" where

$n = 1, 2 \ldots\ldots \;\&\; m = 1, 2, \ldots\ldots, 2^n - 1$, then

Then open sets H_i will have the following properties :

(i) $A \subset H_i \subset \overline{H}_i \subset H_1 \;\forall\; t \in T$.

(ii) Given $t_1, t_2, \in T$ s.t.

$t_1 < t_2 \Rightarrow A \subset Ht_1 \subset \overline{H}_{t1} \subset Ht_2 \subset \overline{H}_{t2} \subset H_1$.

Construct a function $f : X \to R$ s.t. $f(x) = 0 \;\forall\; x \in H_t$

and $f(x) = \{t : x \notin H_t\}$ otherwise.

In both cases $x \in X$.

$f(x) = \sup \{t : x \notin H_t\} = \sup \{t\} = \sup (T) = 1$.

Thus $f(x) = 0 \;\forall\; x \in H_t$

and $f(x) = 1 \;\forall\; x \notin H_t$ otherwise,

$f(x) = 0 \;\forall\; x \in H_t,\; A \subset H_t \;\forall\; t \in T \Rightarrow f(x) = 0 \;\forall\; x \in A$

$\Rightarrow f(A) = \{0\}$.

$f(x) = 1 \;\forall\; x \notin H_t,\; H_t \subset H_1 \;\forall\; t \in T \Rightarrow f(x) = 1 \;\forall\; x \notin H_1$

$\Rightarrow f(x) = 1 \;\forall\; x \notin X - B$

$\because X - B = H_1$

$\Rightarrow f(x) = 1 \;\forall\; x \in B$

$\Rightarrow f(B) = \{1\}$.

Thus we have shown that $f(A) = \{0\}$, $f(B) = \{1\}$.

Step 2:

(ii) Secondly we shall provethat f is continuous.

Let $a \in R$ be arbitrary, then $[0, a)$ and $(a, 1]$ are open sets in R with usual topology.

Write $G_1 = f^{-1}([0, a))$, $G_2 = f^{-1}((a, 1])$.

Then G_1, G_2 can also be expressed as

$G_1 = \{x \in X : f(x) \in [0, a)\}$

$= \{x \in X : 0 \leq f(x) < a\}$

$= \{x \in X : f(x) < a\}.$

$\because$ According to the construction of f,

$0 \leq f(x) \leq 1 \ \forall \ x \in X.$

$G_2 = \{x \in X : f(x) \in [a, 1]\}$

$= \{x \in X : a < f(x) \leq 1\}$

$= \{x \in X : a < f(x)\}.$

$= \{x \in X : f(x) > a\}.$

Finally, $G_1 = \{x \in X : f(x) < a\}$, $G_2 = \{x \in X : f(x) > a\}$

We claim $G_1 = \bigcup_{t<a} H_t$, $G_2 = \bigcup_{t>a} (\overline{H}_t)'$

any $x \in G_1 \Rightarrow f(x) < a$

$\Leftrightarrow x \in H_t$ for some $t < a$.

This proves the $G_1 = \bigcup_{t<a} H_t$

any $x \in G_2 \Rightarrow f(x) > a \Leftrightarrow x$ is out side of $\overline{H}_t$ for $t > a$

$\Leftrightarrow x \in \bigcup_{t>a} (\overline{H}_t)'$

Hence we get $G_2 = \bigcup_{t>a} (\overline{H}_t)'$

Since an arbitrary union of open sets is an open set and hence

$$\bigcup_{t<a} H_t, \ \bigcup_{t>a} (\overline{H}_t)' \subset X$$

are open *i.e.*, $G_1, G_2 \subset X$ are open, *i.e.*,

$f^{-1}([0, a)), f^{-1}((a, 1)]$ are open in X.

$\therefore$ f is continuous

Theorem 8:

Every metric space is completely normal.

Solution:

We know that regular space is completely normal.

Step 1: (i) To prove (X, d)

Step 2: (ii) To prove (X, d) is regular.

Summary of the Results

Let C = complete.

R = regular, N = Normal,

C_1 = first countable, C_2 = second countable.

We have seen that

(1) $T_5 \Rightarrow T_4 \Rightarrow T_{3\frac{1}{2}}$
$\Rightarrow T_3 \Rightarrow T_2 \Rightarrow T_1 \Rightarrow T_0$

(2) CR = R, CN
$\Rightarrow$ N, N + R $\Rightarrow$ CR

(3) Second countable $\Rightarrow$ separable.

(4) C_1, C_2, T_0, T_1, T_2, T_3,
$T_{3\frac{1}{2}}$ R, CR, CN.

are hereditary properties. Separability and normality are not hereditary.

(5) C_1, C_2, T_0, T_1, T_2, T_3,
$T_{3\frac{1}{2}}$ CR, CN,
T_4, T_5, N are all topological properties.

Theorem 9(a):

Tietze Extension Theorem : *Suppose (X, T) is a topological space. The space X is normal iff every continuous real function f defined on a closed subspace F of X into a closed interval [a, b] has a continuous extension.*

$$f : X \to [a, b].$$

Proof:

(i) Suppose (X, T) is a topological space s.t. every continuous real valued function f : F ® [a, b] has continuous extended function g : X $\to$ [a, b].

where F is a closed subset of X, [a, b] being closed interval.

Let F_1 and F_2 be two closed disjoint subsets of X.

Define a map f : $F_1 \cup F_2 \to [a, b]$

s.t. $f(x) = a$ if $x \in F_1$

and $f(x) = b$ if $x \in F_2$.

This map f is certainly continuous over the subspace $F_1 \cup F_2$.

By assumption, f can be extended to a continuous map

$$g : X \to [a, b] \text{ s.t.}$$

$$g(x) = \begin{cases} a & \text{if } x \in F_1 \\ b & \text{if } x \in F_2 \end{cases}$$

The map g satisfies Urysohn's lemma and hence (X, T) is normal.

(ii) Conversely suppose that (X, T) is a normal space.

Let $f : F \to [a, b]$ be a continuous map. F being a closed subset of X

To prove that $\exists$ a continuous extension of f over X.

For convenience, we take $a = -1$, $b = 1$.

Now we define a map $f_0 : F \to [1, 1]$ s.t

$$f_0(x) = f(x) \ \forall \ x \in F.$$

Suppose A_0 and B_0 are two subsets of F s.t.

$$A_0 = \{x : f_0(x) \le -1/3\},$$

$$B_0 = \{x : f_0(x) \ge 1/3\}$$

Then A_0 and B_0 are non-empty disjoint closed sets in F. A_0 and B_0 are closed in X.

For F is closed in X. Applying general form of Urysohn's lemma, $\exists$ a continuous function

$g_0 : X \to [-1/3, 1/3]$ s.t. $g_0(A_0) = -1/3$, $g_0(B_0) = 1/3$.

Write $f_1 = f_0 - g_0$

Then $|f_1(x)| = |(f_0 - g_0)(x)| = |f_0(x) - g_0(x)| \le 2/3$.

Let $A_1 = \{x : f_1(x) \le (-1/3\ (2/3)\}$.

$B_1 = \{x : f_1(x) \ge 1/3, 2/3\}$.

Then A_1, B_1 are non-empty disjoint closed sets in X and hence $\exists$ a continuous functions s.t.

$g_1 : X \to [-1/3.\ 2/3, 1/3.\ 2/3]$

$g_1(A) = -1/3.\ 2/3$, $g_1(B_1) = 1/3.\ 2/3$.

Again we define a function f_2 on F s.t

$f_2 = f_1 - g_1 = f_0 - g_0 - g_1 = f_0 - (g_0 + g_1)$.

Then $|f_2(x)| = |f_0(x) - (g_0 + g_1)(x)| \le (2/3)^2$.

Continuing this process, we get a sequence of function.

$$\langle f_0, f_1, f_2 \ldots\ldots f_n, \ldots\ldots \rangle$$

defined on F s.t. $\quad |f_n(x)| \le \left(\frac{2}{3}\right)^n$.

and a sequence $\langle g_0, g_1, g_2, \ldots\ldots \rangle$

defined on X s.t. $\left|g_n(x)\right| \leq \frac{1}{3} \cdot \left(\frac{2}{3}\right)^n$

and $f_n = f_0 - (g_0 + g_1 + \ldots + g(n-1))$.

Write $s_n = \sum_{r=0}^{n-1} g_r$

Now s_n can be regarded as partial sums bounded-continuous functions defined on X. Since the space of bounded continuous real valued functions is complete and

$$\left|g_n(x)\right| \leq \frac{1}{2} \cdot \left(\frac{2}{3}\right)^n$$

and $\sum_{n=0}^{\infty} \frac{1}{3}\left(\frac{2}{3}\right)^n = 1$,

the sequence $\langle s_n \rangle$ converges uniformly on X to g (say) when $|g(x)| \leq 1$.

$\left|f_n(x)\right| \leq \left(\frac{2}{3}\right)^n \Rightarrow \langle s_n \rangle$ converges uniformly on F to $f_0 = f$, say.

Hence g = f on F.

Thus g is a continuous extension of f to X which satisfies the given conditions.

Theorem 9(b):

Every metric space is a T_1-space.

Proof:

Let (X, d) be a metric space. Let x, y ∈ X be arbitrary s.t. d(x, y) = 2r. Let T be a metric topology. We know that every open sphere is T open. Then $S_{r(x)}$, $S_{r(y)}$ are open sets s.t.

$x \in S_{r(x)}\ y \notin sS_{r(x)}$

$y \in S_{r(y)}\ x \notin S_{r(y)}$

Hence (X, d) is a T_1-space.

Theorem 9(c):

Every metric space is a T_2-space.

Or. Show that if x, y are distinct points of a metric space X, then \$ disjoint pair of open spheres centered respectively at x and y.

Proof:

Let (X, d) be a metric space. Let t be the metric topology on X corresponding to the metric d.

Steps:

(i) To prove that (X, T) is a T_2-space.

Let x, y $\in$ X be arbitrary s.t. $x \neq y$.

Also let d(x, y) = 3r,

$r > 0 \in R$.

An open sphere is always an open set.

Then (i) $S_{(x,r)}$ and $S_{(y, r)}$ are open sets.

(ii) $x \in S_{(x,r)}$, $y \in S_{(y, r)}$

(iii) $S_{(x,r)} \cap S_{(y, r)} = \varnothing$.

For is $S_{(x,r)} \cap S_{(y, r)} = \varnothing$,

then $\exists$ $p \in S_{(x,r)} \cap S_{(y, r)}$ so

that $p \in S_{(x,r)}$ $p \in S_{(y, r)}$.

This $\Rightarrow d(x, p) < r, d(y, p) < r$

$\Rightarrow d(x, p) + d(p, y) < 2r < 3r$

$\Rightarrow d(x, p) + d(p, y) < 3r$

$\Rightarrow d(x, y) \leq d(x, p) + d(p, y) < 3r$

$\Rightarrow d(x, y) < 3r$.

A contradiction. Since d(x, y) = 3r. Hence the result (iii).

From (i), (ii), (iii), it immediately follows that (X, T) is a T_2-space.

Theorem 10:

Every metric space is normal space.

Proof:

Let (X, T) be a metric space and T be the metric topology on X.

To prove that (X, T) is normal. Let $A \subset X$ and $x \in X$.

Be definition, $d(x, A) = \inf \{d(x, a) : a \in A\}$

This $\Rightarrow d(x, A) = 0$ if $x \in A$ ...(1)

(i) Firstly we shall show that is A is closed, then

$d(x, A) = 0 \Leftrightarrow x \in A$.

Obviously $x \in A \Rightarrow d(x, A) = 0$.

Conversely suppose that A is closed and d(x, A) = 0.

To prove that $x \in A$.

We know that

$\overline{A} = \{x \in X : d(x, A) = 0\}$

But $\overline{A} = A$, since A is closed.

Hence $A = \{x \in X : d(x, A) = 0\}$

This says that

$x \in A \Rightarrow d(x, A) = 0$. Hence the result (i).

Let A and B be pair of disjoint closed sets in X.

Let $G_1 = \{x \in X : d(x, A) < d(x, B)\}$. ...(2)

Let $x \in G_1$ be arbitrary,

then d(x, A) < d(x, B),

Take $d(x, A) = d(x, B) - \eta$, ...(3)

for some n > 0 depending upon x.

Let $a \in A$ be arbitrary.

Let $y \in S_{(x,1/2\eta)}$ be arbitrary then $d(x, y) < \eta/2$.

$$d(y, A) \leq d(y, x) + d(x, a)$$

Taking infimum of both sides, we get

$$d(y, A) \leq d(y, x) + d(x, A). \qquad ...(4)$$

Similarly

$$d(x, B) \leq d(y, x) + d(y, B) \qquad ...(5)$$

Putting the value of d(x, A) from (3) in (4), we get

$$d(y, A) \leq d(y, x) + d(x, B) - \eta.$$

$$\leq d(x, y) + d(x, y) + d(y, B) - \eta \qquad \text{on using (5)}$$

$$= 2d(x, y) - \eta + d(y, B)$$

$$d(y, A) \leq 2d(x, y) - \eta + d(y, B)$$

$\because d(x, y) < \eta/2$, so that $2d(x - y) - \eta < 0$,

i.e., $\because d(y, A) < d(y, B)$

This $\Rightarrow y \in G_1$, according to (2).

Finally,

any y Î $S_{(x,1/2\eta)} \Rightarrow y \in G_1$.

This shows that $S_{(x,1/2\eta)} \subset G_1$.

Thus we have shown that

given any $x \in G_1$

$\Rightarrow \exists\ \eta \in R^+$ s.t. $S_{(x, 1/2\eta)} \subset G_1$.

By definition of an open set in a metric space, the last proves that G_1 is an open set.

Let $p \in A$ be arbitrary, then $d(p, A) = 0$.

$$p \in A,\ A \cap B = \phi \Rightarrow p \notin B$$
$$\Rightarrow d(p, B) > 0$$

Thus if $p \in A$, then

$d(p, A) = 0$ and $d(p, B) > 0$.

This proves that

$$d(p, B) > d(p, A)$$

This $\quad \Rightarrow p \in G_1$.

Thus any $p \in A \Rightarrow p \in G_1$.

This proves that $A \subset G_1$.

Thus we have shown that G_1 is an open set containing A.

If we write $G_2 = \{x \in X : d(x, B) < d(x, A)\}$, ...(6)

then by making parallel arguments, we can prove that G_2 is an open set containing B.

We claim $G_1 \cap G_2 = \phi$. ...(7)

Suppose not.

Then $G_1 \cap G_2 \neq \phi$. so that $\exists$ an element

$x \in G_1 \cap G_2$ which $\Rightarrow x \in G_1$, $x \in G_2$

$$x \in G_1 \Rightarrow d(x, A) < d(x, B) \qquad ...(8)$$

$$x \in G_2 \Rightarrow d(x, B) < d(x, A). \qquad ...(9)$$

(8) and (9) contradict each other.

Hence the result (7).

Given a pair of disjoint closed sets A and B in X, $\exists$ a pair of disjoint open sets G_1, G_2 in X s.t. $A \subset G_1$, $B \subset G_2$.

This proves that (X, T) is a normal space.

Theorem 11:

Every metric space is a regular space.

Proof:

Let (X, d) be a metric space and T be the corresponding metric topology on X.

To prove that (X, T) is a regular space.

Let $x \in X$ be arbitrary. Let $G \subset X$ be open s.t. $x \in G$.

By definition of an open set

$$\exists\, r \in R^{+} \text{ s.t. } S_{r(x)} \subset G$$

Evidently $\overline{S_{\frac{1}{2}r(x)}} = S_{\frac{1}{2}r[x]} \subset G.$

so that $\overline{S_{\frac{1}{2}r(x)}} \subset G.$

$S_{\frac{1}{2}r(x)}$ denotes a closed sphere of radius r/2 with its centre at x.

Thus we have shown that given a neighbourhood G of x, $\exists$ a spherical neighbourhood

$$\overline{S_{\frac{1}{2}r(x)}} \text{ of x s.t } \overline{S_{\frac{1}{2}r(x)}} \subset G.$$

By the proves that (X, T) is a regular space.

SOLVED EXAMPLES

Example 1:

Every metric space is a T_3-space.

Solution:

Let (X, d) be a metric space.

Step:

(i) To prove that (X, d) is a T_1-space.

Let $x, y \in X$ be arbitrary s.t. $x \neq y$.

Again let $d(x, y) = r$ ...(1)

Then $x \in S_{r(x)}$, $y \notin S_{r(x)}$. ...(2)

For $y \notin S_{r(x)} \Rightarrow d(x, y) < r \Rightarrow d(x, y) \geq r$

$\Rightarrow d(x, y) = r$, which is true by; virtue of (1)

Similarly $y \notin S_{r(y)}$, $x \in S_{r(y)}$ ...(3)

Also every open sphere is an open set.

In view of this, (2) and (3) taken together imply that

X is a T_1-space.

Step: (ii) To prove that (X, d) is a regular space.

For this refer step

Thus we have shown that (X, d) is a regular T_1-space and hence by definition (X, d) is a T_3-space.

Example 2(a):

Every metric space is T_4-space.

Solution:

A normal T_1-space is called T_4-space.

The (X, d) is a normal spcae. Again prove as in step (i) of theorem 27 that (X, d) is a T_1-space.

Example 2(b):

Show that an indiscrete space consisting of at least two points is not a Hausdorff space.

Solution:

Let t be an indiscrete topology on a space X consisting of at least two points. Then, be def. of T, T = (∅, X).

This proves that ∃ no pair of non-empty disjoint open sets. Consequently (X, T) is not a Hausdorff space.

Example 2(c):

Show that every discrete space is a Hausdorff space.

Solution:

Let (X, T) be a discrete space and $x, y \in X$ be arbitrary s.t. $x \neq y$.

By def. of discrete space, {x} and {y} are T-open sets.

Obviously $\{x\} \cap \{y\} = \varnothing$,

$x \in \{x\}$, $y \in \{y\}$.

Thus $ disjoint open sets {x} and {y} containing x and y respectively.

Consequently (X, T) is a Hausdorff space.

Example 3:

Prove that (R, S) is a Hausdorff space.

Solution:

Let a, b $\in$ R, be arbitrary s.t. $|a - b| = \varepsilon > 0$.

take $G = \left[a - \frac{\varepsilon}{3}, a + \frac{\varepsilon}{3}\right)$, $H = \left[b - \frac{\varepsilon}{3}, b + \frac{\varepsilon}{3}\right)$

Then a $\in$ G, b $\in$ H s.t. H $\in$ S, $G \cap H = \varnothing$.

This proves that (R, S) is a Hausdorff space.

Example 4:

If T is a cofinite topology on an infinite set X, then show that (X, T) is not a Housdorff space.

Solution:

Let G, H $\in$ T be arbitrary. Then by def. of cofinite topology, X – G and X – H are finite sub sets of X. We claim that $\exists$ no pair of disjoint open sets.

Suppose not *i.e.*, suppose G and H are disjoint open sets so that

$$G \cap H = \varnothing.$$

Taking complements of both sides, $(G \cap H)' = G' \cup H' = X$,

i.e., $\qquad (X - G) \cup (X - H) = X$,

i.e., finite union of finite sets = an infinite set. This is impossible.

Hence $\exists$ no pair of disjoint open sets.

This $\Rightarrow$ (X, T) is not a Hausdorff space.

Example 5:

If T is a cofinite topology on an infinite set X, then (X, T) is a T_1-space.

Solution:

Let T be a cofinite topology on X.

We claim that (X, T) is a T_1-space.

Let x $\in$ X be arbitrary, then {x} is a finite set so that X – {x} is t-open set, by def. of T.

Let x, y $\in$ X be arbitrary s.t. $x \neq y$.

Write G = X – {x}, H = X – {y}.

Then G, H are open subsets of X.

Also $y \in G$, $x \notin G$, $x \in H$, $y \notin H$.

By def. this proves that (X, T) is a T_1-space.

Example 6(a):

Prove that property of being a completely normal space is a topological property.

Solution:

Let f : (X, T) → (Y, U) be a homeomorphism and let (X, T) be completely normal. If we show that (Y,U) is completely regular the result will follow.

Let P, Q be separated subsets of Y so that

$$\overline{P} \cap A = \phi, P \cap A = \phi$$

Evidently $f^{-1}(P), f^{-1}(Q) \subset X$

f is continuous $\Rightarrow \overline{f^{-1}(P)} \subset f^{-1}(\overline{P})$,

and $\overline{f^{-1}(Q)} \subset f^{-1}(\overline{Q})$,

$$\therefore \quad \overline{[f^{-1}(P)]} \cap f^{-1}(\overline{Q}) \subset f^{-1}(P) \cap f^{-1}(\overline{Q})$$

$$= f^{-1}(\overline{P} \cap Q) = f^{-1}(\phi) = \phi.$$

$$\therefore \quad \overline{[f^{-1}(P)]} \cap f^{-1}(\overline{Q}) \subset \phi$$

This $\Rightarrow \overline{[f^{-1}(P)]} \cap f^{-1}(Q) = \phi$

Similarly $f^{-1}(P) \cap \overline{[f^{-1}(Q)]} = \phi$.

This proves that $f^{-1}(P)$, $f^{-1}(Q)$ are separated subsets of X. Also X is completely normal. Therefore ∃ G,H ∈ T s.t.

$$f^{-1}(P) \subset G, f^{-1}(Q) \subset H, G \cap H = \phi.$$

$\therefore f[f^{-1}(P)] \subset f(G), f[f^{-1}(Q)] \subset f(H)] \subset f(H), f(G \cap H) = \phi$

or $\quad p \subset f(G), Q \subset f(H), f(G) \cap f(H) = \phi.$

where $\quad f(G), f(H) \in U$

and P, Q are separated sets.

[For f is homeomorphism $\Rightarrow$ f is open]

This proves that (Y, U) is completely normal.

Example 6(b):

Prove that in a T_1-space all finite sets are closed.

Solution:

Let (X, T) be a T_1-space.

To prove that $\{x\}$ is closed $\forall\ x \in X$.

Now write (ii) part of the proof of the above Theorem 4.

Let A be an arbitrary finite subset of X.

Then $A = \cup \{\{x\} : x \in A\}$

$=$ finite union of closed sets $=$ closed set.

$\therefore$ A is a closed set.

Example 6(c):

If (X, T) be a topological space, then show that the following statements are equivalent :

(a) (X, T) is a T_1-space.

(b) every singleton subset of X is closed.

(c) every finite subset of X is closed.

(d) the intersection of all members (nbds) of an arbitrary point of X is a singleton.

Solution:

I. (a) $\Leftrightarrow$ (b). Proves an in Theorem 4.

II. (b) $\Leftrightarrow$ (c) This follows from the fact that finite union of closed sets is closed.

III. (a) $\Leftrightarrow$ (d)

Firstly we shall prove (a) $\Leftrightarrow$ (d)

Let $x, y \in X$ be arbitrary s.t. $x \neq y$.

Write $N = \cap \{N_x : N_x$ is a nbd of $x\}$

Evidently $x \in N$.

X is T_1-sapce $\Rightarrow \exists$ a nbd N_x of x s.t. $x \in N_x$, $y \notin N_x$

This is true for every nbd N_x of x.

$\Rightarrow \exists$ nbd N of x s.t. $x \in N$, $y \notin N$.

Since y is arbitrary and hence no point of X other than x belong to N.

Consequently $N = \{x\}$.

We now prove (d) $\Rightarrow$ (a).

We assume (d). Let $x, y \in X$ s.t.$x \neq y$.

By assumption, the intersection of all nbds of x is $\{x\}$. Hence there must be a nbd of x which does not contain y otherwise the intersection of all nbds of x would not be $\{x\}$ but $\{x, y\}$. Similarly there must be a nbd of y which does not contain x. It follows that the space is T_1-space.

Example 7:

a topological space (X, T) is a T_1-space iff T contains the cofinite topology on X.

Solution:

Let (X, T) be a T_1-space.

To prove that T contains cofinite topology on X, we have to show that T contains subsets A of X s.t. X – A is finite.

Here we shall make use of the fact that

X is T_1-space $\Rightarrow \{x\}$ is closed $\forall\ x \in X$.

$\Rightarrow X - \{x\}$ is open subset of $X \Rightarrow X - \{x\} \in T$

Thus $X - \{x\} \in T \Rightarrow X - (X - \{x\}) = \{x\}$ = finite set

This is true $\forall\ x \in X$.

Hence, by def. T contains cofinite topology on X.

Conversely suppose that T contains cofinite topology on X.

To prove that (X, T) is T_1-space.

$\{x\}$ is a finite subset of X.

Also T contains cofinite topology.

Consequently $X - \{x\} \in T$ so that

$\{x\}$ is closed $\forall\ x \in X$.

This $\Rightarrow$ (X, T) is T_1-space.

Example 8:

Each singleton set in a Hausdorff space is closed.

Solution:

Let (X, T) be a Hausdorff space so that X is a T_2-space

X is a T_2-space $\Rightarrow$ X is a T_2-space.

$\Rightarrow \{x\}$ is closed $\forall\ x \in X$.

Example 9:

Show that every finite T_1-space is a discrete space.

Solution:

Let (X, T) be a finite T_1-space.

To prove that (X, T) is a discrete space.

Let $A \subset X$ be arbitrary.

X is finite, $A \subset X \Rightarrow$ A is a finite set ..(1)

X is a T_1-space $\Rightarrow \{x\}$ is closed for each $x \in X$

$A = \cup \{\{x\} : x \in A\}$

= A finite union of closed sets = A closed set.

$\therefore$ A is closed set. ...(2)

X is finite, $A \subset X \Rightarrow$ A is a finite set

$\Rightarrow X - A$ is closed in accordance with (1) and (2).

$\Rightarrow$ A is open.

Finally, and $A \subset X \Rightarrow$ A is open as well as closed.

This proves that (X, T) is a discrete space.

Example 10:

Every finite Hausdorff space is discrete.

Solution:

Let (X, T) be a finite T_2-space.

X is a T_2-space $\Rightarrow$ X is a T_1-space.

$\Rightarrow \{x\}$ is closed subset of X $\forall\, x \in X$

Let $A = \{a_1, a_2, ..., a_n\}$ be a finite subset of X.

Then $A = \{a_1\} \cup \{a_2\} \cup \; ... \cup \{a_n\}$

= finite union of closed set = closed set.

Thus every finite subset of X is closed ...(1)

X is finite, $A \subset X \Rightarrow X - A$ is finite.

$\Rightarrow X - A$ is closed, by (1) $\Rightarrow$ A is open.

Thus every subset of X is closed as well as open. Consequently X is a discrete space.

Example 11:

Let F be a closed subset of a completely regular space (X, T) and $x_0 \in F'$, then prove that there exists a continuous map $f : X \rightarrow [0, 1]$ s.t. $f(x_0) = 1$, $f(F) = \{0\}$.

Solution:

Suppose (X, T) is a completely regular space and F a closed subset of X s.t. $x_0 \in F'$ so that $x_0 \notin F$.

By def. of completely regular space.

$\exists$ a continuous map $g : X \to [0, 1]$ s.t. $g(x_0) = 0$, $g(F) = \{1\}$.

Now we define a map $f : X \to [0, 1]$ s.t. $f(x) = 1 - g(x)$

Then $f(x_0) = 1 - g(x_0) = 1 - 0 = 1$.

$\because$ $f(x) = 1 - g(x) \ \forall \ x \in X$

$\therefore$ $f(x) = 1 - g(x) \ \forall \ x \in F = 1 - 1 = 0 \ \forall \ x \in F$

$\therefore$ $f(x) = 0 \ \forall \ x \in F$

This $\Rightarrow f(F) = \{0\}$

g is continuous $\Rightarrow$ 1– g is continuous $\Rightarrow$ f is continuous

Finally, we have a continuous map $f : X \to [0, 1]$

s.t. $f(x_0) = 1$ and $f(F) = \{0\}$.

Example 12(a):

Prove that a toplogical space (X, T) is completely regular iff for every $x \in X$ and every open set G containing x, $\exists$ a continuous map

$f : X \to [0, 1]$ s.t. $f(x) = 0$ and $f(y) = 1 \ \forall y \in G'$.

Solution:

Let (X, T) be a toplogical space and $x \in X$ be arbitrary. Let G be an open set s.t. $x \in G$ so that $x \in G'$ and G' is a closed set.

(i) Let X be completely regular.

To prove that $\exists$ a continuous map

$f : X \to [0, 1]$ s.t. $f(x) = 0$ and $f(y) = 1 \ \forall \ y \in G'$.

By def. of completely regular space,

$\exists$ continuous map $f : X \to [0, 1]$ s.t.

$f(x) = 0$ and $f(G') = \{1\}$, *i.e.*,

$f(x) = 0$ and $f(y) = 1 \ \forall \ y \in G'$.

(ii) Conversely suppose that $\exists$ a continuous map

$f : X \to [0, 1]$ s.t. $f(x) = 0$ and $f(y) = 1 \ \forall \ y \in G'$.

To prove that X is completely regular.

Let F be a closed subset of X and $x \in X$ s.t. $x \notin F$.

Then F' = X – F – F is open set with $x \in F'$. So we can take F' in place of G in the given condition so that ∃ a continuous map

$f : X \to [0, 1]$ s.t.

$f(x) = 0$ and $f(y) = 1 \; \forall \; y \in (F')$

This ⇒ X is completely regular.

Example 12(b):

Every completely normal space is normal.

Solution:

Let (X, T) be completely normal space.

To prove that (X, T) is normal.

Let A, B be disjoint closed sets, then $\overline{A} = A$, $\overline{B} = B$.

$\therefore\; A \cap B = \phi \Rightarrow \overline{A} \cap B = \phi,\; A \cap \overline{B} = \phi.$

⇒ A and B are separated sets in X.

Also X is completely normal.

⇒ ∃ disjoint open sets G, H s.t. $A \subset G$, $B \subset H$

Given a pair of disjoint closed sets, A, B, ∃ disjoint open sets G, H s.t. $A \subset G$, $B \subset H$. This proves that X is normal.

Problems related to hereditary property

Example 12(c):

Every subspace of a T_0-space is a T_0-space , that is to say, the property of a space being a T_0-space is a hereditary property.

Solution:

Let (X, T) be a T_0-space and (Y, U) a subspace of (X, T)

If we show that (Y, U) is a T_0-space, we can conclude the result.

Let x, y ∈ y be arbitrary s.t. $y \neq x$. For $Y \subset X$.

By the property of a T_0-space,

$\exists\; G \in T$ s.t. $x \in G$, $y \notin G$.

Consequently $\exists\; G \cap Y \in U$ s.t. $x \in G \cap Y$, $y \notin G \cap Y$.

This ⇒ (Y, U) is T_0-space.

Example 13:

Prove that the T_1-axiom of separation is a hereditary property.

Solution:

Let (X, T) be a T_1-space of (Y, U) a subspace of (X, T).

If we show that (Y, U), is T_1-space, we can conclude the result.

Let $x, y \in Y$ be arbitrary s.t. $x \neq y$.

Then $\quad x, y \notin X$ s.t. $x \neq y$. $\because Y \subset X$.

By the property of a T_1-space.

$\exists$ G, H $\in$ T s.t. $x \in G, y \notin G; y \in H, x \notin H$.

Consequently

$x \in G \cap Y, y \in H \cap Y, y \notin G \cap Y, x \notin H \cap Y$

Take $G \cap Y = G_1$, $H \cap Y = H_1$ and observe that

$G, H \in T \Rightarrow G \cap Y, H \cap Y \in U$

$\Rightarrow G_1, H_1 \in U$.

Given a pair of distinct points $x, y \in Y$.

$\exists\ G_1, H_1 \in U$ s.t. $x \in G_1, y \in H_1 : x \notin H_1, y \notin G_1$.

This proves that (Y, U) is a T_1-space.

Example 14:

Show that the property of a space being a Hausdorff space is a hereditary property.

Or. Every subspace of a T_2-space is a T_2-space.

Solution:

Let (X, T) be a Hausdorff space and (Y, U) be a sub-space.

If we show that (Y, U) is a T_2-space then we can conclude the result.

Let a pair of elements $y_1, y_2 \in Y$ s.t. $y_1 \neq y_2$.

Then $y_1, y_2 \in X$ s.t. $y_1 \neq y_2$. For $Y \subset X$.

By the property of a T_2-space.

$\exists$ disjoint sets $G_1, G_2 \in T$ s.t. $y_1 \in G_1, y_2 \in G_2$.

$G_1, G_2 \in T \Rightarrow \exists\ H_1, H_2 \in U$ s.t.

$H_1 = G_1 \cap Y, H_2 = G_2 \cap Y$

$H_1 \cap H_2 = (G_1 \cap Y) \cap (G_2 \cap Y) = (G_1 \cap G_2)\ y$

$= \varnothing \cap Y = \varnothing.$

$y_1 \in Y, y_1 \in G_1 \Rightarrow y_1 \in G_1 \cap y = H_1$.

$y_2 \in Y, y_2 \in G_2 \Rightarrow y_2 \in G_2 \cap y = H_2$.

Thus we have shown that

given pair of elements $y_2, y_2 \in Y$ s.t. $y_1 \neq y_2$.

$\exists$ disjoint sets $H_1, H_2 \in U$ s.t. $y_1 \in H_1, y_2 \in H_2$.

This proves that (Y, U) is a T_2-space.

Example 15:

Show that the property of a space being regular is hereditary property.

Solution:

Let (Y, U) be a sub-space of a regular space (X, T). We claim that the property of regularity is hereditary property. If we show that (Y, U) is regular, the result will follow.

Let F be a U-closed set and $p \in Y$ s.t. $p \notin F$.

Let $\overline{F}^T$ =closure of F w.r.t. the topology T.

and $\overline{F}^U$ =closure of F w.r.t. the topology U.

We know that $\overline{F}^U = \overline{F}^T \cap Y$.

Since F is a U-closed set $\Rightarrow F = \overline{F}^U \Rightarrow F = \overline{F}^T \cap Y$.

$p \notin F \Rightarrow p \notin \overline{F}^T \cap Y \Rightarrow p \notin \overline{F}^T$ or $p \notin Y$.

$\Rightarrow p \notin \overline{F}^T$ For $p \in Y$.

$\overline{F}^T$ is a T-closed set.

$\because$ closure of any set is always closed.

By the regularity of (X, T),

given a closed set $\overline{F}^T$ and a point $p \in X$.

s.t. $p \notin \overline{F}^T$; $\exists$ disjoint sets $G, H \in T$ with $p \in G, \overline{F}^T \subset H$

Consequently, $F = \overline{F}^T \cap Y \subset H \cap Y, p \in G \cap y$.

$(G \cap Y) \cap (H \cap Y) = (G \cap H) \cap (Y \cap Y) = \varnothing \cap y = \varnothing$.

Thus we have shown that

given a U-closed set F and a point $p \in Y$ s.t. $p \notin F$, we are able to find out the disjoint open sets $G \cap Y$, $H \cap Y$ in Y s.t.

$$p \in G \cap Y, F \subset H \cap Y$$

This proves that (Y, U) is regular. Hence the proposition.

Example 16(a):

A closed sub-space of a normal space is a normal space.

Solution:

Let (X, T) be a topological space which is normal and (Y, U) a closed sub-space of (X, T) so that symbol Y is closed in X.

To prove that Y is a normal space.

Let $F_1, F_2 \subset Y$ be disjoint sets which are closed in Y.

Y is closed in X, a subset F of Y is closed in Y iff F is closed in X.

$\therefore$ F_1 and F_2 are disjoint closed sets in X.

By the property of normal space (X, T).

$\exists\, G_1, G_2 \in T$ s.t. $F_1 \subset G_1, F_2 \subset G_2, G_1 \cap G_2 = \varnothing$.

$F_1 \subset G_1 \Rightarrow F_1 \cap Y \subset G_1 \cap Y$

$\Rightarrow F_1 = F_1 \cap Y \subset G_1 \cap Y \quad \because F_1 \cap Y$

$\Rightarrow F_1 \subset G_1 \cap Y.$

similarly $F_2 \subset G_2 \Rightarrow F_2 \subset G_2 \cap y$.

By definition of relative topology,

$G_1, G_2 \in T \Rightarrow Y \cap G_1, Y \cap G_2 \in U$

$(G_1 \cap Y) \cap (G_2 \cap Y) = (Y \cap Y) \cap (G_1 \cap G_2) = Y \cap \varnothing = \varnothing.$

Finally given a pair of disjoint closed sets F_1, F_2 in Y $\exists$ disjoint sets.

$G_1 \cap Y, G_2 \cap Y \in U$ s.t. $F_1 \subset G_1 \cap Y$ $F_2 \subset G_2 \cap Y$.

This proves that (Y, U) is a normal space.

Example 16(b):

Prove that the property of being a T_3-space is hereditary.

Solution:

Let (X, T) be a T_3-space so that (X, T) is regular and T_1-space.

Let (Y, U) be a subspace of (X, T).

If we show that (Y, U) is regular and T_1-space, the result will follow.

(i) To prove that (Y, U) is T_1-space.

(ii) To prove that (Y, U) is regular.

Example 17:

Prove that the property of being a T_4-space is hereditary.

Solution:

Let (X, T) be a T_4-space so that (X, T) is normal and T_1-space.

Also let (Y, U) be a subspace of (X, T).

If we show that (Y, U) is normal and T_1-space, the result will follow.

(i) To prove that (Y, U) is a T_1-space.

(ii) To prove that (Y, U) is normal.

Suppose P and Q are disjoint closed subsets of Y so that $P, Q \subset X$. For $Y \subset X$.

Let $p \in P$, $q \in Q$, then $p \neq q$. For $P \cap Q = \varnothing$.

(X, T) is T_1-space $\Rightarrow$ {p} and {q} are disjoint closed subsets of X.

Also X is normal

$\Rightarrow$ there exist T-open sets G_p, H_q s.t.

$\{p\} \subset G_p$, $\{q\} \subset H_q$, $G_p \cap H_q = \varnothing$.

This is true $\forall\ p \in P$ and $\forall\ q \in Q$

$\Rightarrow P \subset \cup \{G_p : p \in P\} = G$, $Q \subset \cup \{H_q : q \in Q\} = H$,

$\Rightarrow P \subset G$, $Q \subset H$, $G \cap H = \varnothing$

$\Rightarrow P \cap Y \subset G \cap Y$, $Q \cap Y \subset H \cap Y$,

$(G \cap Y) \cap (H \cap Y) = \varnothing$

$\Rightarrow P \subset G \cap Y$, $Q \subset H \cap Y$, $(G \cap Y) \cap (H \cap Y) = \varnothing$

Also $G \cap Y$, $H \cap Y$, $\in U$. For $G, H \in T$

$\Rightarrow$ (Y, U) is normal.

Example 18(a):

Prove that the property of being a completely normal space is a hereditary property.

Solution:

Let (X, T) be completely normal space and (Y, U) be a subspace of (X, T).

If we show that (Y, U) is ;completely normal space, the result will follow.

Let P and Q be separated subsets of Y, then

$$P \cap (U - \text{closure of } Q) = \varnothing \quad \text{...(1)}$$

$$Q \cap (U - \text{closure of } P) = \varnothing \quad \text{...(2)}$$

Let $\overline{P}$ = T-closure of P, = $\overline{Q}$ T-closure of Q.

Recall that $\overline{P}$ = (U – closure of P) $\cap$ Y

$\overline{P} \cap Q$ = (U – closure of P) $\cap Y \cap Q$

$= (U - \text{closure of } P) \cap Q$. For $Q \cap Y = Q \cap Y = Q$
$= \varnothing$, by (2)

Similarly $P \cap \overline{Q} = \varnothing$,

$\overline{P} \cap \overline{Q} = \varnothing$, $P \cap \overline{Q} = \varnothing \Rightarrow$ P and Q are separated subsets of (X, T).

Also (X, T) is completely normal

$\Rightarrow \exists\, G, H \in T$ s.t. $P \subset G$, $Q \subset H$, $G \cap H = \varnothing$

$\Rightarrow G \cap Y, H \cap Y \in U$ s.t. $P \cap Y \subset G \cap Y$,

$Q \cap Y \subset H \cap Y$, $(G \cap Y) \cap (H \cap Y) = \varnothing$

$\Rightarrow G \cap Y, H \cap Y \in U$ s.t. $P \subset G \cap Y$,

$Q \subset H \cap Y$, $(G \cap Y) \cap (H \cap Y) = \varnothing$

$\Rightarrow$ (Y, U) is completely normal space.

Example 18(b):

Prove that completely regularity is a hereditary property.

Solution:

Let (X, T) be completely regular space and (Y, U) a subspace of it. If we show that (Y, U) is completely regular space, the result will follow.

Let $y \in Y$ and F be a U-closed subset of Y s.t. $y \notin F$.

$$y \in Y,\ Y \subset X \Rightarrow y \in X.$$

F is U-closed $\Rightarrow \exists$ T-closed set F_1 s.t. $F = F_1 \cap Y$

$$y \notin F \Rightarrow y \notin F_1 \cap Y \Rightarrow y \notin F_1.$$

Now $y \in X$ and F_1 is T-closed set s.t. $y \notin F_1$.

By property of completely regularity, $\exists$ continuous map

$f : X \to [0, 1]$ s.t. $f(y) = 0$, $f[F_1] = \{1\}$.

Let g denote restriction of f to Y, then

$g : X \to [0, 1]$ is a continuous map

s.t. $g(u) = f(u)\ \forall\ u \in Y$.

This $\Rightarrow g(y) = f(y) = 0$, $y \in Y \Rightarrow g(y) = 0$

Again $g(u) = f(u) \quad \forall\ u \in F \subset Y$ ...(1)

But $f(u) = 1 \quad \forall\ u \in F_1$. Also $F \subset F_1$

$\therefore f(u) = 1 \quad \forall\ u \in F$ ...(2)

By (1) & (2), $g(u) = 1\ \forall\ u \in F$ or $g(F) = \{1\}$.

Finally, $g : Y \to [0, 1]$ is a continuous map

s.t. $g(y) = 0$, $g(F) = \{1\}$.

This proves that (Y, U) is completely regular space.

Problems related to Topological Property

Example 18(c):

To prove that completely regularity is a topological property.

Solution:

Let : $(X, Y) \to (Y, U)$ be a homeomorphism and let (X, T) be a completely regular space.

If we show that (Y, U) is a completely regular, the result will follow.

Let $y \in Y$ be arbitrary and $F \subset Y$ be U-closed set s.t. $y \notin F$. Since f is one-one onto and hence $\exists\ x \in X$ s.t. $f(x) = y$, then $x = f^{-1}(y)$.

If is continuous $\Rightarrow f^{-1}(F)$ is T-closed set.

$$y \notin F \Rightarrow f^{-1}(y) \notin f^{-1}(F) \Rightarrow x \notin f^{-1}(F)$$

Now $f^{-1}(F)$ is T-closed set s.t. $x \notin f^{-1}(F)$. By definition of completely regularity, $\exists$ continuous map $g : X \to [0, 1]$ s.t.

$$g : X \to [0, 1] = \{1\}$$

$$g(x) = 0 \text{ and } g[f^{-1}(F)] = \{1\}$$

or $$g[f^{-1}(y)] = \text{ and } (gf^{-1})(F) = \{1\}$$

$$f^{-1} : Y \to X,\ g : X \to [0, 1] \text{ both are continuous}$$

$$\Rightarrow gf^{-1} : Y \to [0, 1] \text{ is continuous.}$$

Now $$gf^{-1} : Y \to [0, 1] \text{ is continuous map s.t.}$$

$$(gf^{-1})(y) = 0 \text{ and } (gf^{-1})(F) = \{1\}$$

This $\Rightarrow$ (Y, U) is completely regular.

Example 18(d):

If f : (X, T) is a continuous map and (X, T) is a Hausdorff space, then show that

$$A = x \in X : f(x) = x\} \text{ is closed.}$$

Solution:

Suppose $f : (X, T) \to (X, T)$ is a continuous map and (X, T) is a Hausdorff space. Also suppose that

$$A = x \in X : f(x) = x\}.$$

To prove that A is closed, we have to prove that X – A is open.

$$X - A = \{x \in X : f(x) \neq x\}.$$

Let $p \in X - A$ be arbitrary, then $f(p) \neq p$,

$$p, f(p) \in X \text{ s.t. } p \neq f(p).$$

X is T_2-space.

By def. T_2-space, $\exists$ G, H $\in$ T s.t.

$$p \in G, f(p) \in H, G \cap H = \varnothing.$$

$f(p) \in H$, f is continuous $\Rightarrow p \in f^{-1}(H)$, $f^{-1}(H)$ is open.

Write $V = G \cap f^{-1}(H)$ = finite intersection of open set = open set.

$p \in G, p \in f^{-1}(H) \Rightarrow p \in G \cap f^{-1}(H) \Rightarrow p \in V$

any $y \in V \Rightarrow y \in G, y \in f^{-1}(H)$

$\Rightarrow y \in G, f(y) \in H = \varnothing.$

$\Rightarrow y \in X - A$, by construction of A,

$\therefore$ $V \subset X - A$

any $p \in X - A \Rightarrow \exists, V \in T$ s.t. $p \in V \subset X - A$.

This says that p is an interious point of X – A. But p is arbitrary. Hence every point of X – A is its interior point.

Consequently X – A is open.

Example 19:

Let (X, T) be a T_1-space and let B_p be a local base at $p \in X$. Let $q \in X$ s.t. $p \neq q$, then show that $\exists$ at least one member $B \in B_p$. s.t. $q \notin B$.

Solution:

Let (X, T) be a T_1-space and let B_p be a local base at $p \in X$. Let $q \in X$ s.t. $p \neq q$.

To prove that \$ at least one member $B \in B_p$. s.t. $q \notin B$.

By hypothesis,

$p, q \in X$ s.t. $p \neq q$ and X is a T_1-space.

$\therefore \exists$ G, H $\in$ T s.t. $P \in G, q \notin G$;

$$q \in H, p \notin H.$$

By definition of local base,

given any $G \in T$ s.t. $p \in G \Rightarrow \exists B \in B_p$ s.t. $p \in B \subset G$.

Now $B \subset G, q \notin G \Rightarrow q \notin B$.

Finally we have shown that

$\exists$ B $\in$ B_p s.t. p $\in$ B and q $\notin$ B.

Example 20:

If F_1 and F_2 are T-closed disjoint subsets of a normal space (X, T), then there exists a continuous map g of X into [0, 1] such that

$$g(x) = \begin{cases} 0 & \text{if } x \in F_1 \\ 1 & \text{if } x \in F_2 \end{cases}$$

i.e., $f(F_1) = \{0\}$ *and* $g(F_2) = \{1\}$.

Solution:

Here write the proof of step II of the previous Theorem 25.

Example 21:

if F_1 and F_2 are T-closed disjoint subsets of a normal space (X, T) and [a, b] is any closed interval on the real line, then there exists a continuous map.

f of x into [a, b] such that

$$f(x) = \begin{cases} a & \text{if } x \in F_1 \\ b & \text{if } x \in F_2 \end{cases}$$

i.e., $f(F_1) = \{a\}, f(F_2) = \{b\}$.

This problem is known as general form of Urysohn's lemma.

Solution:

Let F_1 and F_2 be disjoint closed subsets of (X, T).

To prove that $\exists$ a continuous map

$$f : X \to [a, b] \text{ s.t. } f(F_1) = \{a\}, f(F_2) = \{b\}.$$

By Urysohn's lemma, $\exists$ a continuous map

$$g : X \to [0, 1] \text{ s.t. } g(F_1) = \{0\}, g(F_2) = \{1\}.$$

Define a map h : [0, 1] $\to$ [a, b] s.t.

$$h(x) = \frac{(b-a)x}{1-0} + a$$

i.e., $h(x) = x(b - a) + a.$

[This is obtained by writing the equation of the straight line joining (0, a) and (1, b) and then putting y = h(x)].

Evidently h(0) = a, h(1) = b – a + a = b.

Also h is continuous.

Write $f = hg.$

$$g : X \to [0, 1], h : [0, 1] \to [a, b]$$

$$\Rightarrow hg : X \to [a, b] \Rightarrow f : X \to [a, b]$$

Product of continuous functions is continuous.

Therefore $f(F_1) = (hg)(F_1) = h[g(F_1)] = h(\{0\}) = \{a\}$

$$f(F_2) = (hg)(F_2) = h[g(F_2)] = h(\{1\}) = \{b\}.$$

Thus $\exists$ a continuous map

$f : X \to [a, b]$ s.t. $f(F_1) = \{a\}, f(F_2) = \{b\}.$

Example 22:

Prove that a normal space is completely regular iff it is regular.

Solution:

(i) Let (X, T) be a normal space and regular.

To prove that X is completely regular.

Let $x \in X$ be arbitrary and F a closed subset of X, s.t. $x \notin F$ so that $x \in F'$

Now F' is an open nbd of x. By property of regular space $\exists$ open set G s.t.

$x \in G \subset \overline{G} \subset F'$ (Refer theorem 24, page 294)

$\overline{G} \subset F' \Rightarrow \overline{G} \subset F = \varnothing.$

Now $\overline{G}$ and F are disjoint closed subsets of X.

Applying Urysohn's lemma, $\exists$ a continuous map

$$f : X \to [0, 1] \text{ s.t. } f(\overline{G}) = \{0\} \text{ and } f(F) = \{1\}$$

$$f(\overline{G}) = \{0\} \Rightarrow f(y) = 0 \ \forall \ y \in \overline{G}$$

But $\quad x \in \overline{G} \Rightarrow f(x) = 0$

Finally, given $x \in X$ and a closed set $F \subset X$ s.t.

$x \notin F$, $\exists$ continuous map $f : X \to [0, 1]$ s.t.

$f(x) = 0$ and $f(F) = \{1\}.$

This proves that X is completely regular.

(ii) Suppose (X, T) is completely regular space.

To prove that X is regular.

Example 23:

Prove that every T_4-space is a Tychonoff space.

Solution:

Let (X, T) be a T_4-space so that

(i) X is T_1-space. (ii) X is normal.

To prove that (X, T) is a Tychonoff space, we have to show that

(iii) x is T_1-space. (iv) X is completely regular.

Evidently (i) $\Rightarrow$ (iii)

Let $x \in X$ be arbitrary and F a closed subset of X s.t.

$x \notin F$ so that $\{x\} \cap F = \phi$.

X is T_1-space $\Rightarrow$ $\{x\}$ is closed.

Now $\{x\}$ and F are disjoint closed subsets of X.

Applying Urysohn's lemma $\exists$ a continuous map

$$f : X \to [0, 1] \text{ s.t. } f(\{x\}) = \{0\} \text{ and } f(F) = \{1\}$$

i.e., $f(x) = 0 \;\; f(F) = \{1\}$.

Finally, given $x \in X$ and a closed set $F \subset X$ s.t.

$x \notin F$, $\exists$ continuous map $f ; X \to [0, 1]$

s.t. $f(x) = 0$

and $f(F) = \{1\}$.

This says that X is completely regular.

EXERCISES

1. Prove that for any set X, there is a unique smallest topology T s.t. (x, T) is a T_1-space.
2. Define a T_1-space and a Hausdorff space. Prove that each subspace of a Hausdorff space is Hausdorff.
3. Every closed subspace of a product of closed intervals is a compact Hausdroff space. Show, conversely, that every compact Hausdorff space is homeomorphic to a closed subspace of such a product.
4. Prove the following generalization of the Tietze extension theorem. If X is a normal space, F a closed subspace of X, and f a continuous mapping of F into a completely regular space Y, then f can be extended continuously to a mapping f' of X into a compact Hausdorff space Z which contains F as a subspace.

5. We know that every metric space is normal, and also that a normal space, if second countable, is metrizable. Give an example of a normal space which is not metrizable. This shows that metrizable spaces cannot be characterized among topological spaces by the property of normality.
6. Among normal spaces, second countability implies metrizability. Give an example of a metric space which is not second countable. This shows that metrizable spaces cannot be characterized among normal spaces by the property of second countability.
7. Show that a compact Hausdorff space is metrizable $\Leftrightarrow$ it is second countable.
8. State and prove a generalization of Tietze's theorem which relates to functions whose values lie in R^n.
9. Justify the assertion in the last paragraph of the text that the function defined there cannot be extended continuously to X.
10. Compact Hausdorff space X is completely regular and that therefore $\mathcal{C}(X, R)$ separates points. Use this to prove that the weak topology generated by $\mathcal{C}(X, R)$ equals the given topology. Show further that if +S is any subset of $\mathcal{C}(X, R)$ which also separates points, then the weak topology generated by S also equals the given topology.
11. Let X be a completely regular space, and show from the definition that the weak topology generated by $\mathcal{C}(X, R)$ equals the given topology.
12. Show that the topological space defined is not a T_1-space.
13. Show that the topological space defined is a T_1-space but not a Hausdorff space.
14. Show that any finite T_1-space is discrete.
1. Show that a closed subspace of a normal space is normal.
15. Let X be a T_1-space, and show that X is normal $\Leftrightarrow$ each neighbourhood of a closed set F contains the closure of some neighbourhood of F.
16. In the text we used Urysohn's lemma as a tool to prove Tietze's theorem. Reverse this process, and deduce Urysohn's lemma from Tietze's theorem.
17. If X is completely regular, show that every bounded continuous complex function defined on X has a unique extension to a bounded continuous complex function defined on $\beta(X)$.
18. If X is a T_1-space with at least two points, show that an open base which contains X as a member remains an open base if X is dropped.

19. Let X be a topological space, Y a Hausdorff space, and A a subspace of X. Show that a continuous mapping of A into Y has at most one continuous extension to a mapping of A into Y, is a special case of this statement.
20. If f is a continuous mapping of a topological space X into a Hausdorff space Y, prove that the graph of f is a closed subset of the product X × Y.
21. Let X be any non-empty set, and prove that in the lattice of all topologies on X each chain has at most one compact Hausdorff topology as a member. (It is interesting to speculate about whether a compact Hausdorff topology can be defined on an arbitrary nonempty set.)
20. State a normal space. State and prove Urysohn's lemma.
21. (a) Define a T_0-space. Show that a T_1-space is always T_0-space.

    (b) Define a T_2-space.

    Give an example of a T_1-space which in not T_2.

    (c) In a T_1-sapce, prove that the derived set of each set is closed.
22. If X is a T_1-space with at least two points, show that an open base which contains X as a member remains an open if X is dropped.
23. Let X be a normal space and A, B be a disjoint closed subsets of X. Prove that ∃ a continuous real function f on X to [0, 1] such that f(A) = 0 and f(B) = 1.
24. (a) State and prove the *Urysohn's lemma.*

    Or

    State and prove the *Tiet Z* extension theorem.

    (b) Show that a space is completely regular T_1 iff it is homeomorphic to a subspace of a base.
25. Define a normal topological space.
26. Let X be a topological space, Y a Hausdorff space and A a subspace of X. Show that a continuous map of A into Y has almost one continuous extension to a map of $\overline{A}$ into Y.
27. Show that every; convergent sequence in a Hausdorff space has a unique limit.
28. (a) Define a normal space and show that a closed subspace of a normal space is normal.

    (b) Prove that a completely regular space is regular.

29. Define a normal topological space. Prove that a compact Hausdorff space is normal.

30. Let X be an arbitrary topological space and $\{x_n)$ a sequence of points in X. This sequence is said to be *convergent* if there exists a point x in X such that for each neighbourhood G of x a positive integer n_0 can be found with the property that x_n is in G for all n ≥ no. The point x is called a *limit* of the sequence, and we say that x_n *converges* to x (and symbolize this by $x_n \to x$).

    (a) If X is a Hausdorff space, show that every convergent sequence in X has a unique limit.

    (b) Show that if $f : X \to Y$ is a continuous mapping of one topological space into another, then $x_n \to x$ in $X \Rightarrow f(x_n) \to f(x)$ in Y. Prove that the converse of this is true if each point in X has a countable open base.

CONVERGENCE

We shall be mainly concerned with the concept of a generalized sequence called 'net'. The nation of a filter and ultrafilter will also be given in some detail. It is felt that filters are more covenient in general topology, nets in analysis. But before we take up the question of the generalization of the notion of convergence of a sequence, we define some terms concerning sequences,

Recall that a sequence in a set X is a mapping of N into X. If s is a sequence in X, the image s (n) of n ∈ N is usually denoted by s_n. It is customary to denote the sequences by the classical symbol s_{n-1}, or simply by $< s_n >$. A sequence $< s_n >$ is said to be eventually in a set A if and only if there exists a positive integer m such that $n \geq m \Rightarrow s_n \in A$. A sequence $< s_n >$ in a topological space X is said to converge to point s0 in X if and only if it is eventually in every neighbourhood of s_0 and we then say s_0 is a limit point of the sequence and write $s_n \to s_0$. A point s_0 in X is said to be a cluster point of a sequence $< s_n >$ in X if and only if $< s_n >$ is frequently in every neighbourhood of s_0, that is every neighbourhood of s_0 contains sn, for finitely many n. A sequence t in a space X is said to be a subsequence of another sequence s in X iff there exists a mapping ϕ of N into N such that $s_o\ \phi = t$ and $i < j \Rightarrow \phi\ (i) < \phi\ (j)$ for all i, j ∈ N.

Definition : *Let $\mathfrak{J}'$ and $\mathfrak{J}$ be two topologies on a set X. Then $\mathfrak{J}$ is said to be sequentially stronger than $\mathfrak{J}'$ iff sn → s0 in $\mathfrak{J} \Rightarrow$ sn → in $\mathfrak{J}$.. Also $\mathfrak{J}$ and $\mathfrak{J}'$ are said to be sequentially equivalent iff each is sequentially stronger than the other.*

Theorem 1:

Let (X, $\mathfrak{I}$) be a Lindelof space such that every sequence in X has a cluster point. Then X is compact.

Proof:

We have to show that every $\mathfrak{I}$-open cover of X has a finite subcover. Since X is Lindelof, we may assume that the open cover is countable. Let $c = \{G_n : n \in N\}$ be a countable open cover of X. We proceed by induction. Let $H_0 = G_0$ and for each n in N let H_0 be the first member of the sequence of G′s which is not contained in the union $H_0 \cup H_1 \cup ... \cup H_{n-1}$. If such a choice is impossible at any stage, then the sets already chosen will form a finite subcover of C.

In the contrary situation, it is possible to pick a point h_n in H_n for each n in N such that $h_n \notin H_m$ for m < n. We thus obtain a sequence $h < n_n : n \in N >$ satisfying the above condition. By hypothesis this sequence must have a cluster point, say x_0. Since $\{H_n : n \in N\}$ covers, $x_0 \in H_n$ for some $n \in N$. and since x_0 is a cluster point of the sequence h, there exists some i > n such that $h_i \in H_n$. But this is a contradiction. Hence c must have a finite subcover and consequently X is compact.

Theorem 2(a):

Let (X, $\mathfrak{I}$) be a first axiom space. Then a point x_0 of X in an accumulation point of a subset Y of X if and only there exists a sequence in Y – $\{x_0\}$ which converges to x_0.

Proof :

The 'if' part : Let $s = < s_n : n \in N >$ be a sequence in $Y - \{x_0\}$ which converges to x_0 let V be a $\mathfrak{I}$-nhd of x_0. Since s is eventually in V there exists $n_0 \in N$, such that for every $n \in N$, $n \geq n_0 \Rightarrow s_n \in V$. In particular, $s_{n_0} \in V$. Hence $s_{n_0} \in V \cap (Y - \{x_0\})$. This implies that every nhd of x_0 contains a point of Y distinct from x_0. Hence, x_0 is an accumulation point of Y.

The 'only if' part : Assume that x_0 is an accumulation point of a subset Y of X. Since the space is first axiom, there exists a nested local base at x_0. Let $B(x_0) = \{B_n : n \in N\}$ be this nested local base at x_0 so that $B_n + 1 \subset B_n$ for all $n \in N$. Since x_0 is an accumulation point of Y, every nhd of x_0 must intersect $Y - \{x_0\}$. Hence by the axiom of choice for each $n \in N$, we may choose a point s_n in $B_n \cap (Y - \{x_0\}$, this obtaining a sequence $s = < s_n : n \in N >$ in $Y - \{x_0\}$. We now show that s converges to x_0. Let V be any $\mathfrak{I}$-nhd of x=. Then by the definition of local base $B(x_0)$, there exists $n_0 \in N$, such that $B_{n_0} \subset V$ and since $B_{n+1} \subset B_n$, we have $B_n \subset V \vee n \geq n_0$.

But $s_0 \in B_n$. It follows that $s_n \in V \vee n \geq n_0$. This implies that the sequence s is eventually in V and consequently s converges to x_0.

Theorem 2(b):

Let (X, $\mathfrak{I}$) be a first axiom space and let $Y \subset X$. Then a point x_0 of X is a adherent point of Y. that is, $s_0 \in Y$ iff there exists a sequence in Y which converges to x_0.

Proof:

(i) This sequence converges to the point b since only $\mathfrak{I}$-nhd of b are {a, b} and X both of which contain all the terms of the sequence. Similarly it converges to c since the only $\mathfrak{I}$-nhd of c which contains all the terms of the sequence is X.

But a is not a limit point of this sequence since {a} is a $\mathfrak{I}$-nhd of a such that the sequence are b and c only.

(ii) Since b, c are limit point of this sequence they are certainly cluster points. The point a is also a cluster point of this sequence since the nhds of a are {a}, {a, b} and X each of which contains an infinite number of terms of the sequence. Hence a, b, c are the cluster points of this sequence.

(iii) Since a, b, c are limit point of this sequence, they are certainly cluster points.

Theorem 2(c):

Let $\mathfrak{I}$ and $\mathfrak{I}'$ be topologies for a set X with $\mathfrak{I}$ first axiom. Then $\mathfrak{I} \supset \mathfrak{I}'$ if and only if $\mathfrak{I}$ is sequentially stronger than $\mathfrak{I}'$. and $s_n \to s_0$ in $\mathfrak{I}$

Proof:

Suppose $\mathfrak{I} \supset \mathfrak{I}'$ and sn $\to s_0$ in $\mathfrak{I}$. Let V be a $\mathfrak{I}'$-nhd of s0. Then V is a $\mathfrak{I}$-nhd of s_0 since $\mathfrak{I} \supset \mathfrak{I}'$. Since $s_n \to s_0$ in $\mathfrak{I}$, $s_n \in V$ eventually. Hence $s_r \to s_n$ in $\mathfrak{I}'$ and so $\mathfrak{I}$ is sequentially stronger than $\mathfrak{I}'$. Conversely, let $\mathfrak{I}$ be sequentially stronger than $\mathfrak{I}'$.

It follows from (1.6) and (1.7) that $CI_{\mathfrak{I}}$ (S) $\subset CI_{\mathfrak{I}}'$ (S) for all $S \subset X$. Let $G \in \mathfrak{I}'$. Then G′. Then G′ is $\mathfrak{I}'$-closed and so $CI_{\mathfrak{I}}'(G') = G$. Hence $CI_{\mathfrak{I}}$ $(G') = G'$. But $G' \subset CI_{\mathfrak{I}}$ (G′) always. Thus $CI_{\mathfrak{I}}$ (G′) $(G') = G'$ and therefore G′ in $\mathfrak{I}$-closed, that is, $G \in \mathfrak{I}$. Hence $\mathfrak{I}' \subset \mathfrak{I}$.

Theorem 3(a):

Let (X, $\mathfrak{I}$) be a first axiom space and let x_0 be a cluster point of a sequence $s = < s_n : n \in N >$ in X. Then there exists a subsequence of s which converses to x_0.

Proof:

Since (X, $\Im$) is a first axiom space, there exists a nested local base B (x_0) = $\{B_n : n \in N\}$ at x_0 so that $B_{n+1} \subset B_n$. Since x_0 is a cluster point of the sequence s, for each B_i in B (x_0) an each n in N, there exists an m in N such that $m \geq n$ and $s_n \in B_i$.

We now define sequence ϕ in N as follows: Let ϕ (1) be the least positive integer such that s (ϕ (2)) $\in B_i$; ϕ (2) be the least positive integer greater than ϕ (1) such that s (ϕ (2)) $\in B_2$ and so on. Then s o ϕ = {s (ϕ (n)) : n $\in$ N} : n $\in$ N} is a subsequence of s.

It will now be shown that s o ϕ converges to x_0. Let V be any $\Im$-nhd of x_0. Then by the definition of local base B (x_0), there exists n_0 in N such that $B_{n_0} \subset V$ and since $B_{n+1} \subset B$. we have $B_n \subset V$ for all $n \geq n_0$. But s (ϕ (n)) $\in B_{n'}$. It follows is the $n \geq n_0$. Thus implies that the sequence s o ϕ is eventually in V and consequently s o ϕ converges to x_0.

Theorem 3(b):

Let (X, $\Im$) be a first axiom space and let $Y \subset X$. Then Y is $\Im$-open iff each sequence converging to a point of Y is eventually in Y.

Proof:

The 'if part': Let every sequence converging to a point of Y be eventually in Y and suppose , if possible, Y is not open. Then there exists a sequence s = < s_0 : n $\in$ N > in X – Y which converges to a point $y_0 \in$ Y. Such a sequence surely fails to be eventually in Y. But this is contradiction. Hence Y must be open.

The 'only if part': Let Y be open and let s be a sequence which converges to a point y_0 of Y. Then Y is a $\Im$-open nhd of y_0. Hence by the definition of convergence of sequence, s must be eventually in Y.

ULTRAFILTERS

Definition : *A filter F on a set X is said to be an ultrafilter (or a maximal filter) on X iff F is not properly contained in any other filter on X that is, iff there exists no filter on X strictly finer that F.*

Thus F is an ultrafilter on X iff for every filter F′ on X, F′ ⊃ F implies F′ = F.

It follows that an ultrafilter on X is a maximal element of the collection of all filters on X partially ordered by the inclusion relation ⊂.

Definition : *A filter base on a set X is called an ultrafilter base iff it is a base of an ultrafilter.*

Example:

The filters F_5, F_6 and F_7 defined in example (4.5) are ultrafilters on {a, b, c} since there are no filters strictly finer that these filters.

EXISTENCE OF ULTRAFILTERS

Theorem 1:

Let A be a non-empty collection of subsets of a set X such the A has the FIP. Then there exists an ultrafilter F containing A.

Proof:

Let M = {C : C is a collection of subsets of X with the FIP such that C ⊃ A}. Then M is non-empty since A ∈ M. Also M is partially ordered by the inclusion relation ⊂. Further it is easy to see that every linearly ordered subset of M has an upper bound. Therefore by Zorn's lemma there exists a maximal element F of M. We now show that F is filter on X. Since F ⊃ A and A is non-empty, it follows that F is also non-empty. Further.

[F1] : Since F has the FIP, no member of F can be empty. Hence ∅ ∅ F.

[F2] : Let t A ∈ F and B ⊃ A. Since F has the FIP, it follows that {B} ∪ F also has FIP. Also {B} ∪ F ⊃ F. Hence {B} ∪ F ∈ M. But since F maximal. B ∈ F.

Let A ∈ F and B ∈ F. Since F has thew FIP, it follows the {A ∩ B} ∩ F also has FIP. Also {B} ∪ F ⊃ F. Hence {A ∩ B} ∪ F ∈ M. But since F is maximal, A ∩ B ∈ F. Hence F is the desired ultrafilter.

Theorem 2(a):

Every filter on a set X is contained in an ultrafilter on X.

Proof:

Let F be any filter on X and let M be the class of all filters on X containing F. Then m is non-empty since F ∈ M. Also is partially ordered by the inclusion relation ⊂. Now let H be a linearly ordered subset of M. Then by definition of linear ordering for and tow members F_1, F_2 of H, we have either $F_1 \subset F_2$ or $F_2 \subset F_1$. Let $G = \cup \{F_\lambda : F_\lambda \in H\}$. We claim that G is a filter on X. G is evidently non-empty Also.

[F1] : Since each F_λ is a filter, we have $\varnothing \notin F_\lambda \vee F_\lambda \in H$ and so $\varnothing \notin G$.

[F2] : Let A ∈ G and B ⊃ A. Then $A \in F_\lambda$ for some $F_\lambda \in H$ and since F_λ is a filter. $B \in F_\lambda$. It follows that B ∈ G.

[F3] : Let A $\in$ G and B $\supset$ G. Then A $\in F_\lambda$ and B $\in F_\lambda$ for some F_λ $F_\mu \subset F_\lambda$, F_μ. $\in$ H, Since H is linearly ordered, we have tither $F_\lambda \subset F_\mu$ or $F_\mu \subset F_\lambda$. Hence both A and B belong either to F_λ or to F_μ and so A $\cap$ B belongs either to F_λ or to F_μ. It follows that A $\cap$ B $\subset$ G.

Accordingly G is a filter on X. Further G is finer than every member of H and so G is an upper bound of H. Thus we have shown that M is a non-empty partially ordered set in which every linearly ordered subset has an upper bound. Hence by Zorn's lemma M contains a maximal on X containing F.

A more general form of the above theorem is the following.

Theorem 2(b):

Evey filter base on a set X is contained in an ultrafilter on X.

Proof:

The proof of this theorem is the same as that of the above theorem except that M is new the class of all filters, on X containing a filter base instead of a filter. A still more general theorem is the following.

CHARACTERIZATION OF ULTRAFILTERS

Theorem 1:

A filter F on a set X is an ultrafilter iff F contains all those subsets of X which intersect every member of F.

Proof:

The 'if' part : Let F be a filter on X satisfying the condition and let F′ be a filter finer than F, so that F′ so that F′ so that F′ $\supset$ F. Let F′ $\in$ F′. Then by [F1] and [F3], F′ intersects every member of F′ and since F $\subset$ F′, it follows that F′ also intersects every member of F. Hence by definition of F, F′ $\in$ F. Therefore F′ $\subset$ F. Hence F′ = F. Accordingly F is an ultrafilter on X.

The 'only if' part : Let F be an ultrafilter on X and let A be a subset of X intersecting every member of F. Now consider the collection

$$F' = \{F' : F' \supset A \cap F \text{ for some } F \in F\}.$$

We claim that F′ is a filter on X containing F. Surely F′ contains F since F $\supset$ A $\cap$ F $\forall$ F $\in$ F. Moreover,

[F1] : Since A intersects every member of F, we have

$$A \cap F \neq \varnothing \vee F \in F.$$

Again since $F \supset A \cap F$ F, it follows that

$$F' \neq \varnothing \vee F' \in F' \quad .$$

[F2] : Let F' $\in$ F' and G' $\in$ F $\in$. Then F' $\supset$ A $\cap$ F for some F $\in$ F. Hence G' $\supset$ A $\cap$. It follows that G' $\in$ F'.

[F3] : Let F' $\in$ F' and G' $\in$ F'. Then F' $\supset$ A $\cap$ F and G' $\supset$ A $\cap$ G where F, G $\in$ F and so $F' \cap G' \supset (A \cap F) \cap (A \cap G)$, that is, $F' \cap G' \supset A \cap (F \cap G)$. Since F $\cap$ G $\in$ $\in$ F, we have $F' \cap G' \in F'$.

Thus it is shown that F$'$ is a filter containing F. Since F is an ultrafilter, we have F = F$'$. Now since X $\in$ F and A $\cap$ X = A so that A $\supset$ A $\cap$ X, it follows that A $\in$ F and so also A $\in$ F.

Corollary : *A filter F on an a set X is an ultrafilter on X if A or X belongs to F for all subsets A of X.*

Proof:

Since a filter has the FIP, the result follows from the above theorem.

Another statement of the above theorem.

A subset A of a filter on set X is an ultrafilter on X iff for every subset A of X we have either X $\in$ A or X – A $\in$ A.

Theorem 2:

A filter F on a set X is a ultrafilter if and only if for any two subsets A, B, of X such that A $\cup$ B $\in$ F, we have either A $\in$ F or B $\in$ F.

Proof:

Let F be an ultrafilter on X and let A $\cup$ B $\subset$ F_0. To show that either A $\in$ F or B $\in$ F. Suppose, if possible, that A $\notin$ F. There by [F1], either A or B is non-empty. Assume that B $\neq$ $\varnothing$. Now consider the collection

$$G = \{G : G \cup A \in F\}.$$

Then G is non-void since B $\in$ G. We now claim that G is filter on X strictly finer that F.

[F1] : Since $\varnothing \cup A = A \notin F$, we have $\varnothing \notin G$.

[F2] : Let G $\in$ G and H $\supset$ G. Then G $\cup$ A $\in$ F and since F is a filter H $\cup$ A $\supset$ G $\cup$ A implies H $\cup$ A $\in$ F. Hence by definition of G, H $\in$ G.

[F3] : L Remark (4.2) (ii). But X = (X – A) $\cup$ A. Hence by hypothesis ether A $\in$ A $\in$ F or X – A $\in$ F. Hence by the preceding theorem F is an ultrafilter on X.

Corollary 1 : *If F is an ulirafilter, then for every subset A of X, we have either A $\in$ F or X – A $\in$ F.*

Proof:

Since X ∈ F and A ∪ (X – A) = X, by the above theorem, we have either A ∈ F or A – A ∈ F.

Corollary 2 : *If F is an ultrafilter and $A_1 \cup A_2 \cup ... \cup A_n \in F$, then at least one A ∈ F.*

Proof:

Corollary 3 : *If F is an ultrafilter on X and $A_1, A_2, ... A_n$ are subsets of X such that $\cup (A_i : 1 < i < n)$ covers X, then some $A_i \in F$.*

Proof:

Since $\cup \{A_i : 1 < i < n\} = X \in F$, the result follows from the above theorem.

Theorem 3(a):

A filter F on X is an ultrafilter on X iff A or X – A belongs to F for all subsets A of X.

Proof:

Corollary as above.

Theorem 3(b):

A collection F of non-empty subsets of a set X is an ultrafilter on X if the following conditions are satisfied.

(i) F has the FIP

(ii) For every subset A of X, we have A ∈ F or X – A ∈ F.

Proof:

Let F′ be any filter containing F. Let F′ be any member of F′. Let F′ be any member of F′. Then X – F′ ∉ F′ therefore X – R′ ∉ F. Hence by (ii), F′ ∈ F. Then we have shown that F′ F′ ⇒ F′ ∈ F. Hence F′ ⊂ F. But F ⊂ F′. Therefore F′ = F. Accordingly F is an ultrafilter on X.

Theorem 4:

If F is an ultrafilter on a set X, then F is an ultrafilter base on every superset Y of X.

Proof:

Let F′ be the family of all those subsets A of Y such that A contains a member of F. Then F′ is a filter on Y. Now let B be any subset of Y. Since Y is a superset of X, we have Y ∩ X = X and hence

$$(X \cap B) \cup [X \cap (Y - B)] = X \cap [B \cup (Y - B)]$$
$$= X \cap Y = X \in F.$$

since F is an ultrafilter on X, we have either.

$X \cap B \in F$ or $X \cap (Y - B) \in F'$.

If $X \cap B \in F$, then $B \in F'$, then $Y - B \in F$. Thus we have shown that if B is any subset of Y, then either $B \in F'$ or $Y - B \in F$. Hence F' is an ultrafilter on Y. Hence F is an ultrafilter base on Y.

Theorem 5(a):

Every filter F on a set X is the intersection of all the ultrafilters finer than F.

Proof:

Let $H = \cap$ { : G is an ultrafilter containing F}. To show that H = F. Since $F \subset G \vee G.$, we have $F \subset H$. Conversely let $A \in H$. Then $A \in G$ $\forall$ G.

If possible, let $A \notin F$. Then A contains no member of F by [F2]. Hence every $A \in F$ intersects $X - A$. Then exists a filter F' finer than F and containing $X - A$.

There exists an ultrafilter, say G_0, finer than F. Now $X - A \in F' \Rightarrow A \notin F'$ and since $G_0 \supset F'$, we have $A \notin G_0$. But this contradicts (1). Hence $A \in F$. Thus we have shown that $A \in H \Rightarrow A \in F$. Therefore $H \subset F$. It follows that H = F.

Theorem 5(b):

Let A be a subset of X which has non-empty intersection with every member of an ultrafilter F, then A must belong to F.

Proof:

Let A = {A} $\cup$ F. Evidently, the intersection of any finite number of members of A is non-empty. Hence A generates a filter G containing all l the sets in A, and so, in particular A. But since F is a maximal filter F = G, and hence $A \in F$.

Theorem 5(c):

Let f be mapping of set X onto a set Y. If F is an ultrafilter on X, then G = {f [F] : F $\in$ F} is an ultrafilter on Y.

We have G is a filter base on Y. Let B be any subset of Y, Then we have

$$X = f^{1}[Y] = f^{1}[B \cup (Y - B)] = f^{1}[B] \cup f^{1}[Y - B].$$

Proof:

We know that Also , X $\in$ F. It follows that f^{-1} [B] $\cup$ f^{-1} [Y – B] $\in$ F. Hence f^{-1} [B] $\in$ F or f^{-1} [Y – B] $\in$ F, that is, either f^{-1} [B] = F or f^{-1} [Y – B] = H where F, H are members of F. Hence either f [F] f [f^{-1} [B]] = B or f[H] = f [f^{-1} [Y – B]] = Y – B.

Note that since f is onto, we have f [f^{-1} [Y – B]] = Y – B etc. It follows that either B $\in$ C or Y – B $\in$ C. Then C is an ultrafilter on Y

Definition : *Let $\{f_a : a \in A\}$ be a net in X. Then the filter generated by the filter base B consisting of the series*

$$B_{a_0} = \{f_a : a \geq a_0,\ a_0 \in A\}$$

is called the filter generated by $\{f_a\}$.

Definition : *Let F the filter on X, and let*

$$\Lambda_r = \{(x, F) : x \in F \in F\}.$$

Then Λ_r is directed by the relation $(x_1, F_1) \leq (x_2, F_2)$ iff $F_2 \subset F_1$ and so the map $f : \Lambda_r \Rightarrow X$ defined by f (x, F) = x is an net in X, It is called the net based on F.

Theorem 6:

A net in a set X is ultranet if and only if the filter it generates an ultrafilter.

Proof:

Let $\{F_a : a \in A\}$ be an ultranet in X. Let S $\subset$ X and let be as in. Since $\{f_a\}$ is an ultranet, there is $a_0 \in A$ such that $B_{a0} \subset S$ or $B_{a_0} \subset X - S$.

[$\because$ $\{f_a\}$ is eventually in either S or X – S]

Therefore either S $\in$ F or X – S $\in$ F where F is the filter generated by B.

Conversely. Let the filter F generated by the net $\{f_a : a \in A\}$ be an ultrafilter. Let S $\subset$ X. Then since F is an ultrafilter, either S $\in$ F or X – F $\in$ F. By definition of F, there exists $a_0 \in A$ such that $B_{a_0} \subset S$ or $B_{a_0} \subset X - S$. In either case, $\{f_a\}$ is eventually in either S or X – S. Hence $\{f_a\}$ is an ultranet in X.

CONVERGENCE OF FILTERS

Definition : *Let F be a filter on a set X and let A – X. Then F is said to be eventually in A iff A $\in$ F.*

Definition : *A filter on a set X is said to be frequently in a subset A of X iff A intersects every member of F, that is, $A \cap F \neq \varnothing \;\forall\; S \in F$.*

Remarks:

(i) F is frequently in A if F is eventually in A but non-conversely. For $A \in F \Rightarrow A \cup F = \varnothing \nprec F \in F$.

For the converse, consider the filter F = {{b, c}, X} on X = {a, b, c} and the set A = {b}, The filter F is frequently in A since A intersects every member of F but it is not eventually in A since A $\notin$ F.

(ii) F is frequently in A iff there exists a filter finer than F and containing A.

Definition : *Let (X, $\mathfrak{I}$) be a topological space and let F be a filter on X. Then F is said to $\mathfrak{I}$-converge or simply converge (or to be convergent) to point x $\in$ X iff F is eventually in each neighbourhood of x, that is, iff every nhd of x is a member of F and we say that x is a limit point (or simply a limit) of F and writer F $\to$ x.*

LIMIT POINTS OF A FILTER BASE

Definition : *A filter base B on X is said to converge to a point x $\in$ X iff the filter whose base is B (that is, the filter generated by B) converges to x and we say that x is a limit point of B.*

The set of all limit points of a filter F shall be denoted by lim (F). Likewise lim (B) will denote the set of all limit points of a filter base B.

Theorem 1:

Let $\mathfrak{I}$ and $\mathfrak{I}'$ be topological on a set X. Then $\mathfrak{I}$ is finer than $\mathfrak{I}'$ if and only if F $\to$ x in $\mathfrak{I}$ implies F $\to$ x in $\mathfrak{I}'$ for all filters F in X. Hence $\mathfrak{I}$ = $\mathfrak{I}'$ iff $\mathfrak{I}$ and $\mathfrak{I}'$ have the same convergent filters and corresponding limits.

Proof:

Let $\mathfrak{I}$ be finer than $\mathfrak{I}'$ and let F $\to$ x in $\mathfrak{I}$. Then

$$N'(x) \subset N(x) \subset F.$$

where N′ (x) N (x), denote the nhd filters in $\mathfrak{I}'$ and $\mathfrak{I}$ respectively. Hence F $\to$ x in $\mathfrak{I}'$.

Conversely, suppose that F $\to$ x in $\mathfrak{I}$ implies $\mathfrak{I}$ $\to$ x in $\mathfrak{I}'$ and let S $\subset$ X. If x $\in$ $CI_{\mathfrak{I}}$ (S), then there is a filter base B in S such that B $\to$ x in $\mathfrak{I}$.

Then B $\to$ x in $\mathfrak{I}'$ by hypothesis and so x $\in$ $CI_{\mathfrak{I}'}$ (S). It follows that $CI_{\mathfrak{I}}$ (S) $\subset$ $CI_{\mathfrak{I}}$, (S).

It will now be shown that $\mathfrak{I}' \subset \mathfrak{I}$. Let G $\in$ $\mathfrak{I}'$. Then X – G is $\mathfrak{I}$-closed and hence $CI_{\mathfrak{I}}$ (X – G) $\subset$ $CI_{\mathfrak{I}}$, (X – G) = X – G so that X – G that is $\mathfrak{I}$-closed, that is, G $\in$ $\mathfrak{I}$. Hence $\mathfrak{I}' \subset \mathfrak{I}$ *i.e.* $\mathfrak{I}$ is finer than $\mathfrak{I}'$.

Theorem 2(a):

Let $(X, \mathfrak{I})$ be topological space and let F be a filter on X. Then the following statements are equivalent.

(a) : For converges to a point $x \in X$.

(b) : Each neighbourhood of x belongs to F.

(c) : F is finer than the neighbourhood filter N (x) on X.

(d) : For every nhd N of x, there is $F \in F$ such that $F \subset N$.

Proof:

(a) ⇔ (b) follows immediately. (b) ⇔ (c) is an immediate consequence the definition N (x) and , (b) ⇔ (d) follows from [F*2].

Theorem 2(b):

A filter F on a topological space X converges to a point $x \in X$ every ultrafilter containing F converges to x.

Proof:

If F converges to x, then evidently every ultrafilter containing also converges to x.

Conversely, let every ultrafilter containing x converge to x. Now F is intersection of all ultrafilters′ on X finer than F and so F converges to x by above example.

Theorem 3:

Let $(X, \mathfrak{I})$ be a topological space and let B be a filter base on X. Then B converges to a point $x \in X$ iff every member of the local base B (x) at x contains a member of B.*

Proof:

Let B converge to x. The filter F generated by B also converges to x, every member of B* (x), belongs to F and consequently as shown above, every member of B* (x) must contain a member of B.

Conversely's let every member of B* (x) contain a member of B and let F be the filter generated by B. Then by definition F consists of all those subsets of X which contain a member of B. Hence by [F2] every member of B* (x) belongs to F. Now let N be any nhd of x. Then by the definition of local base, there exists a member of B of B* (x) such $B \subset N$. Hence, $N \in F$ by [F2]. It follows that every nhd of x belongs to F and consequently F converges to x. F converges to x.

Corollary : *B converges to x iff every nhd of x contains a member of B.*

Proof:

Let B converges to x and let N be nhd of x. Then there exists B ∈ B (x) such that B ⊂ N. Hence by the above theorem N must contain a member of B. Conversely, if every nhd of x contains a member of B. Then surely every member of B* (x) must contain a member of B. Hence, by the above theorem, B converges to x.

Theorem 4(a):

Let (X, ℑ) be a topological space and let A ⊂ X. Then a point x in X belongs to the closure A of A iff there exists a filter base on A converging to x.

Proof:

The 'if part': Let B be a filter base on A which converges to x. Then by the preceding theorem for each neighbourhood N of x, there exists a member B of B such that B ⊂ N. Since B is non-empty subset of A, it follows that N contains at least one point of A. Hence x is an adherent point of A and so x ∈ $\overline{A}$.

The 'only if' part : Let x ∈ A. Then every member of the nhd filter N (x) must intersect A. Now consider the collection B = {A ∩ N : N ∈ N (x)]. Then B satisfies [F*1] since by hypothesis

$$A \cap N \neq \varnothing \vee N \in N (x).$$

It also satisfies [F*2] since if A ∩ N and A ∩ M are any two members of B, then (A ∩ N) ∩ (A ∩ M) = A ∩ (N ∩ M) ∈ N. Hence B is a filter base On A. The filter generated by B is evidently the neighbourhood filter N (x). Since N (x) converges to x, the result follows.

Theorem 4(b):

Let A be a subset of a topological space X and let x ∈ X. Then x is an accumulation point of A iff there exists a filter base on A – {x} which converges to x.

Proof:

Same as above theorem.

Theorem 4(c):

Let (X, ℑ) be a topological space and let A ⊂ X. Then A is ℑ-open iff A belongs to every filter which converges to a point of A.

Proof:

Let A be a $\Im$-open set and let F be a filter which converges to a point $x \in A$ [Note that there exists at least one such filter, namely, the nhd filter of x]. Then F contains every nhd of x. Since A is $\Im$-open, it is a nhd of x. Hence $A \in B$.

Conversely, let A belong to every filter which converges to a point A. To show that A is open. Let x be an arbitrary point of A. Then the nhd filter N (x) converges to x. By hypothesis, $A \in N(x)$. Hence A is a $\Im$-nhd of x. It follows that A is $\Im$-open.

Theorem 4(d):

A topological space (X, $\Im$) is Hausdorff iff every convergent filter in X has a unique limit.

Proof:

Let X be Hausdorff and let F be filter on X converging to two distinct points x and y. Then $F \subset N(x)$ and $F \subset N(y)$ where N (x) and N (y) are the nhd filters of x and y respectively. Since the space X is Hausdorff, there exists nhds N of x and M of y such that $N \cap M = \varnothing$. Then $M \in F$ and $N \in F$ and so by [F3], $N \cap M = \varnothing \in F$ and consequently F has a unique limit. [Note that since F is convergent it must have a limit point].

Conversely, let every convergent filter in X have a unique limit and suppose, if possible the space X is not Hausdorff. Then there exists $x, y \in X$ with $x \neq y$ such that

$$N \cap M \neq \varnothing \vee N \in N(x) \text{ and } \vee M \in N(y)$$

Now $N(x) \cap N(y)$ is also a filter on X, and hence it is also a filter base on X. Hence $N(x) \cap N(y)$ generates a filter F which is finer than both N (x) and N (y). Now N (x) converges to x and N (y) converges to y and consequently F converges to both x and y this is a contradiction. Hence (X, $\Im$) is Hausdorff.

Theorem 5:

Let (X, $\Im$) and (Y, V) be two topological spaces and let $x \in X$. Let f be a mapping of X into Y. Then the following statements are equivalent.

(a) : f is continuous at x.

(b) : For every filter base B on X converging to x, the filter base f [B] converges to f (x).

(c) : For every ultrafilter base B on X converging to x the ultrafilter base f [by] on Y converges to f (x).

Proof:

First not that f (B) = {f [B] : B ∈ B}:

(a) ⇔ (b) : Let f be continuous at x and let N be any nhd of f (x). Then f^{-1} [N] is a nhd of x. Let B be any filter base on X converging to x. Then there is a member B of B such that B ⊂ f^{-1} [N]. It follows that f [B] ⊂ f [f^{-1} [N]] ⊂ N. Thus we have shown that every nhd of f (x) contains a member of f (B). Hence f (B) converges to f (x).

Conversely, let (b) hold and let N be a nhd of f (x). Now the nhd filter N (x) converges to x and so by hypothesis, f [N (x)] converges to f (x). Hence there exists a member f [M] of f [N (x)] such that f [M] ⊂ N. Now M ∈ N (x) so that M is a nhd of x. Thus, we have shown that to each nhd N of f (x), there exists a nhd M of x such that f [M] ⊂ N. It follows that f is continuous at x.

(a) ⇔ (c) : Since every ultrafilter base is a filter base. (a) ⇒ (c) follows from (a) ⇒ (b) We now prove (c) ⇒ (a). Suppose, if possible, that f such that f^{-1} [N] is not a nhd of x. Then f^{-1} [N] cannot contain any nhd filter N (x) of x.

Since N (x) has the FIP, N (x) ∪ (X – f^{-1} [N]) has also the FIP. Hence, there exists an ultrafilter F containing N (x) ∪ (X – f^{-1} [N]). Since N (x) converges to x, F surely converges to x.

Now consider the member f {X – f^{-1} [N] of f (F). Evidently, we have N ∩ f [X – F N]] = ∅. Thus, there is a member f [X – f^{-1} [N]] of f (F) which does not intersect the nhd N of f (x). Hence f [F] does not converges to f (x). But this contradicts the hypothesis. It follows that f is continuous at x.

Theorem 6(a):

A filter F on a non-empty product space X = × {X_λ : λ ∈ L) converges to a point x = {x_λ : λ ∈ Λ} of X if and only if the filter π_λ [F] = {π_λ [F] Λ F ∈ F} for every λ ∈ Λ, converges to x_λ = π_λ (x) in X_λ.

Proof:

Let F converge to x ∈ X. Since π_λ is continuous for every λ ∈ Λ, it follows that the filter π_λ [F] converges to π_λ (x) [Note that every filter is a filter base.

Conversely, if F is a filter on X such that filter π_λ [F] converges to a point x_λ in X_λ for every λ, then let x denote the point of X, whose l^{th} coordinate is x. Let N be any nhd of x in X. Then there exists a base member G = × {G_λ λ ∈ Λ} where G_λ is open in X_λ and G_λ = X_λ for all but a finite number of l's, say G_λ = X_λ for λ ≠ λ_1, λ_2, ... λ_n, such that x ∈ G ⊂ N. Now π_{λ_i} [G]

$= G_{\lambda_i}$, which is a nhd of x_{λ_i}, i = 1, 2, ..., n.. By hypothesis, π_{λ_i} [F] converges to x_{λ_i}, π_{λ_i} [F] must contain G_{λ_i} for i = 1, 2, ..., n. Hence $\pi_{\lambda_i}^{-1}$ [G_{λ_i}] contains a member of F. Then by definition of filter, $\pi_{\lambda_i}^{-1}$ [G_{λ_i}] ∈ F for i = 1, 2, ..., n, and therefore $\bigcap_{i=1}^{\pi} \pi_{\lambda_i}^{-1} [G_{\lambda_i}] \in F$. But $\bigcap_{i=1}^{\pi} n_{\lambda_i}^{-1} [G_{\lambda_i}] = G$ and so G ∈ F. Since N ⊃ G, N ∈ F by definition of filter. Since N was any nhd of l in Y, F converges to x.

Theorem 6(b):

If {F_λ : λ ∈ Λ} ισ any collection of closed subsets of a topological space X, then ∩ {F_λ : λ ∈ Λ} is a closed set.

Proof:

Fl is closed " l ∈ L

⇒ F′l is open " l ∈ L

⇒ ∪ {F′l : l ∈ L} is open

[By [T3] of the definition of topology]

⇒ [∩ {Gl : l ∈ L}]′ is open [De-Morgan Law]

⇒ ∩ [Fl : l ∈ L} is closed [By definition of closed set]

Theorem 6(c):

If F1 and F2 be two closed subsets of a topological space X. Then F1 ∪ F2 is a closed set.

Proof:

F1, F2 are closed

⇒ F′1, F′2 are open

⇒ F1′ ∩ F′2 are open

[By **[T2]** of the definition of topology]

⇒ [F1 ∪ G2]′ is open [De Morgan Law]

⇒ F1 ∪ F2 is closed.

Note that if F1, F2,..., Fn be a finite number of closed subsets of X, their union will also be a closed subset of X. This follows at once by the repeated application of the above theorem.

Theorem 7:

A subset of a topological spaces is open if and only if it is a neighbourhood of each of its points.

Proof :

Let a subset G of a topological space be open. Then for every x ∈ G ⊂ G and therefore G is a nhd of each of its points. Conversely let G be a nhd of each of its points, If G = ∅, then it is open.

If G ≠ ∅, then to each x G there exits an open set G_x such that x ∈ G_x ⊂ G. It follows that G = ∪ {G_x : x ∈ G}. Hence G is open, being a union of open sets.

Remark : The above theorem shows that the open sets in a topological space are known whenever neighbourhoods of x are known for every x. This means that two topologies on the same set are identical iff they admit the same neighbourhoods.

For suppose that $\Im$ and $\Im'$ are two topologies for X having the same system of nhds of points in X. Then

G is $\Im$-open ⇔ G contains a $\Im$-nhd of each of its points

⇔ G contains a $\Im'$ nhd of each of its points

⇔ G is $\Im'$ -open.

Theorem 8:

Let X be a non-empty set, and for each x ∈ X, let **N** *(x) be a non-empty collection of subset of X satisfying the following conditions.*

[M1] : N ∈ N (x) ⇒ x ∈ N

[M2] : N ∈ N (x), M ∈ N (x) ⇒ N ∩ M ∈ N (x).

Let $\Im$ consist of the empty set and all those non-empty subset G of X having the property that x ∈ G implies that there exists an N ∈ **N** (x) such that x ∈ N ⊂ G. Then $\Im$ is a topology for X.

Proof:

[T1] : ∅ ∈ $\Im$ be definition. We now show that X ∈ $\Im$. Let x be any arbitrary element of X. Since **N** (x) is non-empty, there is an N ∈ **N** (x) and so x ∈ N by [M1] . Since N is a subset of X, we have x ∈ N ⊂ X. Hence X ∈ $\Im$.

[T2] : Let G_1 ∈ $\Im$ and G_2 ∈ $\Im$

If x ∈ $G_1 \cap G_2$, then x ∈ G_1 and x ∈ G_2. Since G_1 ∈ $\Im$ and G_2 ∈ $\Im$, there exists N ∈ N (x) by [M2]. Hence $G_1 \cap G_2$ ∈ $\Im$.

[T3] : Let G_λ ∈ $\Im$ ∨ λ ∈ Λ. If x ∈ ∪ {G_λ : λ ∈ Λ}, then x ∈ G_{λ_x} for some λ_x ∈ L. Since G_{λ_x} ∈ $\Im$, there exists an N ∈ **N** (x), such that x ∈ N ⊂ G_{λ_x} and consequently

$$x \in N \subset \cap \{G_\lambda : \lambda : \Lambda\}$$

Hence $\cup \{G_\lambda : \lambda : \Lambda\} \in \mathfrak{I}$. It follows that $\mathfrak{I}$ is topology for X.

CLUSTER POINTS OF A FILTER

Definitions : *(i) Let (X, $\mathfrak{I}$) be a topological space and let F be a filter on X. Then a point x $\in$ X is said to be cluster point (or an adherent point) of F iff F is frequently in each neighbourhood of x.*

(ii) *A point x in a topological space X is said to be a cluster point (or an adherent point) of a filter base B on X iff B is frequently in each nhd of x.*

The set of adherent point of F is called the adherence of F and is denote by adh (F).

Theorem 1:

Let B be a filter base on a topological space X. Then adh (B) = $\cap$ {B : B $\in$ B} and thus adh (B) is closed in X.

Proof:

x $\in$ adh (B) $\Leftrightarrow$ x is a cluster point of B.

$\Leftrightarrow$ each nhd of d intersects each member of B.

$\Leftrightarrow$ x $\in$ $\bar{B}$ $\forall$ B $\in$ B.

$\Leftrightarrow$ x $\in$ $\cap$ {$\bar{B}$: B $\in$ B}

Hence adh (B) = $\cap$ {$\bar{B}$: B $\in$ B}. Since the intersection of closed sets is closed, it follows that adh (B) is closed in X.

Theorem 2(a):

Let (X, $\mathfrak{I}$) be a topological space, Then the following statements are equivalent.

(a) : X is a compact space.

(b) : Every collection of closed subsets of X with the FIP has a non-empty intersection..

(c) : Every filter on X has a cluster point,

(d) : Every ultrafilter converges.

Proof:

(a) $\Rightarrow$ (b) : See above theorem.

(b) $\Rightarrow$ (c) : Assume (b) and let F be any filter on X. Then F has the FIP. It follows that the collection $\{\overline{F} : F \in F\}$ also has the FIP.

Therefore by (b). There exists at least one point p $\in \cap \{\overline{F} : F \in F\}$. This implies that p $\in \overline{F}$ for every F $\in$ F.

Hence p is a cluster point of F.

(c) $\Rightarrow$ (d) : Assume (c), that is assume that every filter on X has a cluster point. Then surely every ultrafilter on X has a cluster point which is also a limit point of the filter. Hence every ultrafilter on X converges.

(d) $\Rightarrow$ (a) : Assume that every ultrafilter on X converges. To show that X is compact. Let C be an open cover of X and suppose, if possible, C has finite subcover. The collection A = {X – G : G $\in$ C} has the FIP.

For suppose there is a finite subcollection $\{X - F_i : 1 \leq i \leq n\}$ of A such that $\cap \{X - G_i : 1 \leq i \leq n\} = \varnothing$ Then $X - \cup \{Gi : 1 \leq i \leq n\} = \varnothing$ by De-Morgan Law. This implies that $\cup \{G_i : 1 \leq i \leq n\} = X$ which contradicts our supposition that C has no finite subcover. Thus A must have the FIP. Therefore there exists an ultrafilter F on X containing A. By (d), F converges to some point p $\in$ X. Then p is also a cluster point of F.

Hence p $\in \overline{F}$ for every F $\in$ F. In particular,

$p \in X - G = X - G \; \forall \; G \in C.$ [$\because$ X – G is $\mathfrak{I}$-closed].

This implies $p \in \cap \{X - G : C\}$

$= X - U \{G \in C\}.$

Hence $p \notin \cup \{G : G \in C\}$ But this contradicts the fact that C is a cover of X.

Hence C must have a finite subcover and consequently X is compact.

Theorem 2(b):

Let (X, $\mathfrak{I}$) be a topological space and let F be a filter on X. Then the following statements are equivalent :

(a) : x $\in$ X is a cluster point of F.

(b) : Every neighbourhood of x intersects every member of F.

(c) : x $\in$ F for all F $\in$ F where $\overline{F}$ denotes the $\mathfrak{I}$-closure of F.

(d) : N (x) $\cup$ F has the FIP where N (x) is the nhd filter of x.

(e) : There exists a filter finer than both N (x) and F.

(f) : There exists a filter finer than F and converging to x.

Proof:

(a) $\Leftrightarrow$ (b) : This follows from definition.

(b) $\Leftrightarrow$ (c) : This is an immediate consequence of the properties of closure $\overline{F}$

(b) $\Leftrightarrow$ (d) : This follows from the fact that

(i) both N (x) and F have the FIP.

(ii) finite intersections of F belong to F and

(iii) finite intersections of N (x) belong to N (x).

(d) $\Leftrightarrow$ (e) : By theorem (5.1), there exists a filter on X containing. N (x) $\cup$ F and consequently containing both N (x) and F iff N (x) $\cup$ F has the FIP. Hence (d) $\Leftrightarrow$ (e).

(e) $\Leftrightarrow$ (f) : Let G be the filter which is finer than both F and N (x). Then G is the filter which converges to x since it contains N (x). Conversely let G be the filter which is finer than F and which converges to x. Then G must contain N (x) by definition.

Theorem 3(a):

Let (X, $\mathfrak{J}$) be a topological space and let B be a filter base on X. Then a point x $\in$ X is a cluster point of B iff every member of the local base B (x) at intersects every member of B.*

Proof:

x is cluster point of B.

$\Leftrightarrow$ every of x intersects every member of B.

$\Leftrightarrow$ every member of B* (x) intersects every member of B.

Theorem 3(b):

Let F be an ultrafilter on topological space X and let x $\in$ X. Then F converges to x iff x is a cluster point of B.

Proof:

Let F converges to x. Then every nhd of x belongs to F and so every nhd of x intersects every member of F. It follows that x is a cluster point of F.

Conversely, let x be a cluster point of F. There exists a filter F′ finer then F and converging to x. Since F is an ultrafilter, we have F, F′. It follows that F converges to x.

Theorem 4: (Tychonoff)

A non-empty product space

$X = \times \{X_\lambda\ \lambda \in \Lambda\}$ is compact iff each coordinate space Xl is compact.

Proof :

If X is compact, then since each X_λ is the continuous image of X under the projection π_λ for every $\lambda \in \Lambda$, it follows that each X_λ is compact. Conversely, let X_λ be compact for every $\lambda \in$. Let F be an ultrafilter on X, $\pi_\lambda [F] = \{\pi_\lambda \{F\} : F \in F\}$ is an ultrafilter on X_λ for every $\lambda \in \Lambda$ and hence $\pi_\lambda [F]$ converges to some point of X_λ.

Then F converges.

Hence $X = \times \{X_\lambda : \in \Lambda\}$ is compact

NETS

1. Directed Set
2. Binary Relation Ex.
3. Residual Subset
4. Cofinal Subset.

Directed Sets

Definition : *Let A be a non-empty set. We say that a binary relation $\geq$ on A directs the set A iff*

[D1] : $a \in A \Rightarrow a \geq a$.

[D2] : $a \geq b$ and $b \geq c \Rightarrow a \geq c$ (a, b, c $\in$ A).

[D3] : *Given any two members a and b of A, there exists a member $c \in A$ such that $c \geq a$ and $c \geq b$.*

We say that a follows b in order $\geq$ and that b precedes a iff $a \geq b$.

Definition : *Let $(f, X, A, \geq)$ be net and let $Y \subset X$. Then f is said to be in Y iff $f[A] \subset Y$, that is, iff $f_a \in Y \ \forall\ a \in A$, f is said to be eventually in Y iff there exists a residual subset B of A such that $f[B] \subset Y$, that in iff there exists an element $a_0 \in A$ such that for every $a \in A$, $a \geq a_0 \Rightarrow f_a \in Y$, is said to be frequently in Y iff there exists a confinal subset D of A such that $f[D] \subset Y$, that is , iff for each a in A, there exists $b \geq a$ in A such that $fa \in Y$.*

It is evident that a net f is frequently in Y iff f is not eventually in X – Y.

CONVERGENCE OF A NET IN A TOPOLOGICAL SPACE

Definition : *Let $(X, \mathfrak{J})$ be a topological space and let $(f, X, \geq)$ be a net, in X. We say that f converges to a point $x0 \in X$ relative to $\mathfrak{J}$ iff f is* *eventually in every $\mathfrak{I}$-neighbourhood of x0 that is, iff for each $f_a \geq N$ of x_0, there exists an element $a_0 \in A$ such that $a \in A$, $a \geq a_0 \Rightarrow f_a \in N$.*

Theorem 1(a):

Let g be a mapping of space X into a space Y. Then g is continuous at $x_0 \in X$ iff whenever a net $\{a : a \in A\}$ converges to x_0 in X, then the net $\{g(f_a) : a \in A\}$ converges to $g(x_0)$ in X.

Proof:

Suppose g is continuous at x_0 and $f_a \to x_0$. To show that $g(f_0) \to g(x)$. Let M be any nhd of $g(x_0)$. Since g is continuous at x_0, $g^{-1}[M]$ is a nhd of x_0. Since $f_a \to x_0$ there exists $a_0 \in A$ such that $a \geq a_0 \Rightarrow f_a \in g^{-1}[M]$. Hence $g(f_a) \to g(x_0)$.

Conversely, let $fa \to x_0 \Rightarrow g(x_0)$ and suppose, if possible, g is not continuous at x_0. Then for some nhd M of $g(x_0)$, $g(N) \not\subset M$ for any nhd of x_0. Thus, for each nhd N of x_0, we can choose $x_N \in N$ such that $g((x_N) \notin M$.

Then $\{x_N : N \in N(x_0)\}$

is a net in X since the collection N (x), of all nhds of x_0 is a directed set by the inclusion relation $\subset$ [see (2.3)]. It is easy to see that $xN \to x_0$. For if N_0 if any nhd of x_0 then for every $N \geq N_0$ (*i.e.*, $N \subset N_0$) in $N(x_0)$, we have $x_N \in N \subset N_0$. But $\{g(x_N) : N \in N(x_0)$ is a net in Y which does not converge to $g(x_0)$. For M is a nhd of $g(x_0)$ such that

$$g(x_N) \notin M \text{ for every } N \in N(x_0).$$

Theorem 1(b):

A net $\{f_a : a \in \mathfrak{I}\}$ in a product space

$$X = \times \{X_\lambda : \lambda \in L\}$$

converges to $x \in X$ iff for every $\lambda \in L$, the net $\{\pi_\lambda(f_a) : a \in A\}$ converges to $\pi_\lambda(x)$ in X_λ.

Proof:

If $f_a \to x$ in X, then since π_λ is continuous,

$$\pi_\lambda(f_a) \to \pi_\lambda(x)$$

by the previous theorem, for each $\lambda \in \Lambda$.

Conversely, let $\pi_\lambda(f_a) \to \pi_\lambda(x)$ for every $\lambda \in \Lambda$. Let

$$G = \pi_\lambda^{-1}[G_{\lambda_1}] \cap \ldots \cap \pi_{\lambda_n}^{-1}[G_{\lambda_n}]$$

be any basic nhd of $x \in X$. Then G_{λ_1} is nhd of π_{λ_λ} (x) in X_{λ_i} for i = 1, 2, ... n.

Since π_{λ_i} i (f_a) → π_{λ_i} (x), for each i there exists $a_i \in A$ such that $a \geq a_i$ implies π_{λ_i} (f_a) → π_{λ_i}. Thus if a_0 be picked greater than all a_i i = 1, 2, ... n, we have

$$\pi_{\lambda_i} (f_a) \in G_{\lambda_i}$$

for all $a \geq a_0$. It follows that for $a \geq a_0$.

$$f_a \in f_a \in \bigcup_{i=1}^{n} \pi_{\lambda_i}^{-1} [G_{\lambda_i}] = G.$$

and therefore $f_a \to x$ in X.

Theorem 2:

Let (X, $\mathfrak{I}$) be a topological space and let $Y \subset X$. Then Y is $\mathfrak{I}$-open iff no net X – Y can converges to a point in Y.

Proof:

The 'if' part : Assume that no net in X – Y converges to a point in Y and suppose, if possible, Y in not $\mathfrak{I}$-open. Then there exists a point x_0. In Y such that every nhd of x_0 intersects X – Y. Let N (x_0) be the collection of all $\mathfrak{I}$ nhds of x_0. Then N (x_0) is directed by the inclusions relation ⊂. Since $N \cap (X - Y) = \varnothing \;\forall\; N \in N(x_0)$, by the action of choice, for each nhd $N \in N(x_0)$, we may choose a point x (N) in $N \cap (X - Y)$. Now consider the mapping

$$f : N(x_0) \to X - Y.$$

such that $\quad f(N) = x(N) \;\forall\; n \in n(x).$

Then evidently f is a net in X – Y converging to x_0. To see that f converges to x_0., let V be any $\mathfrak{I}$-nhd of x_0. Then for each $W \in N(x_0)$ such that $W \geq V$ that is $W \subset V$), we have

$$f(W) = x(W) \in W \cap (X - Y) \subset V.$$

Thus there exists a net in X – Y converging to x0. But this is contradiction. Hence Y must be open.

The 'only if' part : Assume that Y is open, and suppose, if possible, that there exists a net (f, X – Y, A, ≥) in X – Y converging to a point $x_0 \in Y$.

Since Y is a $\mathfrak{I}$-nhd of x_0., by the definition of convergence f must be eventually in Y. This means that f can never be in X – Y. But this is a contradiction. Hence no net in X – Y can converge to a point in Y.

Corollary : 1: *A subset A of a topological space is closed if and only if no net in A converges to a point in X – A.*

Proof:

Let Y = X – A. Then we have to show that X is open iff intent in X – Y converges to a point in Y. Now proceed as in the above theorem.

Corollary : 2. *Let (x, $\mathfrak{I}$) be a first axiom space and let $Y \subset X$. Then Y is $\mathfrak{I}$-open iff no sequence in X – Y converges to a point in Y.*

Proof:

The proof of the 'only if' part is the same as that of the above theorem if we replace the word 'net' by the word 'sequence'. The proof of the 'only if' part follows from first countability of X and the 'if part from the above theorem.

Theorem 3(a):

Let (X, $\mathfrak{I}$) be topological space and let $Y \subset X$. If x_0 is a point of X, then $x_0 \in \overline{Y}$ if and only if there exists a net in Y converging to x_0.

Proof : The 'if part' : Let (f, Y, A, ≥) be a net in Y converging to $x_0 \in X$, and let N be any neighbourhood of x_0. Since f is eventually in N, there exists $a_0 \in A$ such for every $a \in A$, $a \geq a0 \Rightarrow f_a \in N$. Now $a_0 \geq a0$.

Therefore $f_{a_0} \in N \cap Y$ so that $N \cap Y \neq \varnothing$. Thus every nhd of x_0 contains a point of Y. It follows that x_0 is an adherent point of Y and consequently $x_0 \in Y$.

The 'only if' part : Let $x_0 \in \overline{Y}$. Then every nhd of x_0 must intersect Y. Let N (x_0) be the collection of all $\mathfrak{I}$-nhd of x_0. Then N (x_0) is directed by the inclusion relation $\subset$. Since

$$N \cap Y \neq \phi \ \forall N \in N(x_0)$$

by the axiom of choice for each N in N (x_0), we may choose a point x (N) in $N \cap Y$. Now consider the mapping

$$f : N(x_0) \to Y : f(N) = x(N) \ \forall N \in N(x_0).$$

Then evidently f is a net in Y converging x_0. To see that f converges to x_0, let V be any nhd of x_0. Then for each W in N (x_0) such that $W \geq V$ (that is, $W \subset V$), we have

$$f(W) = x(W) \in W \cap Y \subset V.$$

Theorem 3(b):

Let (X, $\mathfrak{I}$) be a topological space and let $Y \subset X$. Then a point x_0 of X is an accumulation point of Y if an only if there exists a net in Y – $\{x_0\}$ converging to x_0.

Proof:

The 'if' part : Let (f, Y – $\{x_0\}$, A, ≥) be a net in Y – $\{x_0\}$ converging to x_0 and let N be a $\mathfrak{I}$-nhd of x_0. Since f is eventually in N, there exists a_0 ≥ A such that a ∈ A, a ≥ a_0 ⇒ f_a ∈ N. Now a_0 ≥ a_0. Therefore

$$f_{a_0}\ N \cap (Y - \{x_0\}).$$

This implies that every nhd of x_0 contains a point of Y distinct from x_0. Hence x_0 is an accumulation point of Y.

The 'only if' part : Let x_0 be an accumulation point of Y. Then every nhd of x_0 must intersect Y – $\{x_0\}$. Let N $\{x_0\}$ be collection of all $\mathfrak{I}$-nhds of x_0. Then N (x_0) is directed by the inclusion relation ⊂. Since for each N ∈ N (x_0),

$$N \cap (Y - \{x_0\}) \neq \varnothing$$

we may choose a point x (N) in N ∩ (Y – $\{x_0\}$). Now consider the mapping,

$$f : N(x_0) \to Y - \{x_0\} : f(N) = x(N) \ \forall\ N \in N(x_0).$$

Then f is a net in Y – $\{x_0\}$ converging to x_0 (why ?)

Theorem 3(c):

Let (X, $\mathfrak{I}$) be a topological space and let Y ⊂ X. Then Y is $\mathfrak{I}$-closed if and only if no net in Y converges to a point X – Y.

Proof:

The 'if part' : Assume that no net in Y converges to a point in X–Y and suppose, if possible, Y is not $\mathfrak{I}$-closed. Then there exists an accumulation point x_n of Y such that x_0 ∉ Y, that is, x_0 ∈ X – Y. Then by the preceding to x_0 ∈ X – Y. But this is a contradiction.

The 'only if' part : Let Y be $\mathfrak{I}$-closed and suppose, if possible, there exists a net f in Y converging to a point x_0 ∈ X – Y. Since x_0 ∉ Y, f is also a net in Y – $\{x_0\}$ converging to x_0. Hence by the preceding theorem, x_0 is an accumulation point of Y, since x_0 ∉ Y, Y is not $\mathfrak{I}$-closed. But this is a contradiction. Hence no net in Y can converge to a point in X – Y.

The 'if' part : Assume that every net in X can converges at most one point and suppose, if possible, that the space X is not Hausdorff. Then there exists two distinct points x and Y in X such that every nhd of x intersects every nhd of y. Let N (x) and N (y) be the collection of all $\mathfrak{I}$-nhds of x and y respectively. Then both N (y) and N (y) are directed by the inclusion relation ⊂. Consider the Cartesian product P = N (x) × N (y). In this product we define a relation ≥ as follows. Let (N_1, M_1) iff $N_1 \subset N_2$ and $M_1 \subset M_2$. Then it is easy to see that P is directed by ≥. Since every nhd of x intersects every nhd of y, we have N $\cap$ M = $\varnothing$ $\vee$ (N, M) $\in$ P. Hence by the axiom of choice we may choose a point x (N, M) in N $\cap$ M for all (N, M) $\in$ P.

Now consider the mapping.

$$f : P \to X : f((N, M)) = x(N, M) \;\forall\; (N, M) \in P.$$

Then f is a net in X converging to both x and y. To see this, let U be any nhd of a and let V be any b nhd of y. Then for each (N, M) in P such that (N, M) $\geq$ (U, V) so that N $\subset$ U and M $\subset$ V, we have

$$f((N, M)) = x(N, M) \in N \cap M \subset U \cap V.$$

Hence f (N, M)) $\in$ U and f ((N, M) $\in$ V. It follows that f is eventually both in U and V and consequently f converges so both x and y. But this is a contradiction. Hence X must be a Hausdorff space.

The only 'if part' : Let X be a Hausdorff space and let x, y be two distinct points of X. Then there exists a $\mathfrak{I}$-nhd N of x and a $\mathfrak{I}$-nhd M of y such that N $\cap$ M = $\varnothing$. Since a net cannot be eventually in each of two disjoint sects, it is evident that no net in X can converge to both x and y.

Hence a net in X can now or get to at most one point.

Theorem 4:

Let (X, $\mathfrak{I}$) be a Hausdorff topological space. Then every convergent net has a unique cluster point and this is the unique limit point of the net.

Proof:

We know that in Hausdorff space, every convergent net has a unique limit point. Let p be the unique limit point of a convergent net f in X. Since every limit point is also a cluster point , see ξ_3]. Suppose, if possible, f has another cluster point q distinct from p. Since X is a T_2-space there exist disjoint nhd N_p and N_q of p and q respectively. Since p is the limit of the net f, f is eventually in N_p and since N_4 is disjoint from N_p f cannot be frequently in N_q. But this contradicts our supposition that q is a cluster point of f. Hence f cannot have two distinct cluster points. Accordingly, every convergent net in X has unique cluster point.

ULTRANET

Definition : *A net $\{f_a : a \in A\}$ in a set X is said to be an ultranet (or universal net) iff for each subset S of X, the net is eventually is S or eventually S'.*

It follows from this definition that if an ultranet is frequently in S, then it is eventually in S. In particular, an ultranet in a topological space must converge to each of its cluster points.

Example:

For any direct set A, the map $f : A \to X$, defined by $f(a) = x$ for all $a \in A$ is an ultranet on X, called the trivial ultranet.

Theorem:

Let $\{f_a : a \in A\}$ be an ultranet in X and g a mapping of X into a set Y. Then $\{g(f_a) : a \in A\}$ is an ultranet in Y.

Proof:

Let S be any subset of Y. Then it is easy to see that $g^{-1}[S] = X - g^{-1}[Y - S]$. Since $\{f_a\}$ is an ultranet in X and $g^{-1}[S]$ is a subset of X with $g^{-1}[Y - S]$ as is complement, it follows that {fa} is eventually in either $f^1[S]$ or $f^{-1}[Y - S]$, from which it follows that $\{g(f_a)\}$ is eventually in either S or Y – S. Hence $\{g(f_a)\}$ is an ultranet in Y.

Definition : *Let $(f, X, A, \geq)$ and $(g, X, B, \geq)$ be two nets. Then g is said to be a subnet of f iff there exists a mapping*

$$\phi : B \to A \text{ such that}$$

(i) $g = f \circ f$ and

(ii) for each a in A, there exists an element b in B such that $f(x) \geq a$ for every $x \geq^* b$ in B.

ISOTONE MAP

Definition : *Let $(A \geq)$ and $(B, \geq^*)$ be two directed sets. Then a mapping $y : A \to B$ is said to be isotone iff*

$$x \geq y \Rightarrow \psi(x) \geq^* \psi(y) \; (x, y \in A).$$

Theorem 1:

Let y be an isotone map of directed sets $(B, \geq^*)$ into a directed set $(A, \geq)$ such that $\psi[a]$ is confinal in A. Let $(f, X, A, \geq)$ be a net. Then $f\, 0\, \psi$ is a subset of f.

Proof:

Let $g = f\, 0\, \psi$. Since ψ is a mapping of B into A and f is a mapping of A into X, it if follows that $f\, 0\, \psi = g$ is a mapping of B into X. Hence, g is also a net in X satisfying the condition (i). Again since ψ is an isotone map of B into A, we have

$$x \geq^* y \Rightarrow \psi(x) \geq \psi(y), \; (x, y \in B).$$

Further since $\psi[B]$ is confinal in A, for every $a \in A$, there exists an element $b \in B$ such that

ψ (b) $\geq$ a

From (1) and (2), we see that for every a in A, there exists an element b in B such that ψ (x) $\geq$ a for every x $\geq^*$ b in B. Hence condition (ii) is also satisfied. It follows that f 0 ψ is a subnet of f.

Corollary : *Let (A, $\geq$) be a directed set and B be a confinal subset of A so that B is also directed by $\geq$. Let (f, X, A, $\geq$) be a net. Then the restriction map of f to B is a subnet of f.*

Proof:

Here the identity map

$$I : B \rightarrow A : I(x) = x$$

is an isotone mapping such that [B] = B is confinal in A. Hence by the above theorem f o I is a subnet of f.

Since I is a mapping of B into A and j is a mapping of A into X it follows that f 0 I is a mapping of B into X such that (f 0 I) (x) = f (i (x)) = f (x) for all x $\in$ B. Hence f p I is the restriction map of f to B. It follows that the restriction map of f to B is a subnet of f.

Theorem 2:

Let (X, $\mathfrak{I}$) be a topological space and let (f, X, A, $\geq$) be a net in X. Let c be the collection of subsets of X satisfying the following two conditions.

(i) f is frequently in each member of c.

(ii) It S, T are any two members of c., then there exists a member U of c such that

$$U \subset S \cap T.$$

Then there exists a subset of f which is eventually in each member of c.

Proof:

It is evident from (iii) that c is directed by the inclusion relation $\subset$. Now consider the subset B of the cartesian product A $\times$ C defined by.

$$B = \{(a, U), a \in A, U \in c \text{ and } f(a) \in U\}.$$

We define a binary relation >> in B as follows ;

Let (a, S) and (b, T) be any tow members of B. Then

(a, S) >> (a. T) iff a $\geq$ b and S $\subset$ T.

We now show that B is directed by >>. Hence [D1] and [D2] are obvious. To prove [D3], let (a, S) and (b, T) be two members of B. Then by (iii) there exists U in c such that U $\subset$ S $\cap$ T so that U $\subset$ S and U $\subset$ T. Again by (i) and by the definition of $\geq$, there exists an element c in A, such that $c \geq \geq$ a and $\geq$ b and f (x) $\in$ U. Then (c, U) $\in$ B and such that (c, U) >>

(a, S) and (c, U) >> (B, T). Define a mapping.

$$y : B \rightarrow A : y((a, S)) = S \vee (a, S) \in B).$$

Since (a, S) >> (b, T) $\Rightarrow$ a $\geq$ b, it follows that

Hence ψ is an isotone mapping. Further ψ [B] is cofinal in A. [For let a be any element of A. Then a $\geq$ a. But ψ ((a, S)) = a for all S $\in$ c. Hence y ((A, S)) $\geq$ a. Thus of every element a $\in$ A, there exists an element ψ ((a, S)) of y [B] such that ϕ ((a, S)) $\geq$ a]. It follows by the preceding theorem that the mapping

$$\phi = f\, 0 \vee : B \rightarrow X$$

is a subnet of f.

We now show that this subnet in eventually in each member of c. Let U_0 be an arbitrary member of c. By (i), there exists a member a_0 of A, such that f $(a_0) \in U_0$. Hence by definition of B, $b_0 = (c_0, U_0)$ is a member of B. Now let b = (a, U) be an arbitrary element of B such that b >> b_0. Then a $\geq a_0$ and U $\subset U_0$. We then have

$$\phi(b) = (f\, 0\, \psi)(b) = f(\psi(b))$$

$$= f(\psi(a, U)) = f(a) \in U \subset U_0.$$

Thus, there exists an element $b_0 \in$ B, such that for every b $\in$ B. b >> $b_0 \Rightarrow$ f (b) $\in U_0$. Hence f is eventually in U0.

CLUSTER POINTS OF A NET

Definition : *Let (X, $\mathfrak{I}$) be a topological space and let f be a net in X. Then a point $x_0 \in X$ is called a cluster point of the net f iff f is frequently in ever $\mathfrak{I}$-nhd of x_0.*

Theorem 1:

A topological space (X, $\mathfrak{I}$) is compact if and only if each net in X has a cluster point.

Proof:

The 'if' part : Let every net in X have a cluster point and let F be a collection of $\mathfrak{I}$-closed subsets of X with FIP. Let C = {D : D is the intersection of a finite subcollection of {F}. Since the intersection of every two members of C is a member of C, it is evident that C is directed by the inclusion relation $\subset$. Since each D is non-empty, by the axiom of choice we may choose a point x (D) in D. Now consider the mapping.

$$f : C \rightarrow X : f(D) = x(D) \ \forall\ C.$$

Then f is net in X . By hypothesis, f must have a cluster point, say x_0. Let E be an arbitrary member of C. Then for every $D \geq E$ (*i.e.* $D \subset E$) in c. We have $f(D) = x(D) \in D \subset E$. Hence f is eventually in the closed set. E. It follows that is, $x_0 \in E$. Since E was chosen arbitrary,

$$x_0 \in \cap C \subset \cap F \text{ since } F \subset C.$$

Hence $\cap F \neq \emptyset$ and consequently X is compact.

The 'only if' part : Let $(X, \mathfrak{I})$ be compact and let $(f, Y, A, \geq)$ be a net in X. For each a in A, let $M_a = \{f(x) : x \geq a \text{ in } A\}$. Since A is directed by $\geq$, the collection $\{M_a : a \in A\}$ has the FIP. Hence $\{\overline{M_a} : a \in A\} = \neq \emptyset$. Let $p \in \cap \{\overline{M_a} : a \in A\}$. Then $p \in \{\overline{M_a}\ \forall a \in A$. Hence by the proceeding theorem p is cluster point of f.

Corollary : *A topological space $(X, \mathfrak{I})$ is compact iff each net in X has a subnet which converges to some point in X.*

Proof:

It is an immediate consequence of the above theorem.

Theorem 2(a):

Let $(X, \mathfrak{I})$ be a topological space. A point x_0 in X is a cluster point of a net $(f, X, A, \geq)$ iff there exists a subset $(G, X, B, \geq)$ which converges to x_0.

Proof:

The 'if' part : Assume that f has a subset g which converges to x_0. To prove that x_0 is a cluster point of f. Let N be a $\mathfrak{I}$-nhd of x_0 of x_0 and let a_0 by any element of A. Since g is a subnet of f, there exists a mapping f : B $\rightarrow$ A such that

(i) $g = o\ f\ \varphi$ and

(ii) for each a in A, there exists an element b in B such that $f(x) \geq a$ for every $x \geq^*$ be in B.

Hence be (ii). corresponding to $a_0 \in A$, there exists an element $b_0 \in B$ such that

$$\varphi(x) \geq a_0 \text{ for every } x \geq^* \text{by}.$$

Proof:

The 'if' part : We suppose that for every real valued continuous mapping f of a closed subset F of X into [a, b], there exists a continuous extension of f over X. To show that $(X, \mathfrak{I})$ in normal. Let F_1, F_2 be two closed subsets of X such that $F_1 \cap F_2 = \varnothing$ and let [a, b] be any closed interval. We define a mapping.

$$f : F_1 \cup F_2 \to [a, b],$$

such that f (x) = a if $x \in F_1$ and f (x) = b if $\in F_2$. This mapping is certainly continuous over the subspace $F_1 \cup F_2$. For if H be any closed subset of [a,b]

$$\text{then, } f^{-1}(H) = \begin{cases} F_1 \text{ if } a \in \text{H and } b \notin H \\ F_2 \text{ if } b \in H \text{ and } a \notin H \\ F_1 \cup F_2 \text{ if } a \in H \text{ and } b \in H \\ \varnothing \text{ if } a \notin H \text{ and } b \notin H. \end{cases}$$

Now $(F_1 \cup F_2) \cap F_1 = F_1$ and $(F_1 \cup F_2) \cap F_2 = F_2$ so that F_1 and F_2 are closed if $F_1 \cup F_2$. Of course $F_1 \cup F_2$ and $\varnothing$ are closed in $F_1 \cup F_2$. It follows that f^{-1} [H] in closed in $F_1 \cup F_2$. Hence f is a continuous map over the subspace $F_1 \cup F_2$. Therefore ,y hypothesis f can by extended to a continuous map g over X. This means that there exists a continuous map.

$$g = X \to [a, b]$$

such that g (x) = a if $x \in F_1$ and g (x) = b if $x \in F$. The mapping g now satisfies the condition sated in Urysohn's Lemma and hence $(X, \Im)$ is normal.

The 'only if' part. Let $(X, \Im)$ be a normal space and let f be a real; valued continuous map of the closed subset of the closed subset F into closed interval [a, b] which for numerical convenience, we take to be [–1, 1]. To show that there exists a continuous extension of f over X. We begin by defining a map

$$f_0 : F \to [-1, 1],$$

by setting $f_0(x) = f(x) \vee x \in F$.

Let $\quad G_0 = f_0^{-1}\ G_0 = f_0^{-1}\left[\left[-1, -\frac{1}{3}\right]\right], H_0 = f_0^{-1}\left[\left[\frac{1}{3}, 1\right]\right].$

Since $\left[-1, -\frac{1}{3}\right]$ and $\left[\frac{1}{3}, 1\right]$ are closed in [– 1, 1] and f_0 in continuous, it follows that G_0 and H_0 are closed in F and so also closed in X. Furthermore

$$G_0 \cap H_0 = \left(f_0^{-1}\left[\left[-1, -\frac{1}{3}\right]\right]\right) \cap \left(f_0^{-1}\left[\left[\frac{1}{3}, 1\right]\right]\right)$$

$$= \left[\left(f_0^{-1}\left[\left[-1, -\frac{1}{2}\right]\right]\right) \cap \left[\frac{1}{3}, 1\right]\right] = f_0^{-1}[\varnothing] = \varnothing.$$

Since g converges to x_0, there exists an element $p \geq b_0$ in B such that g (p) $\in$ N. Now let φ (p). Then $q \in A$ and $q \geq a_0$. Also f (q) = f (φ (p))

= (f (φ) (p) ∈ N. Thus, we have shown that for each element a_0 in A, there exists an element q≥ a_0 in A such that f (q) ∈ N. Hence f is frequently in N. It follows that x_0 is a cluster point of f.

The 'only if' part. Let x_0 be a cluster of a net (f, X, A, ≥) and let N (x_0) be the collection of all ℑ-nhds of x_0. If L, M are any two members of (N (x_0), then L ∩ M is also a member of N (x_0). Also since x_0 is a cluster point of f, f is frequently in each member of N (x_0). Hence by the preceding theorem there exists a subnet of g of f which is eventually in each member of N (x_0). This implies that g converges to x.

Theorem 2(b):

Let (X, ℑ) be a topological space and let Y be a ℑ-closed subset of X. If x_0 is a cluster point of a net (f, X, A, ≥) which is eventually in Y, then x_0 ∈ Y.

Proof:

By the previous theorem, there exists a subnet g of f which converges to x_0. Since f is eventually in Y. g may be considered as a net in Y which converges to x_0. Hence, x_0 ∈ Y. Since Y closed, we have Y = Y. Hence x_0∈Y. ▪

Theorem 3:

Let (X., ℑ) be topological space and let (f, X, A, ≥) be a net n X. For each a in A, let M_a = {f (x) : x ≥ a in A}. Then a point P of X is a cluster point of f iff p ∈ M_a for all a ∈ A..

Proof:

The 'if' part : Let P ∈ $\overline{M_a}$ ∀ a ∈ A and suppose , if possible, p is not a cluster point of f. Then there exists a nhd N of p and an element a in A such that f (x) ∉ N for every x ≥ a in A. This implies that N ∩ M_a = ∅. It follows that p ∉ $\overline{M_a}$ for this a. But this is a contradiction. Hence p is a cluster point of f.

The 'only if' part : Let p be a cluster point of f and let N be an arbitrary nhd of p. Then f is frequently in N, that is, for each a ∈ A., there exists x ≥ a in A such that f (x) ∈ N. Hence M_a ∩ N ≠ ∅ for every a ∈ A. Thus each nhd of p intersects M_a for all a ∈ A. It follows that p ∈ $\overline{M_a}$ for every a in A.

FILTERS

Definition : *Let X be any non-empty set. A filter on X is a non-empty family F of subsets of X satisfying the following axioms.*

[F1] : $\varnothing \in$ F.

[F2] : *If F $\in$ F and H $\supset$ F, then H $\in$ F.*

[F3] : *If F $\in$ F and H $\in$ F, then F $\cap$ H = F.*

Remarks:

(i) The axiom [F1] states that no member of F is empty :

(ii) Since F is non-empty, it contains at least one subset, say F, of X and since X $\supset$ X, It follows from [F2] that X $\in$ F.

(iii) The axiom [F3] implies that if F_1, F_2, ... F_n are members of F, then $F_1 \cap F_2 \cap ... \cap F_n$ is also a member of F. Then by $[F_1]$, $F_1 \cap F_2 ... \cap F_n \neq \varnothing$. It follows that F has FIP.

(iv) Since X is the intersection of the empty subfamily of F, it follows from (ii) and (iii) that F is closed for finite intersections.

(v) We observe from [F1] that the power set } (X) is not a filter on X. Thus any filter on X must be a proper subset of P (X).

(vi) If F $\in$ F then X – F $\notin$ F. For if both F and X – F belong to F, then by [F3], F $\supset$ (X, – F) = $\varnothing \in$ F which contradicts [F1].

Definition : *A filter F on X is said to be free iff $\cap$ {F : F $\in$ F} = $\varnothing$ and fixed iff $\cap$ {F : F $\in$ F} $\neq \varnothing$.*

COMPARISON OF FILTERS

Definition : *Let F and F ' be two filters on the same set X, Then F is said to be finer than F' or F' is coarser that F iff F' $\subset$ F. If in addition, F $\neq$F', then F is said to be strictly finer than F' or F' strictly coarser than F.*

Two filters F and F′ are said to be comparable iff F is finer than F′ or F′ is finer than F. The set of all filters on X is directed by relation $\leq$ defined by setting F $\leq$ F′ iff F is coarser than F′.

Example:

Consider the filters defined as in example (4.5). Here F_1 is coarser than all the filters on X. F_5 is finer than F_1, F_2 and F_4 ; F_6 is finer than F_1, F_2 and F_3 ; F_7 is finer than F_2 and F_4, F_6 and F_7 are not comparable etc.

Theorem 1:

Let M = $\{F_\lambda : \lambda \in \Lambda\}$ be a non-empty class of filters on a non-empty set X. Then M has a supermum if and only if for all finite sub-families $\{F_\lambda : \lambda \leq i \leq n\}$ of elements of M and all $F_i \in F_i$ $(1 \leq i \leq n\}$ the intersection

$\cap$ {Fi : 1 $\leq$ i $\leq$ n} is non-empty.

Proof:

The 'if' part : Let the conditions hold and let

$$F = \cup \{F_\lambda : \lambda \in L\}.$$

Then by the conditions of the theorem, it is evident that F, has the FIP. Hence there exists the coarsest filter F′ on X containing F, that is, F′ is the coarsest filter on X which is finer than every member of M. Hence F′ is a supermum of M.

The 'only if' part : Suppose that M has a supermum, say F. The F is the F is the coarsest of all filters on X which are finer then every member of M, that is, F is the coarsest of all filters on X such that $F \cup \{F_\lambda : \lambda \in L\}$. Hence, $\cup \{F_\lambda : \lambda \in L\}$ must have FIP and so the conditions of the theorem hold.

Theorem 2(a):

Let $\{F_\lambda : \lambda \in \Lambda$ be any non-empty family of filters on a non-empty set X. Then the set $F = \cap \{F_\lambda : \lambda \in \Lambda\}$ is also a filter on X.

Proof:

Since X is a member of each F_λ, X is a member of F. Hence F is non-empty. Further.

[F1] : Since $\varnothing \in F_\lambda \ \forall\ l \in L$, it follows that $\varnothing \notin F$.

[F2] : Let $F \in F$ and $H \supset F$. Now $F \in F \Rightarrow F \in F_\lambda \ \forall\ \lambda \in \Lambda$. Since each F_λ is a filter and $F \in F_\lambda$ it follows that $H \in F_\lambda \ \forall\ \lambda \in L$. Hence $H \in F$.

[F3] : Let $F \in F$ and $H \in F$. Then $F \in F_1$ and $H \in F_\lambda \ \forall\ \lambda \in \Lambda$. Since F_λ is a filter on X for all $\lambda \in L$, it follows that $E \cap H \in F_\lambda \ \forall\ \lambda \in L$. Hence, $F \cap H \in F$.

Corollary : *The filter F obtained above is the coarsest filter which contains A, that is, F is the smallest element of the ordered family of all filters on X containing A.*

Proof:

For, if F′ is any filter containing A, then F′ must contain all finite intersection of members of A and their supersets and hence $F' \supset F$.

Definition : *The filter F defined in the above theorem is said to be generated by A and A is said to be a sub-base of F.*

Observe that for A to be a sub-base it is necessary and sufficient that A has the FIP.

Theorem 2(b):

Let F be a filter on a non-empty set X and let $A \subset X$. Then there exists a filter F' finer than F such that $A \in F'$ if and only if $A \cap F \neq \varnothing$ for every $F \in F$.

Proof:

The 'if' part : Let $A \cap F \neq F \neq \varnothing \ \forall \ F \in F$. ...(1)

Let $A = \{A \cap F : F \in F\}$.

Then A has then FIP as shown below :

Let $\{A \cap Fi : 1 \leq : \leq n\}$ be any finite sub-family of A and let $F = \cap \{F_i : 1 \leq i \leq n\}$. Since each F_i is a member of F, F is also a member of F. Hence.

$$\cap \{A \cap F_i : 1 \leq i \leq n\}$$
$$= A \cap [\cap \{F_i : i \leq i \leq n\}]$$
$$= A \cap F \neq \varnothing \text{ by (1)}$$

Then by the preceding theorem, there exists a filter F′ on X containing A. We now show that $F \subset F'$ contains A, it follows that $A \cap F$ is also member of F′. Since $F \supset A \cap F$ and F′ is a filter, it follows that $F \subset F'$. Thus we have shown that

$$F \in F \Rightarrow F \in F'$$

Hence $F \subset F'$, that is, F′ is finer than F.

The 'only if' part : Let F′ be a filter on X such that $A \in F'$ and $F \subset F'$. To prove that $A \cap F \neq \varnothing$ for every $F \in F'$. Let F be an arbitrary member of F. Since $F \subset F'$, we have $F \in F'$. Also $A \in F'$. Since F′ is a filter, we must have $A \cap F \neq \varnothing$.

FILTER BASE

Definition : *Let X be any non-empty set. A filter base on X is a non-empty family B of subsets of X satisfying the following axioms.*

[F*1] : $\varnothing \notin B$

[F*2] : If $F \in B$ and $H \in B$, then there exists a $G \in B$ such that $G \subset F \cap H$.

Theorem 1:

Let A be any non-void family of subsets of a set X. Then there exists a filter base on X containing A iff A has the FIP.

Proof:

The 'if' part : Let A have the FIP and let B = {B : B is the intersection of a finite subfamily of A}. We shall show that B is a filter base on X. Evidently B ⊃ A. Since A is non-void, B is so also.

[F*1] : Since A has the FIP, no member of B can be empty. Hence ∅ ∉ B.

[F*2] : Let F ∈ B and H ∈ B. Since F and H are finite intersections of members of A, F ∩ H must also be a finite intersection of members of A. Hence F ∩ H ∈ B. Since F ∩ H ⊂ F ∩ H. [F*2] is satisfied.

The 'only if' part : Let B be a filter base on X containing A. The by [F*1] no member of B is empty. The intersection of every finite subfamily of B contains a member of B. In particular, the intersection of every finite subfamily of A contains a member of B. Hence A has the FIP.

The filter base B defined in the above theorem is said to be generated by A.

Definition : *Let B be a filter base on a set X. Then the filter consisting of all those subsets of X which contain a member of B is called a filter generated by B.*

[Note that such a filter always exists in view]

To filter bases B_1 and B_2 on a set X are said to be equivalent iff they generate the same filter on X.

Definition : *Let F be a filter on a set X. Then a sub-family B of F is called a base of F iff every member of F contains a member of B.*

Remark : *A collection B of subsets of a set X is a filter base on X iff B is a base of some filter on X. Or equivalently, a collection B of subsets of X is a base of some filter on X iff [F*1] and [F*2] are satisfied.*

If A is a sub-base of a filter F. Then the family B of all finite intersections of members of A is a base for F.

Theorem 2(a):

Let B be a family of subsets of a set X. Then B is filter base on X if and only if the family F consisting of all those subsets of X which contain a member of B is a filter on X.

Proof:

First observe that B is non-empty iff F is non-empty.

The 'if' part : Let F be a filter on X. Then ∅ ∈ F. By definition of F, we have B ∈ F. Hence ∅ ∉ B. Thus [F*1] is satisfied. Again let F ∈ B and H ∈ B. Then surely F ∈ F and H ∈ F. Hence by [F3], F ∩ H ∈ F and so by definition F, there exists a G ∈ B such G ⊂ F ∩ H. Thus we have shown that if F, H are members of B, then there exists a member G of B such that G ⊂ F ∩ H and consequently [F*2] is also satisfied. Hence B is a filter base on X.

The 'only if' part : Let B be a filter base on X and let F be family of all those subsets of X which contain a member of B. Since is a filter base, ∅ ∉ B and hence also ∅ ∉ F. Thus [F1] is satisfied. Now let F ∈ F and H ⊃ F. Then by definition of F, F contains a member of B, say G so that G ⊂ F and hence G ⊂ H. It follows that H ∈ F.

Hence [F2] is satisfied. Finally let F ∈ F and H ∈ F. Then there exists G and K belonging to B such that G ⊂ F and K ⊂ H so that G ⊂ K ⊂ F ∩ H. Since B is a filter base by [F*2], there exists a L ∈ B such that L ⊂ G ∩ K, and hence L ⊂ F ∩ H.

Hence by definition of F, F ∩ H ∩ ⊂ F. This [F3] is also satisfied. Hence F is a filter on X.

Theorem 2(b):

On a set X, a filter F, with base B' is finer that a filter F with base B if and only if every member of B contains a member of B'.

Proof:

The 'if' part : Let every member of B contain a member of B′ and let F be any member of F. Then F contains a member of B, say B′. By hypothesis there exists a member B′ of B′ such that B ⊂ B. Hence, F contains B′ and consequently, F ∈ F′. Thus, we have shown that F ∈ F ⇒ F ∈ F′. Hence F ⊂ F, that is, F′ is finer that F.

The 'only if' part : Assume that F′ is finer that F and let B be any member of B. Then B ∈ F. Since F′ is finer than F we have B′ ∈ F′. Since B′ is a base of F′, there exists a B′ ∈ B′ such that B′ ⊂ B. Thus every member of B contains a member of B′.

Theorem 3:

Two filter basis B and B' on a set X are equivalent if and only if every member of B contains a member of B' and every member of B' contains a member of B.

Proof:

The theorem is an immediate corollary of the preceding theorem. However we give an independent proof below.

Assume that every member of B contains a member of B′ and every member of B′ contains a member of B. Let F and F′ be the filters generated by B and B′ respectively. Let F ∈ F. Then there exists a B ∈ B such that B ∈ F. By hypothesis, there exists a B′ ∈ B′ such that B′ ⊂ B. It follows that B′ ⊂ F. Hence by definition (6.10) F ∈ F′. Thus we have shown that

$$F \in F \Rightarrow F \in F' \text{ and hence } F \in F'.$$

Similarly we can show that F′ ⊂ F. Therefore F = F′. Thus B and B′ generate the same filters and so they are equivalent.

Conversely, assume that B and B′ are equivalent so that they generate the same filter, say F. Let B be an arbitrary member of B. Then B ∈ F. Since B. is also a base of F, there exists a B′ ∈ B′ such that B′ ⊂ B. Thus every member of B contains a member of B′. Similarly it can be shown that every member of B′ contains a member of B.

Theorem 4(a):

Consider the usual topology U on R^2. Let B (x) be the collection of all circular nhds and B' (x) the collection of all square nhds of a point x ∈ R^2. Then B and B' are equivalent since they generate the same filter on R^2, namely, the neighbourhood filter at x.

Proof:

Do yourself.

Theorem 4(b):

Let f be a mapping of a set X into another set Y.

(i) If B is a filter base on X, then the family C = {f [B] : B ∈ B} is a filter base on Y.

(ii) If C is a filter base on Y such that every member of C intersects f [X], then the family D = {f^{-1} [C] : C ∈ c} is a filter base on X.

Proof:

(i) Let f [B_1] ∈ c and f [B_2] ∈ c where B_1, B_2 ∈ B. Then by [F*2], there exists a member B_2 of B such that $B_3 \subset B_1 \cap B_2$. Therefore f [B_3] ⊂ f [$B_1 \cap B_2$]. But f [$B_1 \cap B_2$] ⊂ f [B_1] ∩ f [B_2]. Hence f[B_3] ⊂ f [B_1] ∩ f [B_2]. Thus C satisfies [F*2]. Further B ≠ ∅ ⇒ f [B] ≠ ∅ ∀ B ⊂ B. Hence [F*1] is also satisfied. It follows that C is a filter base on Y.

(ii) Let f^{-1} [C] be any member of D where C ∈ c. Since C is a filter base, C ≠ ∅. Also by definition of c, C ∩ f [X] ≠ ∅. We claim that f^{-1} [C] ≠ ∅. For if f^{-1} [C] = ∅, then either C = ∅ which is a contradiction or no member of C belong to the range set f [X] so that $C \cap f[X] = \varnothing$ which is again a contradiction. Hence $f[C] \neq \varnothing$. Thus [F*1] is satisfied.

Now let f–1 [C] and f–1 [D] be any tow members of D where C, D are members of c. Since c is a filter base, there exists a member E of c such that $E \subset C \cap D$.

But $\quad E \subset C \cap D \Rightarrow f^{-1}[E] \subset f^{-1}[C \cap D]$

$\Rightarrow \quad f^{-1}[E] \subset f^{-1}[C] \cap f^{-1}[D]$.

Also $E \in c \Rightarrow f^{-1}[E] \in D$. Thus {F*2] is also satisfied.

Hence D is a filter base on X.

SOLVED EXAMPLES

Example 1:

Let X = {a, b, c} and let

N (a) = {{a}, {a, b}, {a, c}, Y}

N (b) = {{a, b}, X} N (c) = {{a, c}, X}

(i) Show that N (a), N (b) and N (c) satisfy all the conditions.

(ii) Let $\mathfrak{I}$ be the topology for X induced in the sense by N (a), N (b) and N (c). Find all $\mathfrak{I}$-open subsets of X.

(iii) Find all the $\mathfrak{I}$-nhds of a, b and c and verify that they coincide with N (a), N (b), N (c) respectively,

Solution:

(ii) It is easy to see that the conditions [N0], [N2] and [N3] are satisfied by N (a), N (b), and N (c). The condition [N4] is also satisfied. Take from example, {a, b} ∈ N (a). Then {a} ∈ N (a) is such that {a} ⊂ {a, b} and {a} ∈ N (y) ∨ y ∈ {a}. Note that there is only one y ∈ {a}, namely a. Again take {a, c} ∈ N (c). Then {a, c} ⊂ {a, c} such that {a, c} ∈ N (a) and {a, c} ∈ N (c) so that {a, c} ∈ N (y) ∨ y ∈ {a, c}. The reader is advised to verify [N4] for other members of N (a), N (b) and N (c).

(ii) Now we select the members of $\mathfrak{I}$ as follows.

$G \in \mathfrak{I}$ iff $G \in N(x) \vee x \in G$.

Then $\varnothing \in \mathfrak{I}$ since the statement $\varnothing \in N(x) \vee x \in \varnothing$ is trivially true.

{a ∨ x ∈ {a},} ∈ $\mathfrak{I}$ since {a} ∈ N (a) ∨ x ∈ {a}

{b} ∉ $\mathfrak{I}$ since {b} ∉ N (b).

{c} ∉ $\mathfrak{I}$ since {c} ∉ N (c.)

{a, b} ∈ $\mathfrak{I}$ since {a, b} ∈ N {a} and {a, b} ∈ N {b}.

$\{b, c\} \notin \Im$ since $\{b, c\} \notin N$ (b)

$\{a, c\} \in \Im$ since $\{a, c\} \in N$ (a) and $\{a, c\} \in N \{c\}$.

$X \in \Im$ since $X \in N (x) \vee x \in X$.

Hence $\Im = \{\{\varnothing, \{a\}, \{a, b\}, \{a, c\}, X\}$

(iii) Let N* (a), N* (b), N* (c) denote the $\Im$-nhds of a, b, c respectively. Then

$N^* (a) = \{\{a,\}, \{a, b\}, \{a, c\}, X\}$, $N^* (b) = \{\{a, b\}, X\}$,

$N^* (c) = \{\{a, c\}, X\}$

Here $N^* (a) = N^* (a)$, $N^* (b) = N (b)$ and $N^* (c) = N (c)$.

Another way of generating a topology is given by the following theorem.

Example 2:

Give an example to show that the union of an infinite collection of closed sets in a topological space is not necessarily closed.

Solution:

Let (R, U) be the usual topological space and let fn = [1/n, 1], n $\in$ N so that Fn is a closed interval of R.

$$\cup \{Fn : n \in N\} = \{1\} \cup \left[\frac{1}{2}, 1\right] \cup \left[\frac{1}{4}, 1\right] \ldots =]0, 1[.$$

Since]0, 1] is not closed, it follows that the union of an finite collection of closed sets is not necessarily closed.

Example 3:

Show that every finite subset of R is a U-closed set.

Solution:

We have shown that every singleton {a} is a U-closed subset of R. Since a finite subset of R is the union of a finite number of singletons, it follows every finite subset of R is U closed.

Example 4(a):

Let (X, $\Im$) be a co-finite space. Since $\Im$-open sets are those whose complements are finite, it follows that all finite subsets of X together with X are $\Im$-closed.

Example 4(b):

Give two examples of a proper non-empty subset of a topological space such that it is both open and closed and prove your assertion.

Solution:

Consider the topology

$$\mathfrak{I} = \{\varnothing, \{a\}, \{b\}, \{c\}, \{a, b\}, \{b, c\}, \{c, a\}, X\}$$

on $X = \{a, b, c\}$

Here the sets {a}, {b}, {c}, {a, b}, {b, c}, {c, a} are all both open and closed. The set {a} is closed since it is the complement of {b, c} etc. In fact, in every discrete topological space (X, **D**) consisting of more than one element, every subset of X is both open and closed. As another example consider the topology.

$$\mathfrak{I} = \{\varnothing, \{a\}, \{c, d\}, \{a, c, d\}, \{b, c, d, e\}, X\} \text{ on } Y = \{a, b, c, d, e\}.$$

Here {a} is open. Also {a} is closed since it is the complement of the open set {b, c, d, e}.

Example 4(c):

Give an example of a topological space different from the discrete and indiscrete spaces in which open sets are exactly the same as closed sets.

Solution:

Let X = R (the set of reals) and let $\mathfrak{I}$ consist of the empty set and all those subsets G of R having the property that $x \in G$ implies $-x \in G$. Then $(R, \mathfrak{I})$ is the required space.

We first prove that $\mathfrak{I}$ is actually a topology for R. We have

[T1] : $\varnothing \in \mathfrak{I}$ by definition. Also $R \in \mathfrak{I}$ since

$x \in R \Rightarrow -x \in R$

[T2] : Let $G_1 \in \mathfrak{I}$ and $G_2 \in \mathfrak{I}$ and let $x \in G_1 \cap G_2$. But $x \in G_1 \cap G_2$

$\Rightarrow x \in G_1$ and $x \in G_2$.

$\Rightarrow -x \in G_1$ and $-x \in G_2$ [By def. on $\mathfrak{I}$]

$\Rightarrow -x \in G_1 \cap G_2$

Hence $G_1 \cap G_2 \in \mathfrak{I}$

[T3] : Let $G_1 \in \mathfrak{I}$ for all $\lambda \in L$ let $x \in \cup \{G_\lambda : \lambda \in \Lambda\}$

Now $x \in \cup \{G_\lambda : \lambda \in \Lambda\} \Rightarrow x \in G_\lambda$ for some $\lambda \in \Lambda$

$\Rightarrow -x \in G_\lambda$` [$\because G_1 \in \mathfrak{I}$]

$\Rightarrow -x \in \cup \{G_\lambda : \lambda \in \Lambda\}$.

Hence $\cup \{G_\lambda : \lambda \in \Lambda\} \in \mathfrak{I}$.

It follows that $\Im$ is a topology for R. We now show that every open set is closed. Let G be any open set. We shall show that G is also closed, that is, R – G is $\Im$ open. We have

$$x \in R - G \Rightarrow x \notin G$$

$$\Rightarrow -x \notin G \quad [\because G \in \Im]$$

$$\Rightarrow -x \in -G.$$

It follows that R – G is open and so G is closed.

Similarly it can be shown that every closed set is $\Im$-open.

Example 5:

Let $\Im$ and $\Im'$ be two topological spaces on a set X such that $\Im'$ is coarser then $\Im$ and let x be a $\Im$ cluster point of a filter F on X.

Then show that x is also a $\Im'$ cluster point of F.

Solution:

Let N (x) and N′ (x) be the nhd filters of x relative to $\Im$ and $\Im'$ respectively. Since $\Im' \subset \Im$, it follows that $N'(x) \subset N(x)$. Hence x is a cluster point of F relative to $\Im$.

⇔ every member of N (x) intersects every member of F.

⇔ every member of N′ (x) intersects every member of F.

⇔ x is a cluster point of F relative to $\Im'$.

Example 6:

Show that every discrete space is totally disconnected.

Solution:

Let (X, D) be any discrete space and let x, y be any two distinct points of X. Then A = {x} and B = X – {x} are both non-empty open disjoint sets whose union is X such that $x \in A$ and $y \in B$. It follows that (X, D) is totally disconnected.

Example 7(a):

Given two examples of locally connected spaces which are not connected.

Soution:

(i) iet X be a discrete space containing more than one point. Then X is disconnected. For it $x \in X$, then {x} is a non-empty proper subset of X which is both open and closed in X. X is locally connected.

(ii) Let Y be the union of two disjoint open intervals on the real line, that is, let Y =]a, b[∪] c, d[where $a < b \leq c < d$. Then Y as a subspace of (R, U) is locally connected but not connected as shown below. Since [a, b[and [c, d[are open in R, they are also open in Y.

Thus Y is the union of two non-empty disjoint sets both open in Y and therefore Y is disconnected. To show that Y is locally connected, let x be any point of Y and G_x any open nhd of x in Y. Then we can always find $\varepsilon > 0$ such that $]x - \varepsilon, x + \varepsilon[\subset G_x$. But $]x - \varepsilon, x + \varepsilon[$ is connected in **R** and hence connected in Y.

Thus every open nhd of x in Y contains an open connected nhd of x in Y and therefore Y is locally connected.

Example 7(b):

Let X be a non-empty directed set directed by the relation R. For each a ∈ let

$$s(a) = \{a \in X : x \; R \; a\}$$

Then the collection

$$B = \{s(a) : a \in X\},$$

is a filter base on X.

Solution:

Since the relation R directs the set X, R is reflexive transitive and is such that if a, b are any two members of X, then there exists a member c ∈ X such that x R a and c R b.

We now verify [F*1] and [F*2].

[F*1] : Since X is non-empty, there is at least one x ∈ X and so there is at least one member s (x) in B. Hence B is non-empty. Further since R is reflective, x ∈ s (x) so that s (x) is non-empty for each x ∈ X. Hence $\varnothing \notin$ B.

[F*2] : Let s (a) ∈ B and s (b) ∈ B. Since a, b ∈ X, there exists an element ∈ X such that x R a and c R b. We now show that $s(x) \subset s(a) \cap s(b)$. Let x ∈ s (c). Then x R c. Since c R a and c R b, by transitivity of R, we have x R a and x R b. It follows that x ∈ s (a) and x ∈ s (b). Thus we have shown that

$$x \in s(c) \Rightarrow x \in s(a)$$

and
$$x \in s(c) \Rightarrow x \in s(b).$$

Hence $s(c) \subset s(a)$ and $s(c) \subset s(b)$.

Therefore $s(c) \subset s(a) \cap s(b)$.

It follows that B is a filter base on X.

Example 7(c):

Show that in an indiscrete space X, every filter on X converges to every point of X.

Solution:

Let F be any filter on X and let to be any point of X. The neighbourhood filter of x in this case is {X}. Since X ∈ F, it follows that F converges to x. Since x was an arbitrary point of X. We see that F converges to every point of X.

Example 8:

Let M be the collection of all those filters on a topological space X which converge to the same point x ∈ X. Then the intersection F of all the filters in M also converges to x. Show further that F = N (x).

Solution:

First note that F is actually a filter on X by virtue of theorem. Since all the filters in M converge to x, the neighbourhood filter N (x) is coarser than each filter in M and consequently coarser than F. Hence F also converges to x. We now show that F = N (x). We have shown above that

$$N(x) \subset F. \qquad ...(1)$$

Also sine N (x) converges to x, it is a member of M,

we have

$$F \subset N(x). \qquad ...(2)$$

From (1) and (2), we have

$$F = N(x).$$

Example 9:

Let $B = \{H_n : n \in N\}$ *where* $H_n =]\, n, \infty\, [$ *and let* $F = \{F : F \supset H$ *for some* $n \in N\}$.

Show that F is a filter on R with base B. Further show that F converges to no point of R relative to the usual topology of R.

Solution:

B evidently satisfies [F1] and [F2] but not [F3] but F satisfies [F3] also as the reader can easily see. Hence F is filter with base

B. [Note that [F2] ⇒ [F2]*. But F does not converge to any point x ∈ R. For] x – 1, x + 1 [is a U-nhd of x which does not belong to F.

Example 10:

Show that in a discrete space X, the only convergent filters are the neighbourhood filters.

Solution:

We have the neighbourhood filters are the only ultrafilters on X and so by definition these are the only convergent filters.

Example 11(a):

Show that if a filter F in a topological space X converges to a point $x \in X$, then every filter F finer that F also converges to x.

Solution:

Since F converges to x, F contains the nhd filter N (x) of x, and since F' is finer than F, F' also contains N (x) and consequently F' also converges to x.

Example 11(b):

The neighbourhood filter N (x) of a point x in a topological space X converges to x.

Solution:

Do yourself.

Example 11(c):

Consider the topology $\mathfrak{I} = \{\varnothing, \{a\}, \{a, b\}, X\}$ on $X = \{a, b, c\}$ and the filters $F_1 = \{\{a,\}, \{a, b\}, \{a, c\}, X\}$

$$F_2 = \{\{b\}, \{a, b\} \{b, c\}, X\},, F_3 = \{\{c\}, \{b, c\}, \{a, c\}, X\}$$

and find the limit points of these filters.

Solution:

We have N (a) = {{a},, {a, b}, (a, c), X}, N (b) = {{a, b}, X} and N {c} = {X}. Since F_1 contains N (a), N (b) and N (c), it follows that a, b, c are the limit points of F_1. Again F_2 contains N (b) and N (c) and does not contain N (a), and hence b, c are the limit points of F. Finally F_2 contains N (c) only and so c is the limit point of F_2. shows.

Example 11(d):

Show that in a discrete topological space, every neighbourhood filter is an ultrafilter.

Solution:

In a discrete space X, the neighbourhood filter of a point x ∈ X is the collection of all subsets of X which contain x. Hence by the preceding example, every nhd filter on X is an ultrafilter on X. Note that in this space, no other filter can be an ultrafilter.

Example 12:

Show that the collection F (p) all subsets of a set X which contains a given element p ∈ X is an ultrafilter on X. Show filter that {{p}} is an ultrafilter base on X.

Solution:

Here F (p) = {F : F ⊂ X and p ∈ F}. Let A be any subset of X. Then p ∈ A or p ∈ X – A. Hence either A ∈ F (p) or X – A (p). Also evidently F (p) has the FIP.

Hence F (p) is an ultrafilter on X. Further since {{p}} is a filter base why ?) on X which generates the ultrafilter F (p), it follows that {{p}} is an ultrafilter base on X.

Example 13:

Let F be an ultrafilter on a set X. Show that ∩ {R : F ∈ F} is either empty of a singleton subset of X.

Solution:

Let G = ∩ {F : F ∈ F} and let G ≠ ∅. Then there is at least one element x in X such that x ∈ G. Suppose, if possible there is another element y different from x such that y ∈ G. Since F is an ultrafilter, either {x} ∈ F or X – {X} ∈ F. If {x} ∈ F, then, since y ≠ x, y ∉ {x} and hence y ∉ G, which is contradiction.

Similarly if X – {x} ∈ F, then x ∉ G which is again contradiction. These contradictions arise by supposing x ≠ y. Hence x = y. It follows that G = {x}. That is, either G is empty or a singleton set.

Example 14:

Let (X, ℑ) be a topological space and let N (x) be the neighbourhood filter at x. Then a local bases B (x) at x is a base of the filter N (x).

Solution:

By definition N (x) is the collection of all those subsets of X which contain a member of B (x). Also B (x) is a sub-family of N (x). Hence by the above definition B (x) is a base of the filter N (x).

Example 15(a):

Let (X, $\mathfrak{J}$) be any indiscrete topological space show that every nit in X converges to every point of X.

Solution:

Let (f, X, A, ≥) be a net in X and let x be an arbitrary point of X. Then the only nhd of x is X. Since $f_a \in X$ for all a ∈ A, it follows that f converges to x.

This example show that a net can converges to several different points.

Example 15(b):

Let (X, $\mathfrak{J}$) be a topological space and let x ∈ X. Then the local base B (x) at x is a filter base on X.

Solution:

Since x ∈ B ∀ B ∈ B (x), it follows that no member of B (x) is empty. Hence ∅ ∉ B (x). Thus [F*1] is satisfied. Again let A ∈ B (x) and B ∈ B (x). Then there exists a C ∈ B (x) such that C ⊂ A ∩ B. Hence [F*2] is also satisfied.

Example 16:

Let X = {a, b. c}.

Then {{a}, {a, b}, {a, c}} is a filter base in X.

[To obtain filter base on a non-empty set X take any collection A of subsets of X. The collection B of all finite intersection of A is easily seen to be filter base on X].

Example 3:

The family B = {A} where A is any non-void subset of a n-void set X is a filter base on X.

Solution:

Here B consists of a single member A. Since A is non-void, ∅ ∉ B and hence [F*1] is satisfied. Again since A ⊂ A ∩ A. {F*2] is also satisfied. Hence B is a filter base on X.

Example 17:

Consider the sequence f : N → R defined by setting

f (n) = – 1 when n = 1, 3, 5, 7, ...

and $f(n) = n\backslash 2$ when n = 2, 4, 6, 8...

that is the sequence {– 1, 1, – 1, 2, – 1, 3, – 1, ... }.

Then –1 is a cluster point of f since f is evidently frequently in every nhd of –1. Also it is evident that no other real number can be a cluster point of f.

Hence –1 is the unique cluster point of this sequence. But this sequence fails to converge to –1 since] –2, 1 [is a U-nhd of – 1 in which does not eventually lie.

Example 18:

Let N be directed by ≥ in the usual sence. Consider the mapping,

$$f : N \rightarrow R$$

defined in such a way that f (N) = Q. Then f is a sequence (and hence a net) in R..

Solution:

Do your self

Example 19:

Let N Be the set of all natural members directed by ≥ in the usual sense. Then the mapping.

$$f : N - R : f(n) = n \quad \forall n \in N,$$

is a net in (R, U), This net t has no U cluster points in R. For, let x be any point of R. Then]x – 1, x + 1 [is a U-nhd of x. But evidently f is not frequently in this neighbourhood. Hence x is not a cluster point.

Example 20(a):

Show that the set of all cluster points of a net in a topological space need not be closed.

Solution:

Consider the cofinite topology $\Im$ on N. Then the mapping f : N → N = f (n) = 2n – 1 is a net in N the set of whose cluster points in {1, 3, 5, ... }. But this set is not $\Im$=closed.

Example 20(b):

Let (X, D) be any discrete topological space. Show that a net (f, X, A, ≥) in X converges to a point $x_0 \in X$ iff is eventually in $\{x_0\}$, that is, iff there exists an $a_0 \in A$ such that $f_a = x_0$ for all $a \geq a_0$ in A.

Solution:

Let f converge to x0. Since $\{x_0\}$ is a D-nhd of x_0 f must be eventually in $\{x_0\}$. Conversely, let f be eventually in $\{x_0\}$. Since every nhd of x_0 contains $\{x_0\}$, it follows that f is eventually in every nhd of x_0. Hence f converges to x_0.

Example 20(c):

Let c denote the collection of all finite subsets of a set X. Then c is directed by the relation $\subset$

Solution:

[D1] and [D2] hold in this case also. To prove [D3], we observe that since c consists of finite subsets of X, for any two members A and B of c, $A \cup B$ is also a member of c such that $A \cup B \supset A$ and $A \cup B \supset B$. Hence [D3] is also satisfied.

Example 21:

Let $(X, \mathfrak{I})$ *be any topological space and let* $x \in X$. *Then the collection* $N(x)$ *of all* $\mathfrak{I}$*-nhds of x is directed by the inclusion relation* $\subset$ *For, we have*

[D1] : $A \in N(x) \Rightarrow A \subset A$

[D2] : $A \subset B$ *and* $B \subset C \Rightarrow A \subset C$ $(A, B, C \in N(x))$.

In fact [D1] and [D2] are true for any sets whatsoever.

[D3] : *Let A, B be any two members of N (x). Then* $A \cap B$ *is a also a member of N (x) such that* $A \cap B \subset A$ *and* $A \cap B \subset B$.

Example 22:

Let N be directed by $\geq$ *in the usual sense. Then* $M = N - \{1, 2]$ *is a residual subset of N. Here 3 is the required element* a_0 *of the definition. On the other hand the set* $L = \{1, 3, 5, \ldots\}$ *is not a residual subset of N. However L is a confinal subset of N. Note that* $\geq$ *also directs L.*

Solution:

Let $(A, \geq)$ *be a directed system and let f be an arbitrary mapping of A into a set X. Then f is called a not a in X and is denoted by* $(f, X, A, \geq)$.

Note that a sequence in X is a particular type of a net in which A = N and N is directed by $\geq$ in the usual sense.

We shall denote the image f (a) of $a \in A$ under f by f_a and a net in X will be often denoted by the symbol $(\{f_a : a \in A\}, \geq)$ or simply by

$\{f_a : a \in A\}$.

A net is sometimes called a generalized sequence of Moor. Smith family.

Example 23:

Let X be an infinite set and let F be a filer on X such that $\cap$ {F : F $\in$ F} = $\varnothing$. Show that F is finer than the cofinite filter on X.

Solution:

Let c be the cofinite filter on X. To show that F $\in$ c. Suppose, if possible. $\not<$ c. Then there exists C $\in$ c such that C $\in$ F. Then X– C is finite. Let X – C = $\{x_1, X_2, \ldots x_n\}$. Now $\cap$ {F :$\in$ F $\in$ F} = $\varnothing$. Therefore, there exists $F_1 \in F$ {i = 1, 2, ..., n} such that $x_i \notin F$. Since F is a, G = $\cap$ Fl : 1 $\leq$ i $\leq$ n} $\in$ F. This G will not contain and element of $\{x_1, x_2, \ldots x_n\}$. Hence X – G $\supset$ X – C and consequently G $\in$ C. Since F is filter C $\supset$ G implies C $\in$ F. But this is a contradiction . Hence F must be finer than the cofinite filter.

Example 24(a):

Let F be an arbitrary member of F and suppose, if possible, (X, A) $\cap$ F = $\varnothing$. But then F $\subset$ A and so by [F2], A $\in$ F which is a contradiction. Hence (X – A) $\cap$ F $\neq \varnothing$, that is X – A meets every member of F. Then by the preceding theorem there exists a filter F' finer than F such that X – A $\in$ F', that is A $\notin$ F'.

Solution:

Let M be any non-empty class of filters an a non-empty set X. Then a filter F' on X is said to be supermum (infimum) of M iff F is finer (coarser) than each member of M and if F' is any filter on X which is finer (coarser) than each member of M then F is coarser (finer) than F'.

Thus a supermum of M is the coarsest of all filters on X which are finer than every member of M. Similarly an infimum of M is the finest of all filters on X which are coarser than each member of M.

Example 24(b):

Let $M_1 = \{F_2, F_3, F_5\}$

$M_2 = \{F_1, F_2, F_3, F_5\}$

$M_3 = \{F_3, F_2, F_5\}$

and $M_4 = \{F_2, F_5, F_5\}$,

where F_1, F_2, F_3, F_4, F_5 and F_6 are as defined in example (4.5). Find superma and infima of M_1, M_2, M_3 and M_4

Solution:

Here F_0 is supermum of M_1 since it is the coarser of all filters on X which are liner than each member of M_1 in fact, in this case there is no other filter on X which is filter than each member of M_1. Also F1 is an infimum of M_1, because it is the only filter on X which is coarser than every member of M_1. Observe that in this case, supermum is a member of M_1 whereas infimum is not. M_1 has no supermum since there is no filter on X which is finer than each member of M_2. Also F_1 is an infimum of M_2. Similarly M_3 has no supermum and F_1 is an infimum of M_3. An infimum of M_4 is F_2 since it is coarser than every member of M_4 and finer than F_1 which is the only other filter on X coarser than each member of M_4, M_4, however, has no supermum since there is no filter on X which is finer than every member of M_4 is coarser than every member of M. Supermum of M may or may no exist.

Example 24(c):

Let X be any non-empty set and let $x_0 \in X$. Show that the family F = {F : $x_0 \in F$} is a filter on X called the discrete filter.

Solution:

F is non-empty since $\{x\} \in F$.

[F1] : Since $x_0 \in F \vee F \in F$. It follows that no member of F is empty and so $\varnothing \in F$.

[F2] : Let $F \in F$ and $H \supset F$. Now $F \in F$ implies that $x_0 \in F$. Since $G \supset F$., we have $x_0 \in$ We have $x_0 \in H$. Hence $H \in F$.

[F3] : Let $F \in F$ and $H \in F$. Then $x_0 \in F$ and $x_0 \in H$. Hence $x_0 \in F \cap H$. It follows that $F \cap H \in F$.

Example 25:

Let X be any non-empty set and let F_0 be a non-empty set and let F_0 be a non-empty subset of X. Then the family $F = \{F : F \supset F_0\}$ is a filter on X.

Solution:

Since $F_0 \in F$, F is non-empty. Further

[F1] : Since F_0 is non-empty and each member of F contains F_0 it follows that each member of F is non-empty. Hence $\varnothing \notin F$.

[F2] : Let $F \in F$ and $H \supset F$. Now $F \in F$. Now $F \in F$. Then $F \supset F_0$ and $H \supset F$, it follows that $H \supset F_0$. Hence $H \in F$.

[F3] : Let $F \in F$ and $H \in F$. Then $F \supset F_0$ and $H \supset F_0$. Hence $F \cap H \supset F_0$. It follows that $F \cap H \in F$.

Example 26:

Let $(X, \mathfrak{I})$ be a topological space and let N (x) be the collection of all $\mathfrak{I}$-nhd of a point $x \in X$. Then N (x) is a filter on X called the neighbourhood filter of X.

Solution:

Since $X \in N(x)$, it follows that N (x) is non-empty.

[F1] : Since each member of N (x) contains x, it follows that no member of N (x) is empty. Hence $\varnothing \notin N(x)$.

[F2] : Let $F \in N(x)$ and $H \supset F$. Then F is a $\mathfrak{I}$-nhd of x and since $H \supset F$. H is also a $\mathfrak{I}$-nhd of x. It follows that

$$H \in N(x)$$

[F3] : Let $F \in N(x)$ and $H \in N(x)$. Then, we have $F \cap H \in N(x)$.

Example 27(a):

Let X be any infinite set. Then $F = \{F : X - F \text{ is finite}\}$ is a filter on X called the cofinite filter.

Solution:

Since $X - X = \varnothing$ is finite, $X \in F$ and hence F is non-empty. Further.

[F1] : Since X is infinite and X – F infinite, it follows that F is an infinite set and so no member of F os empty. Hence $\varnothing \in F$.

[F2] : Let $F \in F$ and $H \supset F$. Since X – F is finite, then surely X – H is finite. Hence $H \in F$.

[F3] : Let $F \in F$ and $H \in F$. Then X – F and X – H are both finite. Now

$$X - (F \cap H) = (X - F) \cup (X - H)$$

[De-Morgan law] ...(1)

Since X – F and X – H are finite, it follows from (1) that $X - (F \cap H)$ is finite. Hence $F \cap H \in F$. Thus F is a filter on X, In particular if X = N, then co-finite filter on N is called the Erchet filter.

Example 27(b):

Show that there exists no filter on the empty set $\varnothing$.

Solution:

Suppose, if possible, there exists a filter F on $\varnothing$. Now by (ii) of the above remark $\varnothing \in F$. But by [F1]. $\varnothing \in F$. But this is a contradiction. Hence there exists no filter on $\varnothing$.

Example 27(c):

Let X = {A, b, c} and let

$F_1 = \{X\}$, $F_2 = \{\{a, b\}, X\}$, $F_3 = \{\{b, c\}, X\}$,

$F_4 = \{\{c, \}, X\}$, $F_5 = \{\{a\}, \{a, b\}, \{a, c\}, X\}$,

$F_6 = \{\{b,\}, \{a, b\}, \{b, c\}, X\}$, $F_7 = \{\{c,\}, \{c, d\}, \{b, c\}, X,\}$

$F_8 = \{\{a, b\}, \{b, c\}, X\}$, $F_9 = \{\{a,\}, \{b\}, \{a, b\}, X\}$,

$F_{10} = \{\{\varnothing, \{a\}, \{a, b\}, X\}$.

Solution:

Then F_1, F_2, F_3, F_4, F_5, F_6, F_7 are filters on X as the reader can easily verify, F_8 is not a filter because it does not satisfy [F3], F_9 is not a filter because it does not satisfy [F2] and [F3], F_{10} is not filter since it does not satisfy [F1] and [F2].

Example 28:

Consider the topology

$\mathfrak{I} = \{\phi, \{a\}, \{a, b\}, \{a, c\} ; X\}$

On X = ({a, b, c} and the filters

$F_1 = \{\{b, c\}, X\}$,

$F_2 = \{\{c, a\}, X\}$

$F_5 = \{\{a, b\}, X\}$,

and find the cluster points of these filters,

Solution:

We have

$N(a) = \{\{a,\}, \{a, b\}, \{a, c\}, X\}$

$N(b) = \{\{a, b\}, X\}$ and

$N(c) = \{\{a, c\}, X\}$,

Now a is not a cluster point of F_1 since the nhd {a} does not intersect $\{b, c\} \in F_1$, b is a cluster point of F_1 since every member of N (b) intersects every member of F_1. Similarly, c is a cluster point of F_1. Hence adh (F_1) = {b, c}. Similarly the reader can verify that adh (F_2) = {a, b, c} adh (F_3) = {a, b, c}.

Example 29:

Let x be a cluster point of a filter F on topological space X and let F' be a filter on X coarser than F. Then x is also a cluster point of F'.

Solution:

x is a cluster point of F.

$\Leftrightarrow$ x $\in \overline{F}$ $\forall$ F $\in$ F.

$\Leftrightarrow$ x $\in \overline{G}$ $\forall$ G $\in$ F′.

[Since every member of F′ is a member of F]

$\Leftrightarrow$ x is a cluster point of F′.

Example 30:

Let $X = \{a, b, c, d, e\}$ *and let*

$\mathfrak{J} = \{\varnothing, \{b\}, \{c, d\}, \{b, c, d\}, \{a, b, c, d\}, X\}$.

Find (i) interior (ii) exterior (iii) frontier of the following subsets of X.

(a) $A = \{c\}$, *(b)* $B = \{a, b\}$, *(c),* $C = \{a, b, d\}$, *(d),* $D = \{b, c, d\}$.

Solution:

(a) (i) Since A is not a nhd of c, c $\notin$ A°.

Hence, A° = $\varnothing$

(ii) Now A′ = {a, b, c, d, e}. It is easy to see that b is an interior point of A′ since A′ is a nhd of b. But a, d, e are not the interior points of A′.

Hence ext (A) = {b}.

(iii) Since A° = $\varnothing$ and ext (A) = {b}, if follows that

fr (A) = {a, c, d, e}.

(b) (i) Here b is an interior point of B but a is not.

Hence B° = {b}.

(ii) Now B' = {c, d, e}. Since c, d, $\in$ {c, d} $\subset$ B', it follows that B' is a nhd of x, d. Hence c, d are interior points of B'.

Hence ext (B) = (B')° = {c, d}.

(iii) Since B° = {b} and ext (B) = {c, d}, it follows that

Fr (B) = {a, e}

(c) Here (i) C° = {b} (ii) C' = {c, e}

so that ext C = (C')° = $\varnothing$

and (iii) Fr (C) = {a, c, d, e} as the reader can easily verify.

(d) (i) Here D is an open set so that it is nhd of each of its points and consequently every point of D is its interior point.

Hence D° D = {b, c, d}.

(ii) D' = (a, e). Since there exits no open set G such that $a \in G \subset D'$, D' is not a nhd of a. Hence $a \notin (D')^\circ$. Therefore ext (D) = $(D')^\circ = \varnothing$.

(iii) Since $D^\circ = \{b, c, d\}$ and ext D = $\varnothing$, it follows that

$$Fr(D) = (a, e\}$$

Example 31(a):

Show that the spaces (R, U) and (R, S) are Hausdorff.

Solution:

Let a, b be any two distinct points of R so that $a \neq b$ ans let $|a - b| = \varepsilon$. Then

$$]a - \varepsilon/3, a + \varepsilon/3 [\text{ and}] b - \varepsilon/3, b + \varepsilon/3 [$$

are disjoint U-nhds of a and b respectively.

Also $[a - \varepsilon/3, a + \varepsilon/3$ [and $[b - \varepsilon/3, b + \varepsilon/3$ [are disjoint S-nhds of a and b respectively. Hence (R, U) and (R, S) are both Hausdorff.

Example 31(b):

Consider the cofinite topology on an infinite set X and show that it is not T_2.

Solution:

For this topology no two open sets can be disjoint. For suppose, if possible, that G, H are two open disjoint subsets of X so that $G \cap H = \varnothing$. Then

$$(G \cap H)' = \varnothing'$$

i.e., $G' \cup H' = X$. [De-morgan Law]

But G′ and H′ are finite sets so their union is also finite which is a contradiction since X is infinite. Hence no two distinct points can be separated by open sets. Accordingly the space is not T_2.

Example 31(c):

Find the condition under which a closed (open) set in the subspace Y of a topological space X be closed (open) in the larger space X.

Solution:

Let X = {a, b, c, d, e} and let $\mathfrak{I}$ be a topology on X given by

$$\mathfrak{I} = \{\varnothing, \{a\}, \{a, b\}, \{a, c, d\} \{a, b, c, d\}, \{a, b, e\}, X\}.$$

Let Y = {a, c, e}. Then $\mathfrak{I}_Y = \{A \cap Y : A \in \mathfrak{I}\}$ so that the members of $\mathfrak{I}_Y$ are

$\varnothing \cap Y = \varnothing$, $Y \cap X = Y$, $\{a\} \cap Y = \{a\}$,

$\{a, b\} \cap Y = \{a\}$, $\{a, c, d\} \cap Y = \{a, c\}$,

$\{a, b, c, d\} \cap Y = \{a, c\}$,

$\{a, b, e\} \cap Y = \{a, e\}$.

Thus $\mathfrak{I}_Y = \{\varnothing, \{a\}, \{a, c\}, \{a, e\}, Y\}$

Now consider the subset $A = \{a, e\}$, of Y. Then

$\text{Int}_Y \{A\} = \{a, e\}$

and $\text{Int}_X \{A\} = \{a\}$.

Hence in this case $\mathfrak{I}$- interior of A is a proper subset of $\mathfrak{I}_Y$-interior of A, so that $\text{Int}_X (A) \subseteq \text{Int}_Y (A)$.

Example 31(d):

Consider the usual topology U on R and find (i) interior (ii) exterior and (iii) frontier of each of the following subset of R.

(a) $A = [0, 1[$; (b) $B = [0, 1[$; (c) $C = [0, 1]$;

(d) $D = \{1/n : n \in N\}$; (e) N; (f) Q.

Solution:

(a) (i) Since A is an open set, it is a nhd of each of its points and so every point of A is its interior point.

Hence $A^\circ = A =]\,0, 1[$.

(ii) Now $A' =]-\infty, 0] \cup [1, \infty[$.

Here A' is nhd of each of its points except 0 and 1.

Hence ext (A) = (A') = $]-\infty, 0[\cup]\,1, \infty[$.

(iii) Fr (A) = $[A^\circ \cup \text{ext}(A)\,(A)]' = \{0, 1\}$.

Thus the frontier of A consists of two points 0 and only.

(b) Proceeding as in (a), we have

(i) $B^\circ =]0, 1[$.

(ii) Here $B' =]-\infty, 0\,[\cup [1, \infty[$.

Then ext (B) = (B') = $]-\infty, 0\,[\cup]\,1, \infty[$.

(iii) Fr (B) = $\{0, 1\}$.

(c) (i) $C^\circ =]0, 1\,[$.

(ii) Here $C' =]-\infty, 0[\cup]1, \infty[$. Hence

ext (C) = (C') = $]-\infty, 0\,[\cup], \infty[$.

(iii) Fr (C) = {0, 1}

(d) (i) Here D cannot be nhd of any its points 1/n, n = 1, 2, 3, ... since there exits no $\varepsilon > 0$ such that

$]1/n - \varepsilon, 1/n + \varepsilon[\subset D$.

Hence no point of D can be its interior point so that $D^\circ = \varnothing$.

(ii) It is easy to see that D' is a nhd of each of its points except 0. Hence

$$\text{ext}(D) = (D')^\circ = [D \cup \{0\}]'$$

(iii) Fr (D) = $[D^\circ \cup \text{ext}(D)]' = D \cup \{0\}$.

(e) (i) It is easy to see that N is not a nhd of any of its points and so $N^\circ = \varnothing$.

(ii) Since **N** has no limit points we have $\overline{N} = N$ so that N is a closed set. Its complement N' is therefore an open set. Hence

ext (N) = $(N')^\circ = N'$.

(iii) $\quad$ Fr (N) $= [N^\circ \cup \text{ext}(N)]' = [\varnothing \cup \text{ext}(N)]'$
$= [\text{ext}(N)]' = (N')' = N$.

Note that N non-dense since $(\overline{N})^\circ = N^\circ = \varnothing$.

(f) (i) Since Q is not a nhd of any of its points, we have

$Q^\circ = \varnothing$

(ii) Also $Q' = R - Q$ is the set of all irrational points which is also not a nhd of any of its points so that $(Q')^\circ = \varnothing$.

Hence $\quad$ ext (Q) = (Q') = $\varnothing$

(iii) $\quad$ Fr (Q) $= [Q^\circ \cup \text{ext}(Q)]'$
$= [\varnothing \cup \varnothing]' = \varnothing' = R$.

Note that Q is every where dense since $\overline{Q} = R$.

Example 32:

Let (X, d) be a discrete topological space and let A be a subset of X. Find (i) A°, (ii) **ext** *(A), (iii)* **Fr** *(A)..*

Solution:

(i) Since in a discrete space, every set is open, it follows that A is a nhd of each of its points. Hence $A^\circ = A$.

(ii) ext (A) = (A') = A' $\quad$ [$\because$ B' is open (why ?)]

(iii) ext (A) = $[A^\circ \cup \text{ext}(A)]'$

$= [A \cup A'] = X' = \varnothing.$

Example 33:

Consider the usual topological (R,a U) and find the U-closure of the following subsets of R.

(i) A = {1/n : n ∈ N}. (ii) I (iii) Q. (iv)]0, 1 [. (v) [0, 1[.

(vi) [0, 1]. (vii) $B = \left\{2, \frac{3}{2}, \frac{4}{3}, \frac{5}{4}\right\}.$

Solution:

(i) Here D (A) = {0} so that

$$\overline{A} = \{1/n : n \in N\} \cup \{0\}.$$

(ii) Since D (I) = ∅, we have $\overline{I} = I$.

(iii) Since D (Q) = R, we have

$$\overline{Q} = Q \cup D(Q) = Q \cup R = R$$

(iv) $\overline{]0, 1[} = [0, 1]$ (why ?)

(v) $\overline{[0, 1[} = [0, [$. (vi) $\overline{[0, 1]} = [0, 1]$.

(vii) Let] 1 – ε, 1 + ε [be any nhd of 1. Choose a positive integer n such that

$$\frac{1}{n_0} < \varepsilon.$$

Then $1 + \frac{1}{n} \in]1 - \varepsilon\ [\ \forall\ n \geq n_0$.

But $1 + \frac{1}{n} = \frac{n+1}{n} \in B$.

Hence every nhd of 1 contains infinitely many point of B ans so 1 ∈ D (B). It is obvious that no other point can belong to D (B).

Hence D (B) = {1}, ∴ $\overline{B} = B \cup \{1\}$.

Example 34(a):

Let X = {A, b, c} and let

𝔍 = {∅, {a}, {a, b}, {a, c}, X}

Find all limit points of the sets

(i) A = {b, c}, (ii), B = {a, c}

Solution:

(i) a is not a limit point of A since {a} is a $\Im$-nhd of a which contains no points of A. Again b is not a limit point of A since {a, b} is a $\Im$-nhd of b which contains no point of A other than b. Also c is not a limit point of A since {a, c} is a $\Im$-nhd of c which contains no point of A other than c. Hence A has no limit points. Thus D (A) = ∅.

(iii) Here a is not a limit point of B since {a} is a $\Im$-nhd of a which contains no point of B other than a. But b is a limit point of B since $\Im$-open nhds of b are {a, b} and X, each of which contains a point of B other than b. Note that here b is not a point of B.

Also c is a limit point of B since there are only two $\Im$-open nhds of c, namely {a, c} and X each of which contains a point of B other than c. Since b, c are limit points of B, we have

$$D (B) = \{b, c\}.$$

Example 34(b):

Show that the spaces (R, U) and (R, S) are Hausdorff.

Solution:

Let a, b be any two distinct points of R so that a ≠ b ans let | a – b | = ε. Then

$$]a - \varepsilon/3, a + \varepsilon/3 [\text{ and}] b - \varepsilon/3, b + \varepsilon/3 [$$

are disjoint U-nhds of a and b respectively.

Also [a – ε/3, a + ε/3 [and [b – ε/3, b + ε/3 [are disjoint S-nhds of a and b respectively. Hence (R, U) and (R, S) are both Hausdorff.

Example 34(c):

Consider the cofinite topology on an infinite set X and show that it is not T_2.

Solution:

For this topology no two open sets can be disjoint. For suppose, if possible, that G, H are two open disjoint subsets of X so that G ∩ H = ∅. Then

$$(G \cap H)' = \varnothing'$$

i.e., $G' \cup H' = X$. [De-morgan Law]

But G′ and H′ are finite sets so their union is also finite which is a contradiction since X is infinite. Hence no two distinct points can be separated open sets. Accordingly the space is not T_2.

Example 34(c):

Consider the co-countable topology on an uncountable X and show that it is not T_2.

Solution :

The proof is the same as that of the previous example. In this case G′ ∪ H′ is countable whereas X is uncountable.

Example 35:

Let X = {a, b, c, d, e} and let

ℑ = {∅, {b}, {d, e}, {b, d, e}, {a, c, d, e}, X}.

Then ℑ is a topology on X. Consider the subset A = {b, c, d}. Then c is a limit point of A since the ℑ-open nhds of c are {a, c, d, e} and X each of which contains a point of A other than c. But b is not a limit point of A since {b} is a ℑ-open nhd of b which contains no point of A other than b. The reader can easily verify that the other limit points of A are a, c. Hence

D (A) = {a, c, e}.

Isolated points of A are b and d since b, d belong to A but are not limit points of A. The *adherent points* of A are a, b, c, d, e.

Example 36(a):

Consider any indiscrete topological space (X, I). Let A be a subset of X containing two or more points of X. Then every point x ∈ X is a limit point of A since the only open set containing x is X which contains all the points of A and must therefore contains a point of A other than x. (Since A contains more than one point). Hence **D** *(A) = X.*

If A = {p}, consisting of a single point p. Then all the points of X other than p are limit points of A for if x (≠p) ∈ X, then the only open set containing x is X which contains the point p of A which is different from x. Hence **D** *(A) = X – {p} in this case.*

If A = ∅, then evidently no point of X can be a limit point of A ans so **D** *(A) = ∅ in this case.*

Solution:

Do yourself

Example 36(b):

Find the derived set of each of the following subset of R relative to (i) the U-topology. (ii) the I-topology and (iv) the D-topology.

(a)]0, 1[(b) [0, 1[(c)]0, 1]

(d) N i.e. the set of all natural numbers,

(e) the set Q of all rational numbers.

(f) $A = \{1/n : n \in N\}$.

(g) $B = \{-1/n : n \in N\}$.

(h) $C = \{(2n + 1)/n : n \in N\}$.

(i) $D = \{x : x = 0 \text{ or } x \in Q_+ \text{ and } x^2 < 2\}$.

i.e. the set of all non-negative rationals whose square is less than 2.

Solution:

(a) (i) Here all the points of closed interval [0, 1] are **U**-limit points of]0, 1[. Hence

D (]0, 1 [) [0, 1].

Note that 0 is a limit point because any open nhd of 0 must contain an interval of the form]– ε, ε [, ε > 0 which surely contains points of]0, 1[other than 0. As a matter of fact every nhd of 0 contains uncountably many points of]0, 1[ans so 0 is a point of condensation also. Similarly other points of [0, 1] are points of condensation of the set [0, 1[.

(ii) Any S-nhd of 0 must contain an interval of the form [0, ε [, ε > 0. Since [0, e[is S-open nhd of 0, which contains points of]0, 1 [other than 0, it follows that 0 is a S- limit point of [0, 1[. But is not a S-limit point of]0, 1[since [1, 1[[is a S-nhd of 1 which contains no point of]0. 1[.Also all points of]0, 1[are S limit points of]0, 1[as the reader can easily see. Hence D – ( ] = [0, 1 [for the S-topology for R.

(iii) For the indiscrete topology for R every point of R is a limit point of]0, 1[. Hence D ( ] 0, 1 [ ) = R

(iv) For discrete topology for R no point of R can be a limit point of]0, 1. Hence D- ( ]0, 1 [ ) = ∅.

(b) D ([0, 1 [ ) = 0, 1] for the U-topology.

(i) D ([0, 1 [ ) = 0, 1[for the S-topology.

(iii) D ([0, 1 [ ) = R for the I topology.

(iv) D ([0, 1 [ ) = ∅ for the D topology.

(c) As in (b), we have

(i) D ( ]0, 1]) = [0, 1] for the U-topology.

(ii) D ( ]0, 1]) = [0, 1] for the S-topology.

(iii) D ( ]0, 1]) = R for the I-topology.

(iv) D ( ]0, 1]) = ∅ for the D-topology.

(d) (i) The set N has no U-limit point since for each $x \in R$ we can choose $e > 0$ so small that the nhd $]x - \varepsilon, x + \varepsilon[$ contains no natural number other than x. Hence D (N) = ∅.

(ii) As in (i), we shall get D (N) = ∅ for the S-topology

(iii) D (N) R for the I-topology (why ?),

(iv) D (N) = R for the D-topology,

(e) (i) Since rational points are every where dense, every nhd of $x \in R$ must contain infinitely many rational points. Hence every point $x \in R$ is a limit point for the U-topology. Hence

D (Q) = R

(ii) Proceeding as in (i) we shall have D (Q) = R for the S-topology.

(iii) As before

D (Q) = R for the I-topology.

(iv) D (Q) = ∅ for the D-topology.

(f) (i) Here 0 is the only limit point of A for the U-topology for R since for each nhd $]-\varepsilon, \varepsilon[$, $\varepsilon > 0$ of zero, we can choose n so large that the interval [0, /n] is contained in $]-\varepsilon, \varepsilon[$ so that each nhd of 0 contains points of A other than 0. No other point of R can be the limit point of A as the reader can easily see. Hence D (A) = {0}.

(ii) For S -topology also 0 is the only limit point of A.

Hence D (A) = {0}.

(iii) D (A) = R for I-topology.

(iv) D (A) = ∅ for D-topology.

(g) (i) As in (f), we have D (B) = {0} for the U-topology.

(ii) D (B) = ∅ for the S-topology.

Note that here 0 is not a limit point of B for the S-topology. Since [0, 1] is a S-nhd of 0 which contains no point of B.

(iii) D (B) = R for the I-topology

(iv) D (B) = ∅ for the D-topology.

(h) We leave it to the reader to verify that

(i) D (C) = {2} for the U-topology.

(ii) D (C) = {2} for the S-topology.

(iii) D (C) = R for the I-topology.

(iv) D (C) = ∅ for the D-topology.

(1) (i). Here every point of the closed interval $[0, \sqrt{2}]$ is a limit point of D for the U-topology. Hence

$$D(D) = [0, \sqrt{2}].$$

Note that $\sqrt{2}$ is a limit point of D since every nhd of $\sqrt{2}$ must contain infinitely many rational points whose square is less than 2.

(ii) Here $\sqrt{2}$ is not a limit point of D for the S-topology since $[\sqrt{2}, 2[$ is a S-nhd or $\sqrt{2}$ which contains no rational number whose square is less than 2. Also all the points of the interval $[0, \sqrt{2}[$ are the limit points of D for the S-topology. Hence $D(D) = [0, \sqrt{2}$ in this case.

(iii) Here $D(D) = R$ for the I- topology.

(iv) $R(D) = \varnothing$ for the D-topology.

Example 37:

Consider the topology $\mathfrak{J}$ on N which consists of $\varnothing$ and all subsets of N of the form

$$A_n = \{n, n + 1, n + 2 ... \}, n \in N$$

Find (i) the derived set of the set $A = \{3, 12, 28, 35\}$

(ii) those subset G of N for which $D(G) = N$.

(iii) $D(\{1\})$.

Solution:

(i) Here open sets containing any point $m \in N$ are the sets A_n where $n \leq m$. If $m_0 \mathfrak{J} 34$, then every open set containing m_0 also contains $35 \in A$ which is different from m_0.

Hence $m_0 \leq 34$ is a limit point of A. On the other hand, if $m_0 < 34$, the open set

$$A_{m_0} = \{m_0, m_0 + 1, m_0 + 2, ...\}$$

contains no point of A different from m_0. Hence $m_0 > 34$ is not a limit point of A.

$$D(A) = (1, 2, 3, ..., 33, 34\}$$

(ii) Let G be any infinite subset of N. Then G is not bounded from above. Hence if p be any point of N, then every open set containing p must contain points of G other than p.

Hence **D** (G) = N.

Note that if F is a finite subset of N bounded above by n_0 then the open set A_{n_0+1} contains no point of F. Hence $n_0 + 1 \in N$ is not a limit point of F and so $N(F) \neq N$.

(iii) The only nhd of 1 is N which contains n_0 point of {1} other then 1. hence 1 is not a $\mathfrak{I}$-limit point of {1}.

Also if p is any point greater then, then nhd A_p of p contains no point of {1} and so p > 1, is not a limit point of {1},. Hence no point of N can be a limit point of {1}. Thus

$$D(\{1\}) = \varnothing.$$

Example 38:

Let X = {a, b, c} and let

N (a) = {{a}, {a, c}}, N (b) = {{b},

{,b, c}}, N (c) = {X}

(i) Show that N (a), N (b) and N (c) satisfy the conditions [M1], [M2] of the preceding theorem.

(ii) if $\mathfrak{I}$ is the topology for X induced (in the sense of the preceding theorem) by N (a), N (b) and N (c), then all $\mathfrak{I}$ open subsets of X.

(iii) Find all $\mathfrak{I}$ closed subsets of X.

(iv) Is (Y, $\mathfrak{I}$) a door space ?

Solution:

(i) a ∈ {a} and a ∈ {a, c}

Also {a} ∩ {a, c} = {a} ∈ N (a).

Hence N (a) satisfies [M1] and [M2]. Similarly it can be shown that N (b) and N (c) satisfy [M1] and [M2].

(ii) Now we select the members of $\mathfrak{I}$ as follows :

{a} ∈ $\mathfrak{I}$ since {a} ∈ B (a) st. a ∈ {a} ⊂ {a}.

{b} ∈ $\mathfrak{I}$ since {{b} since {b} st. b ∈ {b}.

{c} ∉ $\mathfrak{I}$ since there is no N ∈ N (c) st. c ∈ N ⊂ {c}.

{a, b} ∈ $\mathfrak{I}$ since {a} ∈ N (a) st. ⊂ {a, b} and

{b} ∈ N (b) s.t. {b} ⊂ {a, b}

{b, c} ∉ $\mathfrak{I}$ since c ∈ {b, c} but there is no N ∈ N (c) st. Similarly {c, a}

X ∈ $\mathfrak{I}$ since ∨ x ∈ X. there is an N ∈ N (x).

s.t. x ∈ N ⊂ Y.

Hence $\mathfrak{I}$ = {∅, {a}, {b}, {a, b}, X}

(iii) $\mathfrak{I}$-closed subset of X are.

X, {b, c}, {c, a} {c}, ∅.

(iv) Since every subset of X is either $\Im$-open or $\Im$-closed, it follows that (X, $\Im$) is a door space.

Example 39:

Which of the following subsets of R are U-neighbourhoods of 1 ?

(i) [0, 2[, (ii) [0, 2], (iii) [1, 2],

(iv)]1, 2],(v) [1, 2[, (vi) [0, 2] $- \left\{1\frac{1}{2}\right\}$

(vii) $[0,2]-\left\{\frac{1}{2}\right\}$ *(viii) R.*

Solution:

(*i*) Since]0, 2[is U-open set containing 1, it follows that]0, 2[is U-nhd of 1.

(ii)]0, 2] is U-nhd of 1 since]0, 2[is U-open set such that 1 ∈]0, 2 [⊂] [0, 2].

(iii) [1, 2] is not a U-nhd of 2 since there is no U-open set contained in [1, 2] and containing 1.

(iv)]1, 2] is not a U-nhd of 1 since]1, 2] does not contain 1.

(v) [1, 2[is not a U-nhd of 1 since there exits no U-open set G such that 1 ∈ G ⊂ [1, 2[.

(vi) [0, 2] $- \left\{1\frac{1}{2}\right\}$ is a U-nhd of 1 since [0, $1\frac{1}{4}$ [is a U-open set such that

$$1 \in \text{-}, 1\frac{1}{4} [\subset [0, 2] - \left\{1\frac{1}{2}\right\}.$$

(*vii*) [0, 2] $- \left\{\frac{1}{2}\right\}$ is a U-nhd of 1 since] $- -\frac{3}{4}$, 2 [is a U-open set *i.e,*

$$1 \in \left]\frac{3}{4}, 2\right[\subset [0,2]-\left\{\frac{1}{2}\right\}.$$

Example 40:

Let X be a topological space. If FD is a closed subset of X and x ∈ F′, prove that there exists a nhd N of x such that N ∩ F = ∅.

Solution:

Since F is closed, F$\Im$ is open and F$\Im$ contains a nhd of each its points. Hence there exits a nhd N of, x such that

$$N \subset F' \text{ i.e., } N \cap F = \varnothing.$$

Properties of neighbourhoods : We now prove several essential properties of neighbourhoods which are usually known *Neighbourhood axioms.*

Example 41:

A topological space (X, $\Im$) it said to be a door space iff every subset of X is either open or closed. Give an example of a door space.

Solution:

Let $X = \{a, b, c\}$ and let

$$\Im = \{\varnothing, \{b\}, \{a, b\}, \{b, c\}, X\}.$$

Then closed sets are

$$X, \{a, c\}, \{c\}, \{a\}, \varnothing$$

Hence all the subsets of X are either closed or open and consequently (X, $\Im$) is a door space.

Example 42:

If $a \in R$, show that $\{a\}$ is a closed set in the usual topology for R.

Solution:

$\{a\}' =]-\infty, a[\cup], a, \infty[$. But] $-\infty$ a[and]a, ∞[are U-open sets as the reader can easily show. Hence their union is also U-open. It follows that $\{a\}'$ is U-open. Therefore $\{a\}$ is U closed.

Example 43(a):

Let $a, b \in R$ where $a < b$. Show that the closed interval [a, b] is closed in the usual topology for R.

Solution:

$$[a, b]' = \{x \in R : x < a \text{ or } x > b\}.$$
$$=]-\infty, a\, [\cup]\, b, \infty\, [$$

which is U-open, being the union of two U-open sets.

Hence [a, b] is U closed.

Example 43(b):

Show that the right half-open interval [a, b[is neither U-closed nor U-open. But it is both S-open and S-closed.

Solution:

$$[a, b[' =]-\infty, a\, [\cup\, [b, \infty[. \quad \text{...(1)}$$

Since there exists no U-open interval containing b and contained in [a, b[', is follows that [a, b[' is not U-open. Hence its complement [a, b[is not U-closed. We also see that [a, b[is not U-open since there exist not U-open interval containing a and contained in [a, b[. Thus [a, b[is neither U-closed nor U-open. Similarly the left half-open interval]a, b] is neither U-closed nor U-open.

[a, b[is S-open by definition. Since [$-\infty$, a [and [b, ∞ [are both S open, we see from (1) that [a, b[' is S-open and so [a, b[is S-closed.

Example 43(c):

Give an example of a space which is connected but not locally connected.

Solution:

Consider the subspace Y of the Euclidean plane **R2** defined by $Y = A \cup B$ where

$$A = \{(0, y)\} : y \in [-1, 1]\} \text{ and}$$

$$B = \{(x, y) : x \in]0, 1] \text{ and } y = 1/x\}$$

Now B is the image of the interval]0, 1[under the continuous mapping f defined by f (x) = x sin (1/x). Since]0, 1] is connected, B is also connected. It is easy to see that every nhd of a point $x \in A$ contains an infinite number of points of B. Hence every point of A is a limit point of B. Also no point outside $A \cup B$ can be a limit point of B. Hence $\overline{B} = A \cup B = Y$.

It then follows that $\overline{B}$ i.e., Y is connected. However, X is not locally connected; since it can be easily shown that each point $x \in A$ has nhd which does not contain any connected nhd of x.

Example 44:

Show that the space (R, S) is totally disconnected.

Solution:

Let x, y be any two distinct real numbers and let $x < y$. Let A =]$-\infty$, y[and B = [y, ∞[. Then A, being U-open is also S-open since S is finer than U. Also B is S-open. Since $x < y$, we have $x \in A$. Also $y \in B$. Further $A \cup B = R$ and $A \cap B = \varnothing$. Hence $A \cup B$ is a disconnection of R with $x \in A$ and $y \in B$. It follows that (R, S) is totally disconnected.

Example 45(a):

Show th*at the set Q of all rational numbers with the relativized usual topology is totally disconnected.*

Solution:

Let x, y be any two distinct rational numbers and let x < y. Then there exists an irrational number p such that x < p < y. Let A = Q ∩] –∞, p [and B = Q ∩], ∞[.

Then A ∪ B = (Q ∩) –∞,

p [) ∪ (Q ∩]p, ∞[) = Q

Also A ∩ B = ∅. Hence A ∪ B is a disconnection of Q with x ∈ A and y ∈ B. It follows that Q is totally disconnected.

Example 45(b):

Show that the set of all irrational numbers with the relativized usual topology is totally disconnected.

Solution:

Solution of this problem is on the same lines as that of the preceding example since between any two irrational numbers there is a rational number.

Example 46:

Explain how 'neighbourhood' can be used as primitive notion to define a topology on a set.

Solution:

Do yourself.

Example 47:

Let X be a non-empty set and for each point x ∈ X, let N (*x*) *denote the collection of subset of X containing x. Show that N (x) satisfies all the conditions and determine the induced topology on X.*

Solution:

[N0] Since Y is non-empty, there exists a point x ∈ X. Since X contains x, we have

X ∈ N (x) and so, N (x) ≠ ∅,

[N1] : By hypothesis, x belongs to each member of N (x), that is N ∈ N (x) ⇒ x ∈ N.

[N2] : Let N ∈N (x) and let M ⊃ N. Then by definition of N (x), x ∈ N. Since M ⊃ N, we have x ∈ M. Hence M ∈ N (x). Thus we have shown that

N ∈ N (x), M ⊃ N

$\Rightarrow$ $M \in N(x)$

[N3] : Let M, N $\in$ N (x). Then x $\in$ M and x $\in$N and so x $\in$ M $\cap$ N. Hence M $\cap$ N $\in$ N (x).

[N4] : Let N $\in$ N (x). By definition of N (x), N has the property that N $\in$ N (x) for every x $\in$ N and so [N4] is also satisfied.

Now we fine dht topology $\mathfrak{I}$. A subset G of X is opening the induced topology iff G $\in$ N (x) for every x $\in$ G.

Since by def. of N (x) every subset of X has this property, the induced topology of X is the discrete topology.

Example 48:

Let X = N and let $\mathfrak{I}$ consists of $\varnothing$ and all those subsets of N whose complements are finite. Finite (i) the limit points, (ii) the cluster points of the sequences.

$$< 1, 3, 5... > \textit{ and } < 1, 2, 1, 2, ... > \text{ in N.}$$

Solution:

(i) Consider the sequence < 1, 3, 5, ... >. Let V be any $\mathfrak{I}$-nhd of 1. Then by definition of $\mathfrak{I}$, V contains all points of N except a finite number and since all the terms of this sequence are distinct, it follows that the sequence is eventually in V.. Hence 1 is a limit point of this sequence. Similarly 3, 5, 7... are limit points of this sequence.

Now consider the sequence < 1, 2, 1, 2, 1, 2, ... >. The set N – {2} is a $\mathfrak{I}$-open nhd of 1. This sequence is however not eventually in this neighbourhood. Hence 1 is not a limit of this sequence. Similarly 2 is also not a limit point of this sequence.

(ii). Since 1, 3, 5, ... are the limit points of the sequence < 1, 3, 5, ... >, they are certainly cluster points of this sequence. Now consider the sequence < 1, 2, 1, 2, ... >. Let V be any $\mathfrak{I}$-nhd of 1. Then this sequence is frequently in V for it contains an infinite number of terms of this sequence namely, s_1, s_3, s_5, ... Similarly 2 is a cluster point this sequence.

EXERCISES

1. Let X be an infinite set and F the filter on X generated by the filter base consisting of all complements of finite sets. To which points does F converge if X is given the cofinite topology ?
2. Show that a filter F on a set X converges to x $\in$ X iff the net based on F converges to x.
3. Let Y be a topological sub-space of a space X, let Y $\in$ Y and let B be a filter base on Y. Show that B $\to$ y in Y iff B $\to$ y in X.

4. Which filters F on a set X will converge to $x \in X$ if (i) X has the descrete topology, (ii) X has the indiscrete topology ?
5. Show that the collection of all discs containing the origin in their interiors is a filter base on R^2.
6. Prove that a filter in a Hausdorff space can converge to at most one point.
7. Let $f : X \to Y$, and let B be a filter base on X. Show that f [B] is a filter base on Y, and if F is a filter on X, then f [F] is a filter on Y if and only in f is onto.
8. Show that a filter F on a set X is an ultrafilter only if every net base on F is an ultranet.
9. Show that a necessary and sufficient condition that a filter F on a set X be an ultrafilter is that given any subset A of X, either $A \in F$ or $X - A \in F$.
10. If B is a filter base, than $\{\overline{B} : B \in B\}$ is also a filter base, and generates a smaller filter.
11. Show that if a filter in a metric space converges, it must converges to a unique limit.

    [**Hint** : Since every metric spaces is a Hausdorff space].
12. Show that a filter F on a space X converges to $x \in X$ iff there exists a filter F′ finer than F such that F′ converges to x.
13. Show that a net $\{f_a : a \in A\}$ in a set X converges to $x \in X$ iff the filter generated by $\{f_a\}$ converges to x.
14. Let Y be a proper subset of a set X and let F be a filter on Y. Prove that F is not a filter on X $\{\because X \notin F]$. But F be a filter base on X,. Also prove that if B is a filter base on Y, then B is also a filter base on X.
15. Show by means of an example that a base for a topology is not a filter base, but a local base is always filter base.
16. Show that if x is an isolated point of V, then

    $\{D : D = N - (x)$ and $N \in N(x)\}$ does not generate a filter on X.
17. Let X be a space and let x be a non-isolated point of X. Show that $\{N : N = V - \{x\}\}$ and $V \in N(x)$, generates a filter on X which contains the filter N (x) of all nhds of x in X.
18. Exhibit a sequence $< x_n >$ in a set X and a subset of $< x_n >$ which is not a sequence.

19. Show that every subnet of an ultranet is an ultranet.
20. Show that a space X is compact iff every filter on X has a cluster point.
21. Show that a filter F on a space X has x as a cluster point iff there is a filter finer than F which con verges to x.
22. Show that every net has a subnet which is an ultranet.
23. If every subnet of a net $\{f_a : a \in \Lambda\}$ in a space X has a subnet covering in $x_0 \in X$ (fa) converges to x_0.
24. Show that a mapping g of a space X into a space Y is continuous at $x_0 \in X$ iff f or every net {f, X, A, ≠} converging to x_0, the composition map g o f : A → Y converges to g (x).
25. Let $\{f_a : a \in A\}$ be a net in X. Let $S_1, S_2, \ldots S_n$ be linitely many subset of X and suppose that for each i = 1, 2, ... n. $f_a \in S$, eventually.

    Then $f_a \in \bigcup_{i=1}^{n} S$ eventually.

    [**Hint :** For each i, choose $a_i \in A$ such that $a \geq a$, implies $f_a \in S_i$. Since A is a directed set, there exists $a_0 \in A$ satisfying $a_0 \geq a_i$ for each i. Then $a \geq a_0$ implies $a \geq a_i$ and therefore $f_a \in S_i$ each i]
26. If a sequence s = < s_n: n ∈ N > in a topological space X has a subsequence s o ϕ = {s (ϕ (n) : n ∈ N} converging to a point x_0 in X.
27. Let X be a T_1-space. Show that F ⊂ X is closed iff whenever < sn > is contained in F and $s_n \to s_0$., then $s_0 \in F$.

    Let $\Im$ be the countable topology on an uncountable set X. Show that a sequence s = < s_n > in X is convergent iff it is eventually constant, that is, $s_n \to s_0$ iff $s_n = s_0$ eventually.

    [**Hint :** $s_n = s_0$ eventually ⇒ $s_n \to s_0$ is clear. Now suppose that s is sequence in X, $s_0 \in X$, and it is false that $s_n = s_0$ eventually. Let F = $\{s_n : s_n \neq s_0\}$. Then F = $\{s_n : s_n \neq s_0\}$. Then F is countable and hence F′ is a nhd of s_0. But it is false that sn ∈ F′ eventually. Hence < s_n > does not converge to s_0].
28. Give an example of a non-compact space X such that each sequence in X has cluster point.

2

COMPACTNESS AND CONNECTEDNESS

INTRODUCTION TO COMPACTNESS

The property we have in mind is expressed by the *Heine-Borel theorem*, which asserts the following: if X is a closed and bounded subset of the real line R, then any class of open subsets of R whose union contains X has a finite subclass whose union also contains X. If we regard X as a topological space in its own right, as a subspace of R, then this theorem can be thought of as saying that any class of open subsets of X whose union is X has a finite subclass whose union is also X. As is often the case with crucial theorems in analysis, the conclusion of the Heine-Borel theorem is converted into a definition in topology. This definition singles out for special attention what are called *compact topological spaces*. Like many other notions in topology, the concept of compactness for a topological space is an abstraction of an important property possessed by certain sets of real numbers. The Heine-Borel theorem has a number of profound and far-reaching applications in analysis.

Many of these guarantee that continuous functions defined on closed and bounded sets of real numbers are well behaved. For instance, any such function is automatically bounded and uniformly continuous. In contrast to this satisfying behaviour, we note that the function f defined on the open unit interval (0, 1) by f(x) = 1/x is neither bounded nor uniformly continuous.

Definitions:

(1) A subset of A of a topological space X is said to be compact if every open cover of A has finite subcover, that is, iff for every collection $\{G_\lambda : \lambda \in \Lambda\}$ *of open sets for which*

$$A \subset \cup \{G_\lambda : \lambda \in \Lambda\},$$

there exists finitely many sets $G_{\lambda_1}, \ldots G_{\lambda_n}$ *among the* G_λ*'s such that*

$$A \subset G_{\lambda_1} \cup \ldots \cup G_{\lambda_n}.$$

In particular the space X is said to be compact iff for every collection $\{G_\lambda : \lambda \in \Lambda\} \cup \phi$ *open sets for which*

$$X = \cup \{G_\lambda : \lambda \in \Lambda\}$$

there exists finitely many sets $G_{\lambda_1} ... G_{\lambda_n}$ *among the* G_λ*'s such that there*

$$X = G_{\lambda_1} \cup ... \cup G_{\lambda_n}.$$

It is easy to see that a finite union of compact sets is compact.

(2) By an open cover (or covering) of a subset A of a topological space X, we mean a collection **C** $= (G_\lambda : \lambda \in \Lambda\}$ *of open subsets of X such that* $A \cup \{G_\lambda : \lambda \in \Lambda\}$*. We then say that C covers A. In particular. C is said to be an open cover of the space X iff* $X = \cup \{G_\lambda : \lambda \in \Lambda\}$*. By a subcover of an open cover C of A, we mean subcollection C' of C covers A. An open cover of A is said to be finite if it consists of finite number of open sets.*

Example:

Let $C_1 = \{] = n, n [: n \in N\}$,

$C_2 = \{] - 3n, 3n [:n \in N\}$ and

$C_3 = \{] 2n - 1, 2n + 1 [,] 2n, 2n + 2 [:n \in N\}$

Then C_1, C_2, C_3 *are all U-open covers of R. Also* C_2 *is subcover of* C_2.

COMPACTNESS IN R

Theorem 1:

Every closed and bounded interval on **R** *is compact where* **R** *has the usual topology* **U**.

Proof :

Let $I_1 = [a, b]$ $(a < b)$ be a closed and bounded interval on R. To show that [a, b] is compact. It suffices to consider U-open covers of [a, b] instead of relativized U-open covers. Suppose I_1 is not compact. Then there exists an open cover **C** $= \{G_\lambda : \lambda \in \Lambda\}$ of I_2 which has no finite subcover. Bisect I_1 into two closed intervals I_{11} and I_{22}. At least one of these must have no finite subcover. We relabel this interval as I2. Bisect I_2 and obtain I_3 (as we obtained I_2) which has no finite subcover. Continuing in this manner, we obtain a nested sequence of closed intervals $< _{In} >$ such that their length $| I_n |$ $\to 0$ as n $\to$ and such that for every n, In has no finite subcover. By Cantor's intersection theorem, there exists a real number $x \in \cap (I_n : n \in N\}$. Then $x \in I_n \subset I_1 \subset \cap \{G_\lambda : \lambda \in \Lambda\}$ so that $x \in$ Gl for some $\lambda \in \Lambda$. Since G_λ is U-open there exists $\varepsilon > 0$ such that

$$] x - \varepsilon, x + \varepsilon [\subset G_\lambda.$$

Now we can take n so large that $I_n \subset]\, x - \varepsilon, x + \varepsilon\, [\subset G_\lambda$, whence I_n is covered by a single member G_λ of **C**. But this contradicts the fact that **C** has no finite subcover of I_1. Hence [a, b] must be compact.

Theorem 2(a) (Heine-Borel Theorem):

A subset of **R** *is compact if and only if A is bounded and closed.*

Proof :

Let A be bounded and closed. Since A is bounded, there exists a, b ∈ **R**; a < b such that A is closed in **R**, it follows that A is compact. Conversely, let A be compact. The collection **C** = $\{G_a : a \in A\}$ where Ga =] a – 1, a + 1[is an open cover of A. Since A is compact, there exist finite number of points $_{al,}$ ... an in A such that

$$A \subset G_{a_1} \cup \ldots \cup G_{a_n}$$

Let $\quad M_0 = \max\{a_1, \ldots, a_n\}$

and $\quad m_0 = \min\{a_1, \ldots, a_n\}$

Then $\quad G_{a_1} \cup \ldots \cup G_{a_n} \subset [m_0 - 1, M_0 + 1]$

and so $\quad A \subset [m_0 - 1, M_0 + 1]$.

Hence, A is bounded. Further since A is a compact subset of Hausdorff space **R.** A is closed. Thus the theorem is completely established.

Theorem 2(b):

The space (**R, U**) *is not compact where U denotes the usual topology on* **R.**

Proof :

Let **C** = $\{G_n : n \in N\}$ where $G_n =]-n, n[$. Then it is clear that **C**, is a **U-open cover of R.** If $\{N_{n_1}, \ldots N_{n_2}, \ldots N_{x_1} N_{n_k}\}$ is any finite subcollection of **C**, let $n_0 = \max\{n_1, n_2, \ldots, n_k\}$. Then $n_0 \in G_{n_i}$, i = 1, ... k. Since $n_0 \in$ **R**, we see that no finite subcollection of **C** can cover **R.** Hence **R** is not compact.

Theorem 3:

Show that Cantor's set Γ is Compact.

Proof :

(i) Recall that $\Gamma = \bigcap_{n=1}^{\infty} F_n$ where $F_0 = [0, 1]$,

$$F_1 = \left[0, \frac{1}{3}\right] \cup \left[\frac{2}{3}, 1\right], F_2 = \left[0, \frac{1}{9}\right] \cup \left[\frac{2}{9}, \frac{1}{3}\right] \cup \left[\frac{2}{3}, \frac{7}{9}\right]\left[\frac{8}{9}, 1\right]$$

and in general F_n is the union of 2^n disjoint closed intervals (each of length 1/3).

Each F_n is closed being a finite union of closed intervals. It follows that Γ is closed. Also Γ is bounded since $\Gamma \subset [0, 1]$. Hence Γ is compact by the previous theorem.

COMPACTNESS IN R^N

Lemma : *if* $< In >$ *is a sequence of closed intervals in* **R** *such that* $In \supset In - 1$ (n = 1, 2, 3, ...), *then* $\cap \{In : n \in N\} \neq \varnothing$.

Proof :

Let $I_n = [a_n, b_n]$ and suppose A is the set of all a_n. Then A is non-empty and bounded above (by b_1). Hence A must have the least upper bound, say u, If m and n are positive integers, then

$$a_n \leq a_{m+n} \leq b_{m+n} \leq b_m.$$

so that $u \leq bm$ for each m, Also obviously $am \leq u$ for each m,

It follows that $u \in I_m$ for m = 1, 2, 3, ...

Hence $\cap \{I_n : n \in N\} \neq \varnothing$.

Lemma : *Let k be a positive integer. If* $< I_n >$ *is a sequence of k-cells in* $\mathbf{R}^n$ *such that*

$$In \supset I_n + 1 \text{ (n = 1, 2, 3, ...) then } \cap [I_n : n \in N\} \neq \varnothing.$$

Proof :

First note that a k-cell I_n consists of all points $x = \{x_1 \, x_2, ... \, x_k\}$ such that

$$a_n, i \leq x_i \leq b_n, i \text{ (}1 \leq k; n = 1, 2, 3,...)$$

Put $I_{n,i} = [a_{n,i}, b_{n,i}]$. For each i, the sequence $< I_n, i >$ satisfies conditions of the previous theorem. Hence there exists real numbers x_i $(1 \leq i \leq k)$ such that

$$a_{n,i} \leq x_i \leq b_{n,i} \text{ (}1 \leq i \leq k, n = 1, 2, 3, ...)$$

If we set $x^* = \left(x_1^* ... x_k^*\right)$,

we see that $x^* \in I_n$ for n = 1, 2, 3, ...

so that $\cap \{I_n : n \in N\} \neq \varnothing$.

Theorem 1:

(Generalized Heine-Borel Theorem). *A subset of* ***Rn*** *is compact is and only f it is closed and bounded.*

Proof :

The 'if part' : Let A be a closed and bounded subset of $\mathbf{R}^n$. Then A $\subset$ I for some k-cell I. I is compact. Thus A is a closed subset of a compact set I and so by theorem, A is compact.

The 'only if' part : Let A be a compact subset of $\mathbf{R}^n$. To show that A is closed and bounded. Since $\mathbf{R}^n$ is Hausdorff, by theorem, A is closed. So we need only show that A is bounded. Suppose A is bounded. Then A contains points xn with

$$|x_n| > n \ (n = 1, 2, 3, ...)$$

The set consisting of these points x_n is infinite and evidently has no limit point in $\mathbf{R}^n$ and hence has none in A. But this contradicts theorem. Hence A must be bounded.

Theorem 2(a): (The Generalized Heine-Borel Theorem):

Every closed and bounded subspace of R^n is compact.

Proof:

A closed and bounded subspace of R^n is a closed subspace of some closed rectangle, it suffices to show that any closed rectangle is compact as a subspace of R". Let X = $P_{i=1}^{n}$ $[a_i,b_i]$ be a closed rectangle in R^n. Each coordinate space $[a_i,b_i]$ is compact by the classical Heine-Borel theorem, so by Tychonoff's theorem it suffices to show that the product topology on X is the same as its relative topology as a subspace of R^n. It is easy to see that the open rectangles in R^n form an open base for its usual topology, that is, for its metric topology, and from this it follows that the product topology on R^n is the same as its usual topology. The relative topology on X is the weak topology generated by its n projections onto the coordinate spaces $[a_i, b_i]$; but this is the product topology on X, so the proof is complete.

The n-dimensional Euclidean space R^n is the most important example of a type of topological space which is of great significance in modern analysis, especially in the theory of integration. A topological space is said to be *locally compact* if each of its points has a neighbourhood with compact closure. It is easy to see by the above theorem that R^n actually is locally compact, because any open sphere centered on any point is a neighbourhood of the point whose closure, being a closed and bounded subspace of R^n, is compact. It is trivial that any compact space is locally compact, for the full space is a neighbourhood with compact closure of every point in the space. We return to the study of locally compact spaces, where we give a more detailed analysis of their structure and properties.

Theorem 2(b):

Every k-cell is compact.

Proof :

Let I be a k-cell consisting of all points $\mathbf{x} = (x_1, ..., x_k)$ such that ai ≤ $x_i \leq b_i$ $(1 \leq i \leq k)$

Let $\delta = \left[\sum_{i=1}^{k} (b_i - a_i)\right]^{1/2}$. Then $\mathbf{x} \in I, y \in I$ implies $|x - y| < \delta$. Suppose, if possible, I is not compact

$$C = [G_\lambda : \lambda \in \Lambda\}$$

of I which contains no finite subcover. Put $c_i = (a + b_i)/2$. Then the intervals $[a_i, c_i]$ and $[c_i, b_i]$ determine 2^k k-cells Q_i whose union is I. At least one of these sets Qi, cell it I_1 cannot be so covered by any finite sub-collection of **C** (otherwise I could be son covered). We next divide I_1 and continue the precess ad infinitum. We thus obtain a sequence $< I_n >$ of k-cells with the following properties

(i) $I \supset I_1 \supset I_2 ...$

(ii) In is not covered by any finite subcollection of **C**.

(iii) $\mathbf{x} \in I_n, y \in I_n \Rightarrow |\mathbf{x} - \mathbf{y}| \leq 2^{-n} \delta$

By (i) and lemma (1.21), there is a point $\mathbf{x}^* \in G_\lambda$ for some l ∈ Λ. Since G_λ is open, there exists r > 0 such that $|\mathbf{x} - \mathbf{y}^*| < r$ implies that $y \in G_\lambda$. Choose n so large that $2^{-n} \delta < r$. Then (iii) implies that $I_n \subset G_\lambda$ which contradicts (ii). The proof is now complete.

UNIFORM CONTINUITY AND COMPACTNESS

Theorem 1:

Let f be a continuous mapping of compact Measure Theory (X, d) into a Measure Theory (Y, ρ). Then f is uniformly continuous on X.

Proof :

Let ε > 0 be given. Since f is continuous, to each p ∈ X, there exists $\delta_p > 0$ such that

$$q \in X, d(p, q) < \delta_p \Rightarrow \rho(f(p), f(q)) < \varepsilon/2. \quad ...(1)$$

The collection $\left\{S\left(p, \frac{1}{2}\delta_p\right) : p \in X\right\}$ of open spheres is evidently and open cover of X. Since X is compact, there exists a finite set of points $p_1, ... p_n$ such that

$$X = S\left(p_1, \frac{1}{2}\delta_{p_1}\right) \cup \ldots \cup S\left(p_n, \frac{1}{2}\delta_{p_2}\right) \qquad \ldots(2)$$

Let $\delta = 1/2 \min \{\delta_{p_1}, \ldots, dpn\}$. Then $\delta > 0$ since the minimum of a finite set positive numbers is positive,. [Note that this is one point where the finiteness of the covering inherent in the definition of compactness is essential. The glb of an infinite set of positive numbers may very well be zero.].

Now let x, y be any two points of X such that d (x, y) < δ. By (2), there exists an integer m, ≤ n, such that $x \in S\left(p_m, \frac{1}{2}\delta p_m\right)$ so that

$$d(x, p_m) > \frac{1}{2}\delta_{p_m} \qquad \ldots(3)$$

Also $d(y, p_m) \leq d(y, x) = d(x, p_m) < \delta + \frac{1}{4}\delta_{p_m}$

$$\leq \frac{1}{2}\delta_{p_m} + \frac{1}{2}\delta_{p_m} \qquad \text{[See the choice of d]}$$

$$= \delta_{p_m} \qquad \ldots(4)$$

Finally, we have

$$\rho(f(x), f(y) \leq \rho(f(x), f(p_m)) + \rho\ (f(p_m), f(y))$$

[By triangle inequality]

$$< \varepsilon/2 + \varepsilon + \varepsilon/2 = \varepsilon \text{ by (1), (3) and (4).}$$

Thus we have shown that for every ε > 0, there exists a δ > 0 such that

$$x, y\ \varepsilon\ X,\ d(x, y) < \delta \Rightarrow \rho(f(x), f(y)) < \varepsilon.$$

Hence f is uniformly continuous on X.

CONTINUITY AND LOCAL COMPACTNESS

Theorem 1:

Let f be a mapping of a locally compact space X onto a Hausdorff space Y. If f is both continuous and open, then Y is also locally compact.

Proof :

Let y = Y. Since f is onto Y, there exists x ∈ X such that f (x) = y. By local compactness of X, there exists an open nhd N of x such that $\bar{N}$ is compact. But then

$$x \in N \subset \bar{N} \Rightarrow Y = f(x) \in f[N] \subset f[\bar{N}].$$

Since f is open map, f [$\bar{N}$] is open Y and so f [$\bar{N}$] is a nhd of y. Since f is continuous and $\bar{N}$ is compact, if follows that f [$\bar{N}$] is compact in Y. Moreover, since Y is Hausdorff, f [$\bar{N}$] is closed in Y so that

$$\overline{f[N]} = f[\bar{N}].$$

Thus we have shown that every point y in Y has a nhd whose closure is compact in Y and so y is locally compact.

COMPACT SPACES

Let X be a topological space. A class $\{G_i\}$ of open subsets of X is said to be an *open cover* of X if each point in X belongs to at least one G_i, that is, if $\cup_i G_i = X$. A subclass of an open cover which is itself an open cover is called a subcover. A *compact space* is a topological space in which every open cover has a finite subcover. A *compact subspace* of a topological space is a subspace which is compact as a topological space in its own right. We begin by proving two simple but widely used theorems.

Theorem 1:

A topological space is compact ⇔ every class of closed sets with empty intersection has a finite subclass with empty intersection.

Proof:

This is a direct consequence of the fact that a class of open sets is an open cover ⇔ the class of all their complements has empty intersection.

A class of subsets of a non-empty set is said to have the *finite intersection property* if every finite subclass has non-empty intersection. This concept enables us to express Theorem C as follows.

Theorem 2(a):

A topological space is compact ⇔ every class of closed sets with the finite intersection property has non-empty intersection.

Let X be a topological space. An open cover of X whose sets are all in some given open base is called a basic *open cover*, and if they all lie in some given open subbase, it is called a *subbasic open cover*. We observe the trivial fact that if X is compact, then every basic open cover has a finite subcover. Our next theorem asserts that compactness not only implies this property, but is also implied by it.

Theorem 2(b):

A topological space is compact if every basic open cover has a finite subcover.

Proof:

Let $\{G_i\}$ be an open cover and $\{B_i\}$ an open base. Each G_i is the union of certain B_j's, and the totality of all such B_j's is clearly a basic open cover.

By our hypothesis, this class of B_j's has a finite subcover. For each set in this finite subcover we can select a G_i which contains it. The class of G_i's which arises in this way is evidently a finite subcover of the original open cover.

We go one more step in this direction and prove a similar (and much deeper) theorem relating to subbasic open covers. The proof is rather difficult, and we introduce the following concepts in an effort to make it as simple as possible. They also make some of its applications considerably easier to handle. Let X be a topological space. A class of closed subsets of X is called a *closed base* if the class of all complements of its sets is an open base, and a *closed subbase* if the class of all complements is an open subbase. Since the class of all finite intersections of sets in an open subbase is an open base, it follows that the class of all finite unions of sets in a closed subbase is a closed base. This is called the closed base *generated* by the closed subbase.

Theorem 3(a):

A topological space is compact if every subbasic open cover has a finite subcover, or equivalently, if every class of subbasic closed sets with the finite intersection property has non-empty intersection.

Proof:

The equivalence of the stated conditions is an easy consequence of Theorems C and D. Consider a closed subbase for our space, and let $\{B_i\}$ be its generated closed base, that is, the class of all finite unions of its sets. We assume that every class of subbasic closed sets with the finite intersection property has non-empty intersection, and we prove from this that every class of B_i's with the finite intersection property also has non-empty intersection. By Theorem E, this will suffice to prove our theorem.

Let $\{B_i)$ be a class of B_i's with the finite intersection property. We must show that $\cap_i$, B, is non-empty. We use Zorn's lemma to show that $\{B_i\}$ is contained in some class $\{B_k\}$ of B_i's which is maximal with respect to having the finite intersection property, in the sense that $\{B_k\}$ has this property and any class of B_i's which properly contains $\{B_k\}$ fails to have this property. The argument runs as follows. Consider the family of all classes of B_i's which contain $\{B_j\}$ and have the finite intersection property. This is a partially ordered set with respect to class inclusion. If we consider a chain in this partially ordered set, the union of all classes in it is a class of Bi's which contains every member of the chain and has the finite intersection property, as we see from the fact that every finite class of its sets is contained in some member of the chain, and that member has the finite intersection property. We conclude that every chain in our partially ordered set has an upper bound, so Zorn's lemma guarantees that the partially ordered set has a maximal element. This argument yields the existence of a class $\{B_k\}$ with the properties stated above. Since $\cap_i B_k \subseteq \cap_j B_i$, it now suffices to show that $\cap_k B_k$ is is non-empty.

Each B_k is a finite union of sets in our closed subbase, for instance, $B_1 = S_1 \cup S_2 \dots \cap S_n$. It now suffices to show that at least one of the sets $S_1, S_2, \dots, S_n$ belongs to the class $\{B_k\}$. For if we obtain such a set for each B_k, the resulting class of subbasic closed sets will have the finite intersection property (since it is contained in $\{B_k\}$), and therefore, by our hypothesis relating to the subbasic closed sets, it will have non-empty intersection; and since this non-empty intersection will be a subset of $\cap_k B_k$, we shall know that $\cap_k B_k$ is itself non-empty.

We finish the proof by showing that at least one of the sets $S_1, S_2, \dots, S_n$ does in fact belong to the class $\{B_k\}$. We assume that each of these sets is not in this class, and we deduce a contradiction from this assumption. Since S_1 is a subbasic closed set, it is also a basic closed set; and since it is not in the class $\{B_k\}$, the class $\{B_k, S_1\}$ is a class of B_i's which properly contains $\{B_k, S_1\}$. By the maximality property of $\{B_k\}$, the class $\{B_k, S_1\}$ lacks the finite intersection property, so S_1 is disjoint from the intersection of some finite class of B_k's. If we do this for each of the sets $S_1, S_2, \dots, S_n$, we see that B_k—the union of these sets—is disjoint from the intersection of the total finite class of all the B_k's which arise in this way. This contradicts the finite intersection property for the class $\{B_k\}$ and completes the proof.

The great power of this theorem can be surmised from the complexity of its proof. It is really a tool, and we illustrate the manner in which it can be used by applying it to give a simple proof of the classical Heine-Borel theorem stated in the introduction to this chapter.

Theorem 3(b): (The Heine-Borel Theorem):

Every closed and bounded subspace of the real line is compact.

Proof:

A closed and bounded subspace of the real line is a closed subspace of some closed interval [a, b], and by Theorem A it suffices to show that [a, b] is compact. If a = b, this is clear, so we may assume that a < b. We know that the class of all intervals of the form [a, d) and (c, b], where c and d are any real numbers such that a < c < b and a < d < b, is an open subbase for [a, b]; therefore the class of all [a, c]'s and all [d, b]'s is a closed subbase. Let $S = \{[a, c_i], [d_j, b]\}$ be a class of these subbasic closed sets with the finite intersection property. It suffices by Theorem 6 to show that the intersection of all sets in S is non-empty. We may assume that S is non-empty. If S contains only intervals of the type $[a, c_j]$, or only intervals of the type $[d_j, b]$, then the intersection clearly contains a or b. We may thus assume that S contains intervals of both types. We now define d by $d = \sup \{d_j\}$, and we complete the proof by showing that $d \leq c_i$ for every i. Suppose that $c_{i_0} < d$ for some i_0. Then by the definition of d there exists a d_{i_0} such that $c_{i_0} < d_{j_0}$. Since $[a, c_{i_0}] \cap [d_{j_0}, b] = \phi$, this contradicts the finite intersection property for S and concludes the proof.

The reader should understand that there are elementary proofs of the Heine-Borel theorem which do not use Theorem F or anything like it. Theorem F will render us its major service in connection with the proof of the vital Tychonoff theorem.

Theorem 4:

Any closed subspace of a compact space is compact.

Proof:

Let Y be a closed subspace of a compact space X, and let $\{G_i\}$ be an open cover of Y. Each G_i, being open in the relative topology on Y, is the intersection with Y of an open subset H_i of X. Since Y is closed, the class composed of Y' and all the H_i's is an open cover of X, and since X is compact, this open cover has a finitc subcover. If Y' occurs in this subcover, we discard it. What remains is a finite class of H_i's whose union contains X. Our conclusion that Y is compact now follows from the fact that the corresponding G_i's form a finite subcover of the original open cover of Y.

Theorem 5:

Any continuous image of a compact space is compact.

Proof:

Let $f: X \rightarrow Y$ be a continuous mapping of a compact space X into an arbitrary topological space Y. We must show that f(X) is a compact subspace of Y. Let $\{G_i\}$ be an open cover of f(X). As in the above proof, each G_i is the intersection with f(X) of an open subset H_i of Y. It is clear that $\{f^{-1}(H_i)\}$ is an open cover of X, and by the compactness of X it has a finite subcover. The union of the finite class of H_i's of which these are the inverse images clearly contains f(X), so the class of corresponding G_i's is a finite subcover of the original open cover of f{X}, and f(X) is compact.

It is sometimes quite difficult to prove that a given topological spaee is compact by appealing directly to the definition. The following theorems give several equivalent forms of compactness which are often easier to apply.

TYCHONOFF'S THEOREM AND LOCALLY COMPACT SPACES

The main theorem of this section, to the effect that any product of compact spaces is compact, is perhaps the roost important single theorem of general topology. We shall use it repeatedly throughout the rest of this book, and the reader will come to see that its commanding position is due largely to the fact that in the higher levels of our subject many spaces constructed for special purposes turn out to be closed subspaces of products of compact spaces. Such a subspace is necessarily compact, and since compact spaces are so pleasant to work with, this makes the resulting theory much cleaner and smoother than would otherwise be the case.

Theorem 1: (The Generalized Heine-Borel Theorem):

Every closed and bounded subspace of R^n is compact.

Proof:

A closed and bounded subspace of R^n is a closed subspace of some closed rectangle, it suffices to show that any closed rectangle is compact as a subspace of R". Let $X = P_{i=1}^{n} [a_i,b_i]$ be a closed rectangle in R^n. Each coordinate space $[a_i,b_i]$ is compact by the classical Heine-Borel theorem, so by Tychonoff's theorem it suffices to show that the product topology on X is the same as its relative topology as a subspace of R^n. It is easy to see that the open rectangles in R^n form an open base for its usual topology, that is, for its metric topology, and from this it follows that the product topology on R^n is the same as its usual topology. The relative topology on X is the weak topology generated by its n projections onto the coordinate spaces $[a_i, b_i]$; but this is the product topology on X, so the proof is complete.

The n-dimensional Euclidean space R^n is the most important example of a type of topological space which is of great significance in modern analysis, especially in the theory of integration. A topological space is said to be *locally compact* if each of its points has a neighbourhood with compact closure. It is easy to see by the above theorem that R^n actually is locally compact, because any open sphere centered on any point is a neighbourhood of the point whose closure, being a closed and bounded subspace of R^n, is compact. It is trivial that any compact space is locally compact, for the full space is a neighbourhood with compact closure of every point in the space. We return to the study of locally compact spaces, where we give a more detailed analysis of their structure and properties.

Theorem 2: (Tychonoff's Theorem):

The product of any non-empty class of compact spaces is compact.

Proof:

Let $\{X_i\}$ be a non-empty class of compact spaces, and form the product $X = P_i X_i$. Let $\{F_j\}$ be a non-empty subclass of the defining closed subbase for the product topology on X. This means that each F_j is a product of the form $F_j = P_i F_{ij}$, where F_{ij} is a closed subset of X_i which equals X_i for all i's but one. We assume that the class $\{F_j\}$ has the finite intersection property, and we conclude the proof by showing that $\cap_j F_j$ is non-empty. For a given fixed i, $\{F_{ij}\}$ is a class of closed subsets of X_i with the finite intersection property; and by the assumed compactness of X_i, there exists a point x_i in X_i which belongs to $\cap_j F_{ij}$. If we do this for each i, we obtain a point $x = \{x_i\}$ in X which is in $\cap_j F_j$.

As our first application of Tychonoff's theorem, we prove an extension of the classical Heine-Borel theorem. We prepare the way for this proof by defining what we mean by open and closed rectangles in the n-dimensional Euclidean space R^n. If (a_i, b_i) is a bounded open interval on the real line for each i = 1,2, ..., n, then the subset of R^n defined by

$$P_{i=1}^{n} (a_i, b_i) = \{(x_1, x_2, ..., x_n): a_i < x_i < b_i \text{ for each } i\}$$

is called an open rectangle in R^n. A closed rectangle is defined similarly, as a product of n closed intervals.

PRODUCTS OF SPACES

There are two main techniques for making new topological spaces out of old ones. The first of these, and the simplest, is to form subspaces of some given space. The second is to multiply together a number of given spaces. Our purpose in this section is to describe the way in which the latter process is carried out.

We defined what is meant by the product $P_i X_i$ of an arbitrary non-empty class of sets. We also defined the projection p_i of this product onto its the coordinate set X_i. The reader should make certain that these concepts are firmly in mind. If each coordinate set is a topological space, then there is a standard method of defining a topology on the product. It is difficult to exaggerate the importance of this definition, and we examine it with great care in the following discussion.

Let us begin by recalling of open rectangles and open strips in the Euclidean plane R^2. We observed there that the open rectangles form an open base for the topology of R^2, and also that the open strips form an open subbase for this topology whose generated open base consists of all open rectangles, all open strips, the empty set, and the full space. The topology of the Euclidean plane is of course defined in terms of a metric. If we wish, however, we can ignore this fact and regard the topology of R^2 as generated in the sense of the class of all open strips. This situation provides the motivation for the more general ideas we now develop.

Let X_1 and X_2 be topological spaces, and form the product $X = X_1 \times X_2$ of the two sets X_1 and X_2. Consider the class S of all subsets of X of the form $G_1 \times X_2$ and $X_1 \times G_2$, where G_1 and G_2 are open subsets of X_1 and X_2, respectively. The topology on X generated by this class in the sense is called the *product topology* on X. The product topology therefore has S as an open subbase; in fact, it is defined by the requirement that S be an open subbase. The open base generated by S, that is, the class of all finite intersections of its sets, is clearly the class of all sets of the form $G_1 \times G_2$, and the open sets in X are all unions of these sets. There are two projections p_1 and p_2 of X onto its coordinate spaces X_1 and X_2, and by definition they carry a typical element (x_1, x_2) of X to x_1 and x_2, respectively. We note that S is precisely the class of all inverse images in X of all open subsets of X_1 and X_2 under these projections: $G_1 \times X_2 = p_1^{-1}(G_1)$ and $X_1 \times G_2 = p_2^{-1}(G_2)$. The product topology is thus a topology on the product with respect to which both projections are continuous mappings, and it is evidently the weakest such topology. The product topology can be regarded as the weak topology generated by the projections. With the above concepts to guide the way, one more step carries us to the product topology in its full generality. Let $\{X_i\}$ be any non-empty class of topological spaces, and consider the product $X = P_i X_i$ of the sets X_i.

A typical element x in X is an array $x = \{x_i\}$ of points in the coordinate spaces, where each x_i belongs to the corresponding space X_i; and for each index i, the projection p_i is defined by $p_i(x) = x_i$. We now define the *product topology* on X to be the weak topology generated by the set of all projections. This means the *product topology* is that generated by the class S of all inverse images in X of open sets in the X_i's, that is, the class S of all subsets of X of the form $S = p_i^{-1}(G_i)$, where i is any index and G_i is any open subset of X_i. It is easy to see that S can also be described as the class of all products of the form $S = P_i G_i$, where G_i is an open subset of X_i, which equals X_i for all its but one. The class S is called the defining open subbase for the product topology; and the class of all complements of sets in S—namely, the class of all products of the form $P_i F_i$, where F_i is a closed subset of X_i which equals X_i for all i's but one—is called the *defining closed subbase*. The open base generated by S, that is, the class of all finite intersections of its sets, is called the *defining open base* for the product topology; and this is evidently the class of all products of the form $P_i G_i$, where G_i is an open subset of X_i which equals X_i, for all but a finite number of i's. It should be clearly understood that an unrestricted product of open sets in the coordinate spaces need not be open in the product topology. A convenient way of thinking about this defining open base is that a typical one of its sets consists of all points $x = \{x_i\}$ in the product such that the z'th coordinate x_i is required to lie in an open subset G_i of X_i for a finite number of i's, all other coordinates being unrestricted.

When the product of a non-empty class of topological spaces is equipped with the product topology defined in the above paragraph, it is called a *product space*, or more simply, the *product* of the *spaces* involved.' It should be clear and these definitions that all projections of a product space onto its coordinate spaces are automatically both continuous and open.

We conclude this section by analyzing an example which we hope will increase the reader's capacity to "see" the structure of product spaces. Let the index set I consist of all real numbers i in the closed unit interval [0, 1]. I is to be considered as a set without any structure. Now let each index i have attached to it a topological space X_i, and let every X_i be a replica of the closed unit interval [0,1] with its usual topology. The index set I, and each vertical cross section represents the coordinate space X_i attached to the index at its base. An element of the product space X is an array of points, one of which lies in each X_i. Such an element is essentially a function—if we identify a function with its graph—defined on the set I, with values in the closed unit interval. We can now visualize as follows a typical set in the defining open base for the product topology. We choose a finite set of indices, say $\{i_1, i_2, i_3\}$, and for each of these we choose an open set on the vertical segment above it. Our basic open set then consists of all functions in X whose graphs cross each of these three vertical segments within the given open set on that segment. In the figure, f belongs to our basic open set, but g does not. The product topology on any product space can be visualized in a similar way. All one has to do is imagine the coordinate spaces as fibers, each attached to a specific element of the index set. The resulting mental image of the product space will then look something like a bundle of fibers, or perhaps a bed of reeds growing in a pond.

LEBESGUE NUMBERS FOR COVERS

Definition : *Let (X, d) be a Measure Theory and*

$$C = \{G_\lambda : \lambda \in \Lambda\}$$

be an open cover of X, A real number l > 0 is called a Lebesgue Number for **C** *iff every subset of X with diameter less than is contained in at least one* G_λ.

Theorem 7:

A compact subset of a Measure Theory is closed and bounded.

Proof :

Let (X, d) be a Measure Theory and let Y be a subset of X such that the subspace (Y, d) is compact. If Y is finite, then surely is closed and bounded. So we may suppose that Y is infinite. The Suppose Y is sequentially compact. We now prove that Y is bounded. Suppose the contrary to be true. Then we can always find pairs of points of Y at arbitrarily large distance apart. We use this property to construct a sequence of points of Y which contains no convergent subsequence. We start with any $y_1 \in Y$. Then we choose a point y_2 such that $d(y_1, y_2) > 1$. We then select a point y_3 such that

$$d(y_1, y_3) > 1 + d(y_1, y_2)$$

and so on. Thus for every $n \in N$, we choose y_n such that

$$d(y_1, y_n) > 1 + d(y_1, y_{n-1}).$$

It follows that if $m > n$,

$$d(y_1, y_n) > 1 + d(y_1, y_n)$$

Hence $\quad d(y_m, y_n) \geq |d(y_1, y_m) - d(y_1, y_n)| > 1.$

From this, we conclude that the sequence $< y_n >$ contains no convergent subsequence. But this is contradiction since Y is sequentially compact. Hence Y must be bounded.

It remains to show that Y is closed. Let y be any limit point of Y. Then there exists a sequence of points $< y_n >$ of Y which converges to y, and every subsequence of $< y_n >$ converges to y. But since Y is compact subspace, $< y_n >$ has a subsequence which converges to a point of Y. Consequently $y \in Y$. It follows that Y is closed.

Theorem 8:

A Measure Theory is compact iff it is complete and totally bounded,

Proof :

Let (X, d) be a Measure Theory and let it be compact. Then by theorem Y is complete. Again by theorem, X is sequentially compact and then by theorem, X is totally bounded. Thus it is shown that X is complete and totally bounded.

Conversely, let X be complete band totally bounded. We shall show that X is sequentially compact. Let

$$s = < x_1, x_2, ..., x_n, ... >$$

be any arbitrary sequence in X. Since X is totally bounded, there exists an ε-net for X for each ε > 0. Take ε = 1. Then X is contained in the union of finite collection of open spheres of radius 2. One of these spheres, say S (a_1,

1) a ∈ X, must contain a subsequence $s_1 = \{x_1^{(1)}, x_2^{(1)} ..., x_n^{(1)} ...\}$ of s. The distance between any two points of s_1 is less than 2. Similarly with ε = 1/ 2, we obtain a subsequence

$$s_2 = \{x_1, ^{(2)}, x_2^{(2)}, ...\}$$

all of whose points line in and open sphere.

$$S = \left(a_2, \frac{1}{2}\right), a_2 \in X,$$

of radius 1/2. Clearly the distance between any two points of s_2 is less than 1. Proceeding by induction, for ε = 1/k) we obtain a subsequence

$$s_k = \{x_1^{(k)}, x_2^{(k)}, ... \}$$

all of whose points lie in an open sphere S (a, 1/k) so that the distance between any two points of s_k is less then 2/k. We claim that the diagonal sequence

$$d = \{x_1^{(1)}, x_2^{(1)}, ... x_n^{(n)}, ... \}$$

is a Cauchy sequence of s. To show this, we observe that

$$x_n^{(n)} \in S\left(a_n, \frac{1}{n}\right) \subset S\left(a_{n-1}, \frac{1}{n-1}\right) \subset ... \subset S(a_1, 1) \qquad ...(1)$$

Now for each ε > 0,
we choose $n_0 \in N$,
such that $2/n_0 < \varepsilon$.
Therefore, $n > n_0$

$$\Rightarrow x_n (n), x_m (m) \in S (a_{n_0}, 1/n_0) \text{ by } (1)$$

$$\Rightarrow d\left(x_n^{(n)}, x_m^{(m)}\right) < \frac{2}{n_0}$$

$$\Rightarrow d (x_n^{(n)}, x_m^{(m)} < \varepsilon.$$

This shows that d is Cauchy subsequence of s.

Since X is complete, d converges to a point in X.

This it is shown that every sequence in X. has a subsequence which converges to a point in X. Hence X is sequentially compact. It follows that X is compact

Theorem 1(a):

A subset A of a complete Measure Theory (X, d) is compact iff A is closed and totally bounded.

Proof :

Let A be compact. Since A is a compact subset of X, it follows from theorem that A is closed. Also A is compact implies A is sequentially compact. Then by theorem, A is totally bounded.

Conversely, let A be closed and totally bounded. Now a closed subset of a complete Measure Theory is automatically complete and so A is complete and totally bounded. Then by the preceding theorem A is compact.

Corollary : *A matrix space is compact iff it is countably compact. Then from the above theorem, it follows that a subset of a complete Measure Theory is countably compact iff it is closed and totally bounded.*

Theorem 1(b):

Every sequentially compact Measure Theory (X, d) is totally bounded.

Proof :

Suppose X is not totally bounded. Then there exists $\varepsilon > 0$ such that X has no e-net .Let $x_1 \in X$. Then there must exist a point $x_2 \in X$ such that d $(x_1, x_2) \geq \varepsilon$ foe otherwise $\{x_1\}$ would be an ε-net for X. Again there must exists a point x_3 ε X such that d $(x_1, x_2) \geq \varepsilon$ and d $(x_2, x_3) \geq \varepsilon$ for otherwise $\{x_1, x_2\}$ would be an e-net for X. Continuing this process, we get a sequence.

$$\{x_1, x_2, x_3,\dots\}$$

having the property that d (xi, xj) ≥ e for i ≠ j. It follows that the sequence {xn} cannot contain any convergent subsequence. Hence X is not sequentially compact.

Theorem 2(a) (Lebesgue Covering Lemma):

Every open cover of a sequentially compact Measure Theory has a Lebesgue number.

Proof :

Let (X, d) be sequentially compact Measure Theory and let **C** = $\{G_\lambda : \lambda \in \Lambda\}$ be an open cover of X, We want to show that **C** has a Lebesgue number. Suppose the contrary to be true, that is, suppose that **C** has no Lebesgue number. Then for every $n \in N$, there exists a subset C_n of X having the property

$$0 < \delta (C_n) < 1 \, n$$

and $\quad Cn \not\subset G_\lambda \forall \lambda \in \Lambda.$ ...(1)

Choose a point $cn \in C_n$ for each $n \in N$ and consider the sequence $< cn >$. Since X is sequentially compact, it must contain a subsequence.

$$< c_{k_1}, c_{k_2}, c_{k_3}, \dots , >$$

which converges to a point $c \in X$.

Since **C** is a cover of X, there exists an open set $G_{\lambda_0} \in$ **C** such that G_{λ_0}. Since G_{λ_0} is open, is open, there an open sphere S (c, e) with centre c and radius ε such that

$$S(c, \varepsilon) \subset G_{\lambda_0} \quad \text{...(2)}$$

Also since $< c_{k_n} >$ converges to c, there exists $n_0 \in N$ such that

$$k_n \geq n_0 \Rightarrow c_{k_0} \in S(c, \varepsilon/2) \quad \text{...(3)}$$

Choose a positive integer k ≥ n0 such that

$$\frac{1}{k_0} < \frac{\varepsilon}{2}. \quad \text{...(4)}$$

Then $c_{k_0} \in S(c, \varepsilon/2)$ by (3). Also $C_{k_0} \in C_{k_0}$ Hence

$$C_{k_0} \in C_{k_0} \cap S(c, \varepsilon/2) \quad \text{...(5)}$$

Again from (1) and (3),

$$0 < d(C_{k_0}) < \frac{1}{k_0} < \varepsilon/2 \quad \text{...(6)}$$

Now (5) and (6) show that C_{k_0} is a set of diameter < ε/2 and it intersects the open sphere S (c, ε/2). This implies that

$$C_{k_0} \subset S(c, e). \quad \text{...(7)}$$

From (2) and (7)

$C_{k_0} \subset C_{k_0}$. But this contradicts (1).

It follows that **C** has a Lebesgue number.

Theorem 2(b):

The 'only if' part : Let X be a sequentially compact Measure Theory and let A be any infinite subset of X. Let $< x_n >$ be any sequence of distinct points of A. Since X is sequentially compact, this sequence contains a convergent subsequence $< x_{i_n} >$ whose limit is x, say. Then x is a limit point of A. Hence X has BWP.

The 'if part': Conversely, let X has BWP. Let $< x_n >$ be any sequence in X. If the range set

$A = (x_1, x_2, \dots x_n, \dots\}$ is finite then one of the points, say x_{m_0}, is such that $x_i = x_m$ for infinitely many $i \in N$. It follows that $< x_{m_0}, x_{m_0}, \dots >$ is a sub-sequence of $< x_n >$ which converges to $x \in X'$. Now let the range set of A be infinite. Since X has BWP. A has limit point, say x_0. Then it is easy to extract from $< x_n >$ a subsequence which converges to x_0. Hence X is sequentially compact.

Theorem 3:

A metric X is compact *if and only if it is sequentially compact.*

Proof:

The 'only if' part : Let X be compact. Then X has BWP. Hence by the previous theorem, X is sequentially compact,

The 'if part': Conversely, let X be sequentially compact. To show that X is compact, let **C** = {Gλ : λ ∈ Λ} be an open cover of X. We have to show that **C** has a finite subcover. Since X is sequentially compact, the cover **C** has a Lebesgue number l > 0. Further X is totally bounded by theorem (taking ε = i/3), there exists an e-net.

$$A = \{a_1, a_2, \ldots a_n\}$$

where a_i's belong to X so that

$$X = \cup \{S (a_i, \varepsilon) : a_i \in A\} \ldots(1)$$

It follows that

$$\delta (S (a_i, \varepsilon)) \leq 2\varepsilon = \frac{2l}{3} < l, \text{ for } i = 1, 2, \ldots n.$$

By the definition of Lebesgue number, for each i, there exists and $G_{\lambda_i} \in$ **C** such that $S (a_i, \varepsilon) \subset G_{\lambda_i}$.

It follows from (1) and (2) that

$$X = \cup \{G_{\lambda_i} : i = 1, 2, \ldots n\}.$$

Hence X is compact.

Theorem 4:

Let (X, d) be Measure Theory. Then the following statements are equivalent

(a) X is compact

(b) X is countably compact

(c) X has BWP

(d) X is sequentially compact.

Proof :

(a) ⇒ (b) follows from definition (b) ⇒ (c) follows from theorem. (c) ⇒ (d) follows from the 'if part' of theorem.And (d) ⇒ (a) follows from the 'if part' of theorem.

Theorem 5:

Every compact Measure Theory is complete.

Proof :

Let X be a compact Measure Theory and let $< x_n >$ be a Cauchy sequence in X. By theorem (3.6), X is sequentially compact so that the sequence

$< x_n >$ has a subsequence $< x_{n_k} >$ converging to some $x_0 \in X$. We shall show that $< x_n >$ also converges to x_0. Let $\varepsilon < 0$ be given. Since $< x_n >$ is a Cauchy sequence, there exists $n_0 \in N$ such that

$$m, n \geq n_0 \Rightarrow d(x_m, x_n) < \varepsilon/2 \quad ...(1)$$

Again since $< x_{n_k} >$ converges to x_0, there exists $k_0 \in N$ such that

$$k \geq k_0 \Rightarrow d(x_{n_k}, x_0) < \varepsilon/2 \quad ...(2)$$

Choose $k_1 \geq k_0$ such that $k_{k_1} \geq n_0$. Then by (1) and (2), we have

$$n \geq n_{k_1} \Rightarrow d(x_n, x_{n_{k1}}) < \varepsilon/2$$

and $\quad d(x_{n_{k1}}, x_0) < \varepsilon/2.$

Hence by triangle inequality, we obtain

$$d(x_n, x_0) \leq d(x_n, x_{n_{ki}}) + d(x_{n_{ki}}, x_0) < \varepsilon/2 + \varepsilon/2 = \varepsilon$$

for every $n \geq n_{k_1}$. It follows that

$\lim_{n \to \infty x_n = x_0}$ and so X is complete.

Theorem 6:

Every totally bounded Measure Theory is separable.

Proof :

Let (X, d) be a totally bounded Measure Theory and let A_n be a set of points in X which from a 1/n – net. Then since A_n is a finite set, the set of points $\cup \{A_n : n \in X\}$ is countable and dense in X. Hence X is separable.

Corollary : *Every compact space is separable.*

Proof :

Since a compact Measure Theory is totally bounded the result follows from the above theorem.

COMPACTNESS FOR METRIC SPACES

In all candor, we must admit that the intuitive meaning of compactness for topological spaces is somewhat elusive. This concept, however, is so vitally important throughout topology that we consider it worthwhile to devote this and the next section to giving several equivalent forms of compactness for the special case of a Measure Theory. Some of these are quite useful in applications and are perhaps more directly comprehensible than the open cover definition. We hope they will help the reader to achieve a fuller understanding of the geometric significance of compactness.

We begin by recalling the classical *Bolzano-Weierstrass theorem*: if X is a closed and bounded subset of the real line, then every infinite subset of X has a limit point in X. This suggests that we consider the property expressed here as one which a general Measure Theory may or may not possess. A Measure Theory is said to have the *Bolzano-Weierstrass* property if every infinite subset has a limit point. Another property closely allied to this is that of sequential compactness: a Measure Theory is said to be *sequentially compact* if every sequence in it has a convergent subsequence.

Our main purpose in this section is to prove that each of these properties is equivalent to compactness in the case of a Measure Theory. The following is an outline of our procedure: we first prove that these two properties are equivalent to one another; next, that compactness implies the Bolzano-Weierstrass property; and finally, that sequential compactness implies compactness. The first two of these steps are relatively simple, but the last involves several stages.

Theorem 1:

Every sequentially compact Measure Theory is compact.

Proof:

Let X be a sequentially compact Measure Theory, and let $\{G_i\}$ be an open cover. By Theorem 3, this open cover has a Lebesgue number a. We put $\in = a/3$, and use Theorem 4 to find an $\in$-net

$$A = \{a_1, a_2, ..., a_n\}.$$

For each k = 1, 2, ..., n, we have $d(S_\in(a_k) \leq 2\in = 2a/3 < a$. By the definition of a Lebesgue number, for each k we can find a G_{ik} such that $S_\in(a_k) \subseteq G_{ik}$. Since every point of X belongs to one of the $S_\in(G_k)$'S, the class $\{G_{i1}, G_{i2}, ..., G_{in}\}$ is a finite subcover of $\{G_i\}$. X is therefore compact.

Our results so far can be summarized by the statement that if X is a Measure Theory, then the following three conditions are all equivalent to one another:

(1) X is compact;

(2) X is sequentially compact;

(3) X has the Bolzano-Weierstrass property.

Also, of course, we have as by-products the additional information that a compact Measure Theory is totally bounded and that every open cover of a compact Measure Theory has a Lebesgue number. The latter fact has the following useful consequence.

Theorem 2(a):

Any continuous mapping of a compact Measure Theory into a Measure Theory is uniformly continuous.

Proof:

Let f be a continuous mapping of a compact Measure Theory X into a Measure Theory Y, and let d_1 and d_2 be the metrics on X and Y. Let $\in >$ 0 be given. For each point x in X, consider its image f(x) and the open sphere $S_{\in/2}(f(x))$ centered on this image with radius $\in/2$. Since f is continuous, the inverse image of each of these open spheres is an open subset of X, and the class cf all such inverse images is an open cover of X. Since X is compact, Theorem 3 guarantees that this open cover has a Lebesgue number δ. If x and x' are any two points in X for which di(x,x') < δ, then the set {x, x'} is a set with diameter less than δ, both points belong to the inverse image of some one of the above open spheres, both f(x) and f(x') belong to one of these open spheres, and therefore $d_2(f(x), f(x')) < \in$, which shows that f is indeed uniformly continuous.

We continue our study of compact metric spaces in the next section.

Theorem 2(b):

A Measure Theory is sequentially compact ⇔ it has the Bolzano-Weierstrass property.

Proof:

Let X be a Measure Theory, and assume first that X is sequentially compact. We show that an infinite subset A of X has a limit point. Since A is infinite, a sequence $\{x_n\}$ of distinct points can be extracted from A. By our assumption of sequential compactness, this sequence has a subsequence which converges to a point x. a: is a limit point of the set of points of the subsequence, and since this set is a subset of A, x is also a limit point of A.

We now assume that every infinite subset of X has a limit point, and we prove from this that X is sequentially compact. Let $\{x_n\}$ be an arbitrary sequence in X. If $\{x_n\}$ has a point which is infinitely repeated, then it has a constant subsequence, and this subsequence is clearly convergent. If no point of $\{x_n\}$ is infinitely repeated, then the set A of points of this sequence is infinite. By our assumption, the set A has a limit point x, and it is easy to extract from $\{x_n\}$ a subsequence' which converges to x.

Theorem 3(a):

Every compact Measure Theory has the Bolzano-Weierstrass property.

Proof:

Let X be a compact Measure Theory and A an infinite subset of X. We assume that A has no limit point, and from this we deduce a contradiction. By our assumption, each point of X is not a limit point of A, so each point of X is the center of an open sphere which contains no point of A different from its center. The class of all these open spheres is an open cover, and by compactness there exists a finite subcover. Since A is contained in the set of all centers of spheres in this subcover, A is clearly finite. This contradicts the fact that A is infinite, and concludes the proof.

Our next task is to prove that compactness is implied by sequential compactness. We carry this out in several stages, the first of which can be motivated by the following considerations. Let $\{G_i\}$ be an open cover of a Measure Theory X. Then each point x in X belongs to at least one G_i, and since the G_i's are open, each point x is the center of some open sphere which is contained in at least one G_i. If we now move to another point of X, we may be forced to decrease the radius of our open sphere in order to squeeze it into a G_i. Under special circumstances it may not be necessary to take radii below a certain level as we move from point to point over the entire space. The following concept is useful for handling this sort of situation. A real number $a > 0$ is called a *Lebesgue number* for our given open cover $\{G_i\}$ if each subset of X whose diameter is less than a is contained in at least one G_i.

Theorem 3(b) (Lebesgue's Covering Lemma):

In a sequentially compact Measure Theory, every open cover has a Lebesgue number.

Proof:

Let X be a sequentially compact Measure Theory, and let $\{G_i\}$ be an open cover. We say that a subset of X is "big" if it is not contained in any G_i. If there are no big sets, then any positive real number will serve as our Lebesgue number a. We may thus assume that big sets do exist, and we define a' to be the greatest lower bound of their diameters. Clearly, $0 \leq a' \leq +\infty$. It will suffice to show that $a' > 0$; for if $a' = +\infty$, then any positive real number will do for a, and if a' is real, we can take a to be a'. We therefore assume that $a' = 0$, and we deduce a contradiction from this assumption. Since every big set must have at least two points, we infer from $a' = 0$ that for each positive integer n there exists a big set B_n such that $0 < d(B_n) < 1/n$. We now choose a point x_n in each B_n. Since X is sequentially compact, the sequence $\{X_n\}$ has a subsequence which converges to some point x in X. The point x belongs to at least one set G_{i0} in our open cover, and since G_{i0} is open, x is the center of some open sphere $S_r(x)$ contained in G_i. Let $S_{r/2}(x)$ be the concentric open sphere with radius r/2. Since our subsequence of $\{x_n\}$ converges to x, there are infinitely many positive integers n for which x_n is in $S_{r/2}(x)$. Let w_o be one of these positive integers which is so large that $1/n_0 < r/2$. Since $d(B_{n0}) < 1/n_0 < r/2$, $B_{n0} \subseteq S_r(x) \subseteq G_{i0}$. This contradicts the fact that B_{n0}, is a big set, and completes the proof.

The next stage requires the following concepts. Let X be a Measure Theory. If $\in > 0$ is given, a subset A of X is called an $\in$-net if A is finite and $X = \cap_{e \in A} S_\in(a)$, that is, if A is finite and its points are scattered through X in such a way that each point of X is distant by less than $\in$ from at least one point of A. The Measure Theory X is said to be *totally bounded* if it has an $\in$-net for each $\in > 0$. It is clear that if X is totally bounded, then it is also bounded; for if A is an $\in$-net, then the diameter of A is finite (since A is a non-empty finite set) and $d(X) \leq d(A) + 2\in$. Total boundedness is actually a much stronger property than boundedness, as we shall see below.

Theorem 3(c):

Every sequentially compact Measure Theory is totally bounded.

Proof:

Let X be a sequentially compact Measure Theory, and let $\in > 0$ be given. Choose a point di in X and form the open sphere $S_\in(a_1)$. If this open sphere contains every point of X, then the single-element set $\{a_1\}$ is an $\in$-net. If there are points outside of $S_\in(a_1)$, let a_2 be such a point and form the set $S_\in(a_1) \cup S_\in(a_2)$. If this union contains every point of X, then the two-element set $\{a_1,a_2\}$ is an e-net. If we continue in this way, some union of the form $S_\in(a_1) \cup S_\in(a_2) \cup ... \cup S_\in(a_n)$ will necessarily contain every point of X; for if this process could be continued indefinitely, then the sequence $\{a_1, a_2, ..., a_n, ...\}$ would be a sequence with no convergent subsequence, contrary to the assumed sequential compactness of X. We see by this that some finite set of the form $(a_1, a_2, ..., d_n\}$ is an $\in$-net, so X is totally bounded.

We are now in a position to complete this line of thought by proving that compactness is implied by sequential compactness.

ASCOLI'S THEOREM

Our previous characterizations of compactness for a Measure Theory strongly suggest that this property is related to completeness and total boundedness in some way yet to be formulated. We begin by proving a theorem which clarifies this situation.

Theorem 1:

A closed subspace of a complete Measure Theory is compact $\Leftrightarrow$ it is totally bounded.

Proof:

Since a closed subspace of a complete Measure Theory is automatically complete, this is a direct consequence of Theorem 1.

What sort of property is total boundedness? We have seen that it always implies boundedness, the converse of this is true for subspaces of the finite-dimensional Euclidean space R^n. It is false, however, that boundedness implies total boundedness for subspaces of the infinite-dimensional Euclidean space R^∞. In fact, the closed unit sphere in R^∞, defined by $X = \{x:\| \|x\| \leq 1\}$, is not totally bounded, though it is obviously bounded. To see this, it suffices to observe that the sequence $\{x_n\}$ in X defined by

$$x_1 = \{1,0,0, ..., 0, ...\},$$
$$x_2 = \{0,1,0, ..., 0, ...\},$$
$$x_3 = \{0,0,1, ..., 0, ...\},$$

has no convergent subsequence, for the distance from any point of the sequence to any other is $2^{1/2}$. This shows that X is not compact, hence not totally bounded. The following fact, which we may add to the reader's intuition about the relation between boundedness and total boundedness: a Banach space is finite-dimensional $\Leftrightarrow$ every bounded subspace is totally bounded.

We now turn to the problem of characterizing compact subspaces of $\mathcal{C}(X, R)$ or $\mathcal{C}(X, C)$. By Theorem B, we know at once that a closed subspace of $\mathcal{C}(X, R)$ or $\mathcal{C}(X, C)$ is compact $\Leftrightarrow$ it is totally bounded. Unfortunately, however, this information is of little value in most applications to analysis. What is needed is a criterion expressed in terms of the individual functions in the subspace. Furthermore, for most of the applications it suffices to consider only the case in which X is a compact Measure Theory. We describe the relevant concept as follows. Let X be a compact Measure Theory with metric d, and let A be a non-empty set of continuous real or complex functions defined on X. If f is a function in A, then by Theorem 24-F this function is uniformly continuous; that is, for each $\in > 0$, there exists $\delta > 0$ such that $d(x, x') < \delta \Rightarrow |f(x) - f(x')| < \in$. In general, S depends not only on $\in$ but also on the function. A is said to be *equicontinuous* if for each $\in$ a δ can be found which serves at once for all functions f in A, that is, if for each $\in > 0$ there exists $\delta > 0$ such that for every f in A $d(x,x') < \delta \Rightarrow |f(x) - f(x')| < \in$.

Theorem 2(a) (Ascoli's Theorem):

If X is a compact Measure Theory, then a *closed subspace* of $\mathcal{C}(X, R)$ or $\mathcal{C}(X, C)$ is compact $\Leftrightarrow$ it is bounded and equicontinuous.

Proof:

Let d be the metric on X, and let F be a closed subspace of $\mathcal{C}(X, R)$ or $\mathcal{C}(X, C)$.

We first assume that F is compact, and we prove that it is bounded and equicontinuous. F is bounded. We prove that F is equicontinuous as follows. Let $\epsilon > 0$ be given. Since F is compact, and therefore totally bounded, we can find an $(\epsilon/3)$-net $\{f_1, f_2, ..., f_n\}$ in F. Each f_k is uniformly continuous, so for each k = 1, 2, ..., n, there exists $\delta_k > 0$ such that $d(x, x') < \delta_k \Rightarrow |f_k(x) - f_k(x')| < \epsilon/3$. We now define S to be the smallest of the numbers $\delta_1, \delta_2, ..., S_n$. If f is any function in F and f_k is chosen so that $||f - f_k|| < \epsilon/3$, then

$$d(x, x') < \delta \Rightarrow |f(x) - f(x')| \leq |f(x) - f_k(x)| + f_k(x) - f_k(x')\}$$
$$+ |f_k(x')| < \epsilon/3 + \epsilon/3 + \epsilon/3 = \epsilon.$$

This shows that F is equicontinuous.

We now assume that F is bounded and equicontinuous, and we demonstrate that it is compact by showing that every sequence in it has a convergent subsequence. Since F is closed, and therefore complete, it suffices to show that every sequence in it has a Cauchy subsequence. As we proceed, the reader will see that our proof is similar in structure to the last part of the proof of Theorem 1. X has a countable dense subset. Let the points of this subset be arranged in a sequence $\{x_i\} = \{x_2, x_3, ..., x_i, ...\}$, where we start with the subscript 2 for reasons which will become clear below. Now let

$$S_1 = (f_{11}, f_{12}, f_{13}, ...)$$

be an arbitrary sequence in F. Our hypothesis that F is bounded means that there exists a real number K such that $||f|| \leq K$ for every f in F, or equivalently, such that $|f(x)| \leq K$ for every f in F and every x in X. Consider the sequence of numbers $\{f_{1j}(x_2)\}$, j = 1, 2, 3, ..., and observe that since this sequence is bounded, it has a convergent subsequence. Let $S_2 = \{f_{21}\ f_{22}, f_{23}, ...\}$ be a subsequence of S_1 such that $\{f_{2j}, (x_2)\}$ converges. We next consider the sequence of numbers $\{f_{2j}, (x_3)\}$, and in the same way we let $S_3 = \{f_{31}, f_{32}, f_{33}, ...\}$ be a subsequence of S_2 such that $\{f_{3j}(x_3)\}$ converges. If we continue this process, we get an array of sequences of the form

$$S_1 = \{f_{11}, f_{12}, f_{13}, ...\},$$
$$S_2 = \{f_{21}, f_{22}, f_{23}, ...\},$$
$$S_3 = \{f_{31}, f_{32}, f_{33}, ...\},$$

in which each sequence is a subsequence of the one directly above it, and for each i the sequence $S_i = \{f_{i1}, f_{i2}, f_{i3}, ...\}$ has the property that $\{f_{ij}(x_i)\}$ is a convergent sequence of numbers. If we define $f_1, f_2, f_3, ...$ by $f_1 = f_{11}, f_2 =$

f_{22}, $f_3 = f_{33}$, ..., then the sequence $S = \{f_1, f_2, f_3, ...\}$ is the "diagonal" subsequence of +Si. It is clear from this construction that for each point x, in our dense subset of X, the sequence $\{f_n(x_i)\}$ is a convergent sequence of numbers. It remains only to show that S, as a sequence of functions in $\mathcal{C}$(X, R) or $\mathcal{C}$(X, C), is a Cauchy sequence. Let $\in > 0$ be given. Since F is equicontinuous, there exists $\delta > 0$ such that $d\{x,x'\} < \delta \Rightarrow |f_n(x) - f_n(x')| < \in/3$ for all functions f_n in S. We now form the open sphere $S_\delta(x_i)$ with radius δ centered on each of the x_i's. Since the x_i's are dense, these open spheres form an open cover of X, and since X is compact, $X = \bigcup_{i=2}^{i0} S_\delta(x_i)$ for some i_0. It is easy to see that there exists a positive integer n_0 such that $m, n > n_0 \Rightarrow |f_m(x_i) - f_n(x_i)| < \in/3$ for all the points $x_2, x_3, ..., x_{i0}$, Our proof is completed by the remark that if x is an arbitrary point of X, then an i can be found in the set $\{2, 3, ..., i_0\}$ such that $d(x, x_i) < \delta$, and that therefore

$$m, n \geq n_0 \Rightarrow |f_m(x) - f_n(x)| \leq |f_m(x) - f_m(x_i)| + |f_m(x_i) - f_n(x_i)| + |f_n(x_i)| - f_n(x)| < \in/3 + \in/3 + \in/3 = \in.$$

We observe that the total boundedness in Theorem 2 is replaced, in Ascoli's theorem, by the weaker condition of boundedness, and that the resulting deficiency is made up by the additional condition of equicontinuity.

Theorem 2(b):

-A Measure Theory is compact $\Leftrightarrow$ it is complete and totally bounded.

Proof:

Let X be a Measure Theory. The first half of our proof is easy, for if X is compact, then it is totally bounded, and it is complete and the fact that every sequence (and therefore every Cauchy sequence) has a convergent subsequence.

We now assume that X is complete and totally bounded, and we prove that X is compact by showing that every sequence has a convergent subsequence. Since X is complete, it suffices to show that every sequence has a Cauchy subsequence. Consider an arbitrary sequence

$$S_1 = [x_{11}, x_{12}, x_{13}, ...\}.$$

The reason for this notation will soon be clear. Since X is totally bounded, there exists a finite class of open spheres, each with radius 1/2, whose union equals X; and from this we see that S_1 has a subsequence $S_2 = (x_{21}, x_{22}, x_{23} ....\}$ all of whose points lie in some one open sphere of radius %. Another application of the total boundedness of X shows similarly that S_2 has a subsequence $S_3 = \{x_{31}, x_{32}, x_{33}, ...\}$ all of whose points lie in some one open sphere of radius 1/3. We continue forming successive subsequences in this manner, and we let

$$S = \{x_{11}, x_{12}, x_{13}, ...\}$$

be the "diagonal" subsequence of S_1. By the nature of this construction, S is clearly a Cauchy subsequence of S_1, and our proof is complete.

This theorem gives total boundedness a prominent part in determining whether a Measure Theory is compact or not. As we know, many metric spaces occur as closed subspaces of complete metric spaces, and for these we can make the role of total boundedness even more striking.

Theorem 3:

Let Y be a subspace of a topological space X and let $A \subset Y$. Then A is compact relative to X if and only if A is compact relative to Y.

Proof :

Let A be compact relative to X and let $\{V_\lambda : \lambda \in \Lambda\}$ be a collection of sets, open relative to Y, which covers A so that $A \subset \cup \{V_\lambda : \lambda \in \Lambda\}$. Then there exists G_λ, open relative to X, such that $V_\lambda = Y \cap V_\lambda$ for every $\lambda \in \wedge$. It then follows that,

$$A \subset \cup \{G_\lambda : \lambda \in \Lambda\}$$

so that $\{G_\lambda : \lambda \in \Lambda\}$ is an open cover of A relative to X. Since A is compact relative to X, there exists finitely many indices $\lambda_1, ... \lambda_n$ such that

$$A \subset G_{\lambda_1} \cup ... \cup G_{\lambda_n}.$$

Since $A \subset Y$, we have $A \subset Y \cap [G_{\lambda_1} \cup ... \cup G_{\lambda_n}]$

$= (Y \cap G_{\lambda_1}) \cup ... \cup (Y \cap G_{\lambda_1})$ [Distributive law]

Since $Y \cap G_{\lambda_1} = V_{\lambda_i}$ $(i = 1, 2, ..., n)$ we obtain

$$A \subset V_{\lambda_i} \cup ... \cup V_{\lambda_i}$$

This show that A is compact relative Y.

Conversely, let A be compact relative to Y and let $\{G_\lambda : \lambda \in \Lambda\}$ be a collection of open subset of X which cover A so that

$$A \subset \cup \{G_\lambda : \lambda \in \Lambda\} \quad ...(1)$$

Since $A \subset Y$, (1) implies that

$$A \subset Y \cap [\cup \{G_\lambda : \lambda \in \Lambda\}] \, [= \cup \{Y \cap G_\lambda : \lambda \in \Lambda\}\}$$

[Distributive law]

Since $Y \cap Gl$ is open relative to Y, then collection

$$(Y \cap G_\lambda : \lambda \in \Lambda\}$$

is an open cover of A relative to Y. Since A is compact relative to Y, we must have

$$A \subset (Y \cap G_{\lambda_1}) \cup \ldots \cup (Y \cap G_{\lambda_n}) \qquad \text{..(2)}$$

for some choice of finitely many indices, l1, ... λ_n. But (2) implies that

$$A \subset G_{\lambda_1} \cup \ldots \cup G_{\lambda_n}.$$

It follows that A is compact relative to X.

Remark : BY virtue of the above theorem, a subset Y of a topological space X is compact relative to X if and only if Y is compact as a space in its own right. Thus compactness is not a relative property, that is, if Y is compact, it is compact in whatever space it is embedded.

Theorem 4:

Every compact subset A of a Hausdorff space X is closed

Proof :

We shall show that A' is open. Let $p \in A'$. Since X is Hausdorff, for every $q \in A$ there exists open nhds of and q which we denote respectively by M (q) and N (q) $\cap$ N (q) such that M (q) $\cap$ N (q) = $\varnothing$. Now the collection. {N (q) : $q \in A$} is an open cover of A. Since A is compact there exists finite number of points q_i, i = 1, ... such that

$$A \subset \bigcup_{i=1}^{n} N(q_i) \qquad \text{...(1)}$$

Let $\quad M = \bigcap_{i=1}^{n} N(q_i), N = \bigcup_{i=1}^{n} M(q_i).$

Then M is an open nhd of p (being the intersection of a finite number of open nhds of (p). We claim that M $\cap$ N $\varnothing$. We have

$$x \in N \Rightarrow x \in N(q_i) \text{ for some } i$$

$$\Rightarrow x \notin M(q_i) \qquad [\because N(q_i) \cap M(q_i) = \varnothing]$$

$$\Rightarrow x \notin M.$$

Thus M $\cap$ N = $\varnothing$ and since A $\subset$ N, we have A $\cap$ M = $\varnothing$ which implies that M $\subset$ A'. This shows that A' contains a nhd of each of its points and so A' is open *i.e.,* A is closed.

Corollary 1 : *In a Hausdorff space, a point and a compact set not containing it can be separated by open sets. More precisely, if A is a compact subset of a Hausdorff space X and $p \notin A$, then there exists disjoint open sets M and N such that $p \in M$ and $A \subset N$.*

Example 1:

Give an example of a compact space which is not Hausdorff.

Solution:

Consider the topology $\Im = \{\varnothing, \{a\}, \{a, b\}, X\}$ on $X = \{A, b, c\}$. Since X is finite, it is compact. But X is not Hausdorff since a, b are distinct points having no disjoint nhds.

Example 2:

Show by means of an example that a compact subset of a non-Hausdorff topological space need not be closed.

Solution:

Consider any indiscrete space X consisting of more that one point and let A be any proper subset of λ. Then A is not closed since the only closed sets are $\varnothing$ and X. But A is compact since the only open cover of A is $\{X\}$.

Theorem 5(a):

Closed subsets of compact sets are compact

Proof :

Let Y be a compact subset of a topological space X and let F be subset of Y, closed relative to X. To show that F is compact. Let

$$\mathbf{C} = \{G_\lambda : \lambda \in \Lambda\}$$

be an open cover of F. Then the collection

$$\mathbf{D} = \{G_\lambda : \lambda \in \Lambda\} \cup \{X - F\}$$

forms an open cover of Y. Since Y is compact, there is a finite subcollection D' of D which covers Y′ and hence covers F. If X – F is a member of D', we may remove it from D′ and still retain an open finite cover of F.

We have thus shown that a finite subcollection of C covers F. Hence F is compact.

Theorem 5(b):

A topological space X is compact if an only if every basic open cover of X has a finite subcover.

Proof :

Let X be compact. Then every open cover of X has a finite subcover. In particular every basic open cover of X must have a finite subcover.

Conversely, suppose that every basic open cover of X has a finite subcover and let

$$\mathbf{C} = \{G_\lambda : \lambda \in \Lambda\}$$

be any open cover of X, if

$$\mathbf{B} = \{B_\alpha : \alpha \in \Delta\}$$

be any open basic for X, then each Gl is union of some members of B and the totality of all such members of B is evidently a basic open cover of X. By hypothesis this collection of members of B, has a finite subcover say.

$$\{B_{\alpha_i} : i = 1, 2, \ldots n\},$$

For each B_{α_i} in this finite subcover, we can select G_{λ_i} from **C** such that $B_{\alpha_i} \subset G_{\lambda_i}$. It follows that the finite subcollection $\{G_{\lambda_i} : i = 1, 2, \ldots n\}$, which arises in this way is subcover of **C**. Hence X is compact.

Definition : A collection C of sets is said to have to finite intersection property (FIP) or to be finitely common iff the intersection of members of each finite subcollection of C is non-empty.

A collection of sets is called fixed if it has a non-empty intersection and Free if its intersection is empty.

Theorem 6:

A topological space X is compact iff every collection of closed subsets of X with the finite intersection property is fixed, that is, has a non-empty intersection.

Proof :

Let X be compact and let $F = \{F_\lambda : {}_\lambda \in {}_\Lambda)$ be a collection of closed subsets of X with the FIP and suppose, if possible, $\cap \{G_\lambda : \lambda \notin \Lambda\} = \varnothing$. Then means that $\{F'_\lambda\ \lambda \in \Lambda\}$ is open cover of X since Fl's are closed. Since X is compact, we have that $\cup \{F'_{\lambda i} : 1, 2, \ldots n\} = X$ where n is finite and so by De-Morgan law $[\cap F_{\lambda i} : i = 1, 2, \ldots n\}]' = X$ which implies that $\cap \{F_{\lambda i} : i = 1, 2, \ldots n\} = \varnothing$.

But this contradicts the FIP of F. Hence we must have $\cap \{F_\lambda : \lambda \in \Lambda\} \neq \varnothing$.

Conversely let every collection of closed subsets of X with the FIP have a non-empty intersection and let $C = \{G_\lambda : \lambda \in \Lambda\}$ be an open cover of X so that $X = \cup \{G_\lambda : \lambda \in \Lambda\}$ whence taking complements $\varnothing = [\cup \{G_\lambda : \lambda \in \Lambda\}] = \cup \{\lambda'\ \lambda \in \Lambda\}$. Thus $\{G'_\lambda : \lambda \notin \Lambda\}$.

is a collection of closed sets with empty intersection and so by hypothesis this collection does not have the FIP. Hence there exists a finite number of sets G'_{λ_i} $i = 1, 2 \ldots n$ such that

$$\varnothing = \cap \{G'_{\lambda_i} = 1, 2, \ldots n\} = [\cup \{G'_{\lambda_i} : i = 1, 2, \ldots n\}]'$$

[De-Morgan law]

Definition : *A space X is said to have Bolzano Weierstrass Property BPW) If every infinite set in X has limit point. A space with BPW is also said to be Frechet compact.*

Theorem 7:

If A is an infinite subset of a compact space Y then A has limit point in X. In other words, compact spaces have Bolzano Weierstrass Property.

Proof :

Suppose A has no limit point in X. Then for every $x \in X$, there exists an open neighborhood N_x of x which contains no point of A other than (possibly) x. Now the collection

$$(N_x : x \in X\}$$

forms an open cover of X. Since X is compact, there are finitely many points $x_1, \ldots x_n$ in X such that

$$X \subset N_{x_1} \cup \ldots \cup N_{x_n}.$$

Since each Nx contains at most one point of A, (1) shows that A has at most n points and so A is finite which is contradiction. Hence A must have a limit point in X.

COUNTABLE, SEQUENTIAL AND LOCAL COMPACTNESS

Definition : *A topological space X is said to be* **countably compact** *iff every countable open cover of X has a finite subcover.*

It follows from this definition that every compact Measure Theory is countably compact.

Theorem 1:

Every closed subspace of a locally compact space is locally compact.

Proof :

Let Y be a closed of locally compact space X and let $y \in Y$ be arbitrary. Then $y \in X$. Since X is locally compact, there exists an open neighborhood N of y such that $\overline{N}$ is compact. But then $N \cap Y$ is an open neighborhood of y such that $\overline{N \cap Y} \subset \overline{N}$

$[\because N \cap Y \subset N \Rightarrow \overline{N \cap Y} \subset \overline{N}]$.

Thus $\overline{N \cap Y}$ is a closed subset of the compact set $\overline{N}$ and is therefore itself compact.

Also since Y is closed in X, it is easy to see that the closure of N $\cap$ Y in X is the same as its closure in Y. Thus we have shown that every point in Y has a neighborhood in Y whose closure in Y is compact. Hence the subspace Y is locally compact.

Theorem 2:

A Hausdorff space X is locally compact $\Leftrightarrow$ each of its points is an interior point of some compact subspace of X.

Proof :

If X is locally compact, then each x $\in$ X has a nhd N such that $\overline{N}$ is a compact nhd of x and so is an interior point. This every point of X is an interior point of some compact subspace of X. Conversely let every point of X be an interior point of some compact suppose. To show that X is locally compact. Let x $\in$ X be arbitrary. By hypothesis, there exists a compact subspace K of X such that x $\in$ K. Then K is a nhd of x. Since X is Hausdorff, K is a closed subset of X so that $\overline{K}$ = K. Thus every point of X has a nhd whose closure is compact and so X is locally compact.

Theorem 3(a):

*A topological space is countably compact iff every countable collection of closed subsets of X with **FIP** has non-empty intersection.*

Proof :

Proof of this theorem is the same as that of theorem, except that we not take countable open covers instead of arbitrary open covers.

Theorem 3(b):

A topological space X is countably compact iff every sequence in X has a cluster point in X.

Proof :

Let X be countably compact and let $< x_n >$ be any sequence in X. Let $A_n = \{x_k : k \geq n\}$. Then $\{\overline{A}_n : n \in N\}$ is a countable collection of closed subset of X with FIP.

[For example if we take $\overline{A}_1, \overline{A}_{10}, \overline{A}_{18}, \overline{A}_{27}$, then

$\overline{A}_1 \cap \overline{A}_{10} \cap \overline{A}_{18} \cap \overline{A}_{27} = \overline{A}_{27}$ which is non-empty]. By the previous theorem, this countable family of closed sets has a non-empty intersection. Let x_0 be a point of the intersection of the family so that $x_0 \in \overline{A}_n \ \forall n$. Then it can be shown that xn is a cluster point of the sequence $< x_m >$. For, let N be any nhd of x_0 and let m be any positive. Then

$x_0 \in \overline{A}_m$ $[\because x_0 \in \overline{A}_n \ \forall n]$

and so $N \cap A_m \neq \varnothing$ [by def. of adherent point]

Hence $x_p \in N$ for some $p \geq m$. It follows that x_0 is a cluster point. We shall show that the space is countably compact. Let $\{A_n : n \in N\}$ be a countable collection of closed subsets of X with FIP. For each n select a point x_n from $\cap \{A_i : i\ 1, 2, ...\}$

[This is possible as a consequence of FIP].

Let x_0 be a cluster point of $< x_n >$. Let N be any nhd of x_0 and m and positive integer. Then $x_p \in N$ for some $p \geq m$.

But $x_p \in \cap \{A_i : i = 1, 2, ...p\}$ [See the choice of x_n]

Hence $x_p \in A_m$

$A_m \cap N \neq \varnothing$ $[\because p \geq m]$

and so x_0 is an adherent point of Am, that $x_0 \in \overline{A}_m$

But $\overline{A}_m = A_m$ $[\because A_m$ is closed$]$

Hence $x_0 \in A_m$.

Since m was any integer, we have $x_0 \in A_m \ \forall m$.

It follows that $\cap \{A_n : n \in N\} \neq \varnothing$.

Thus it is shown that every countable collection of closed subsets of X with FIP has a non-empty intersection. Therefore, by the previous theorem X is countably compact.

Theorem 4(a):

A countably compact topological space has BWP.

Proof :

Let X be a countably compact space and suppose, if possible, that it does not have *BWP*. Then there exists an infinite set S having no limit point. Let A be a countably infinite subset of S. Then A has no limit point. It follows that A is a closed set. Also for each $a_n \in A$, a_n is not a limit point of A. Hence there exists an open set G_n, such that $a_n \in G_n$ and $G_n \cap A = \{a_n\}$. Then the collection $\{G_n : n \in N\} \cap A'$ is a countable open cover of X. This cover has no finite subcover. For if we remove a single G_n, it will not be a cover of X since then an will not be covered. Hence Xis not countably compact. But this contradicts the hypothesis. Hence X must have *BWP* must

Definition : *A topological space X is said to be sequentially compact iff every sequence in X has a convergent subsequence.*

Theorem 4(b):

Every sequentially compact topological space is countably compact.

Proof :

Let X be a sequentially compact space. If X is finite, then the space is surely countably compact. Let now X be infinite. We prove the contra positive of the statement given in the theorem. Let $\{G_i : i \in N\}$ be a countable open cover of X which has no finite subcover. We now define the sequence $< x_1, x_2, \ldots x_n \ldots >$ as follows.

Let n_1 be the smallest positive integer such that

$$X \cap G_{n_1} \neq \varnothing.$$

Choose $x_1 \in X \cap G_{n_1}$. Now let n_2 be the least positive integer greater than n_1 such that

$$X \cap G_{n_2} \neq \varnothing$$

Choose $x_2 \in (X \cap G_{n_2}) - (\cap G_{n_2})$.

Note that such a point x_2 always exists, for otherwise G_{n_1} will be a cover of X. Continuing this process, we get the sequence.

$$< x_1, x_2 \ldots ,x_n .. >$$

having the property that, for each $i \in N$.

$$x_1 \in X \cap G_{n_i}\ x_i \notin \cup \{X \cap G_k : k = 1, 2, \ldots n - 1\}$$

and $m > n_i - 1$. It can be seen that $< x_n >$ has no convergent sub-sequence in X. For let $x \in X$. Then there exists

$$G_{m_0} \in \{G_n : n \in N\}$$

such that $x \in G_{m_0} \neq \varnothing$, there exists $k_0 \in N$ such that $G_{n_{1k}} = G_{m_0}$. But by the choice of the sequence $< x_1, x_2, \ldots >$, we have

$$i > k_0 \Rightarrow x \notin G_{m0.}$$

Since to G_{m0} is an open set containing x, no subsequence of $< x_0 >$ converges to x. Since x was arbitrary , X is not sequentially compact,

Definition : *A topological space X is said to be locally compact iff every point in X has at least one neighborhood whose closure is compact.*

Theorem 4(c):

Every compact topological space X is locally compact.

Proof :

Let X be a compact space, Since X is both open and closed, it is neighborhood of each of its points such that $\bar{X}$ = X is compact. Hence X is locally compact.

COMPACTNESS IN METRIC SPACES

ε-net and Total Boundedness

Definition : *Let (X, d) be a Measure Theory and let ε > 0 be given. A finite subset A of X is said to be an ε-net for X iff for every x ∈ X, there exists a point a ∈ A*

such that d (a, x) < ε. in other words, A is an e-net for X iff A is finite and

$$X = \cup \{S\ (a, \varepsilon) : a \in A\}$$

A Measure Theory (X, d) is said to be totally bounded iff X has an e-net for every ε > 0.

Theorem 1 (Ascoli's Theorem):

If X is a compact Measure Theory, then a *closed subspace* of $\mathcal{C}$(X, R) or $\mathcal{C}$(X, C) is compact ⇔ it is bounded and equicontinuous.

Proof:

Let d be the metric on X, and let F be a closed subspace of $\mathcal{C}$(X, R) or $\mathcal{C}$(X, C).

We first assume that F is compact, and we prove that it is bounded and equicontinuous. F is bounded. We prove that F is equicontinuous as follows. Let ∈ > 0 be given. Since F is compact, and therefore totally bounded, we can find an (∈/3)-net $\{f_1, f_2, ..., f_n\}$ in F.

Each f_k is uniformly continuous, so for each k = 1, 2, ..., n, there exists $\delta_k > 0$ such that $d(x, x') < \delta_k \Rightarrow |f_k(x) - f_k(x')| < \in/3$. We now define S to be the smallest of the numbers $\delta_1, \delta_2, ..., S_n$. If f is any function in F and f_k is chosen so that $||f - f_k|| < \in/3$, then

$$d(x, x') < \delta \Rightarrow |f(x) - f(x')| \leq |f(x) - f_k(x)| + f_k(x) - f_k(x')\} \\ + |f_k(x')| < \in/3 + \in/3 + \in/3 = \in.$$

This shows that F is equicontinuous.

We now assume that F is bounded and equicontinuous, and we demonstrate that it is compact by showing that every sequence in it has a convergent subsequence. Since F is closed, and therefore complete, it suffices to show that every sequence in it has a Cauchy subsequence. As we proceed, the reader will see that our proof is similar in structure to the last part of the proof of Theorem 1. X has a countable dense subset. Let the points of this subset be arranged in a sequence $\{x_i\} = \{x_2, x_3, ..., x_i, ...\}$, where we start with the subscript 2 for reasons which will become clear below. Now let

$$S_1 = (f_{11}, f_{12}, f_{13}, ...)$$

be an arbitrary sequence in F. Our hypothesis that F is bounded means that there exists a real number K such that $\|f\| \leq K$ for every f in F, or equivalently, such that $|f(x)| \leq K$ for every f in F and every x in X. Consider the sequence of numbers $\{f_{1j}(x_2)\}$, j = 1, 2, 3, ..., and observe that since this sequence is bounded, it has a convergent subsequence. Let $S_2 = \{f_{21}\ f_{22}, f_{23}, ...\}$ be a subsequence of S_1 such that $\{f_{2j}, (x_2)\}$ converges. We next consider the sequence of numbers $\{f_{2j}, (x_3)\}$, and in the same way we let $S_3 = \{f_{31}, f_{32}, f_{33}, ...\}$ be a subsequence of S_2 such that $\{f_{3j}(x_3)\}$ converges. If we continue this process, we get an array of sequences of the form

$$S_1 = \{f_{11}, f_{12}, f_{13}, ...\},$$
$$S_2 = \{f_{21}, f_{22}, f_{23}, ...\},$$
$$S_3 = \{f_{31}, f_{32}, f_{33}, ...\},$$

in which each sequence is a subsequence of the one directly above it, and for each i the sequence $S_i = \{f_{i1}, f_{i2}, f_{i3}, ...\}$ has the property that $\{f_{ij}\ (x_i)\}$ is a convergent sequence of numbers. If we define $f_1, f_2, f_3, ...$ by $f_1 = f_{11}$, $f_2 = f_{22}$, $f_3 = f_{33}$, ..., then the sequence $S = \{f_1, f_2, f_3, ...\}$ is the "diagonal" subsequence of +Si. It is clear from this construction that for each point x, in our dense subset of X, the sequence $\{f_n(x_i)\}$ is a convergent sequence of numbers. It remains only to show that S, as a sequence of functions in $\mathcal{C}$(X, R) or $\mathcal{C}$(X, C), is a Cauchy sequence. Let $\epsilon > 0$ be given. Since F is equicontinuous, there exists $\delta > 0$ such that $d\{x,x'\} < \delta \Rightarrow |f_n(x) - f_n(x')| < \epsilon/3$ for all functions f_n in S. We now form the open sphere $S_\delta(x_i)$ with radius δ centered on each of the x_i's. Since the x_i's are dense, these open spheres form an open cover of X, and since X is compact, $X = \bigcup_{i=2}^{i_0} S_\delta(x_i)$ for some i_0. It is easy to see that there exists a positive integer n_0 such that m, n > n_0 $\Rightarrow |f_m(x_i) - f_n(x_i)| < \epsilon/3$ for all the points $x_2, x_3, ..., x_{i0}$, Our proof is completed by the remark that if x is an arbitrary point of X, then an i can be found in the set $\{2, 3, ..., i_0\}$ such that $d(x, x_i) < \delta$, and that therefore

$$m, n \geq n_0 \Rightarrow |f_m(x) - f_n(x)| \leq |f_m(x) - f_m(x_i)| + |f_m(x_i) - f_n(x_i)|$$
$$+ |f_n(x_i)| - f_n(x)| < \epsilon/3 + \epsilon/3 + \epsilon/3 = \epsilon.$$

We observe that the total boundedness in Theorem 2 is replaced, in Ascoli's theorem, by the weaker condition of boundedness, and that the resulting deficiency is made up by the additional condition of equicontinuity.

Theorem 2: (Ascoli Theorem):

A Measure Theory is compact ⇔ it is complete and totally bounded.

Proof:

Let X be a Measure Theory. The first half of our proof is easy, for if X is compact, then it is totally bounded, and it is complete and the fact that every sequence (and therefore every Cauchy sequence) has a convergent subsequence.

We now assume that X is complete and totally bounded, and we prove that X is compact by showing that every sequence has a convergent subsequence. Since X is complete, it suffices to show that every sequence has a Cauchy subsequence. Consider an arbitrary sequence

$$S_1 = [x_{11}, x_{12}, x_{13}, ...\}.$$

The reason for this notation will soon be clear. Since X is totally bounded, there exists a finite class of open spheres, each with radius 1/2, whose union equals X; and from this we see that S_1 has a subsequence $S_2 = (x_{21}, x_{22}, x_{23} ....\}$ all of whose points lie in some one open sphere of radius %. Another application of the total boundedness of X shows similarly that S_2 has a subsequence $S_3 = \{x_{31}, x_{32}, x_{33}, ...\}$ all of whose points lie in some one open sphere of radius 1/3. We continue forming successive subsequences in this manner, and we let

$$S = \{x_{11}, x_{12}, x_{13}, ...\}$$

be the "diagonal" subsequence of S_1. By the nature of this construction, S is clearly a Cauchy subsequence of S_1, and our proof is complete.

This theorem gives total boundedness a prominent part in determining whether a Measure Theory is compact or not. As we know, many metric spaces occur as closed subspaces of complete metric spaces, and for these we can make the role of total boundedness even more striking.

CONTINUITY AND COMPACTNESS

Theorem 1:

Let X be a compact space and let Y be a Hausdorff space. Then every bijective continuous mapping of X onto Y is a homeomorphism.

Proof :

Let f : X → Y be bijective and continuous. In order that f may be a homemorphis, it suffices to show that f (F) is closed in Y for every closed set F in X. So let F be any closed set in X.

Then F is compact, being a closed subset of a compact space X. Since f is continuous, it follows that f [F] is compact in Y. Since Y is Hausdorff, f (F) is closed in Y.

Theorem 2:

Let f be a continuous mapping of a compact topological space X into a topological space Y. Then f [X] is compact. In other words continuous image of a compact space is compact.

Proof :

Let $\{G_\lambda : \lambda \in \Lambda\}$ be any open cover of f [X]. Since f is continuous, $f^{-1}[G_\lambda]$ is an open set in X. Then collection $\{f^{-1}[G_\lambda] : \lambda \in \Lambda\}$ then forms an open cover of X. Since X is compact, there exists finitely many indices $\lambda_1, \ldots \lambda_n$ such that

$$X = f^{-1}[G_{\lambda_1}] \cup \ldots \cup f^{-1}[G_{\lambda_n}]$$

$$= f^{-1}[G_{\lambda_1} \cup \ldots \cup G_n]$$

so that $f[X] = G_{\lambda_1} \cup \ldots \cup$ Gln.

Hence f [X] is compact.

Corollary : *Let X and Y be topological spaces and let A be a compact subset of X. If f : X → Y is continuous show that f [A] is a compact subset of Y.*

Proof :

We know that the restriction f_A : → Y of f to A defined by fA (x) = f (x) for all x ∈ A is a continuous mapping. Since A is a compact subspace of X, it follows from the above theorem that $f_A[A] = f[A]$ is a compact subset of Y.

BOUNDED MAPPING

Definition : *Let X be any non-empty set. Then a mapping f : X →* R is said to be *bounded* iff f (X) is bounded subset of R.

Theorem 1:

Every real valued continuous function on a compact space X is bounded and attains its infimum and supermum on X in the following sense.

$$\text{if } a = \inf \{f(x) : x \in X\}$$

$$\text{and} \qquad b = \sup \{f(x) : x\ X\}.$$

then there exists points a and b in X such that f (a) = a and f (b) = b.

Proof :

Let f be a real valued continuous function defined on a compact space X. Then by theorem (4.1), f [X] is a compact subset of **R** and therefore by the Heine-Borel theorem f [X] is closed and bounded set of real numbers. Let α = inf f [X] and β = f [X]. Since f [X] is bounded, we have $\alpha, \beta \in$ **R**. Since f [X] is closed, we have $\alpha, \beta \in$ f [X]. Choose a $\in f^{-1}$ [{α}] and b $\in$ f^{-1} [{β}]. Then f (a) = α and f (b) = β, that is, f attains its infimum and supremum on X.

EXERCISES

1. Let X be a Hausdorff space. Prove that intersection of two locally compact subsets of X is locally compact. Hence or otherwise, show that the intersection of an open and a closed set in a locally compact Hausdorff space is locally compact.

2. Generalize the preceding problem by proving that a compact subspace of an arbitrary Measure Theory is closed and bounded. (It should be carefully noted that a closed and bounded subspace of an arbitrary Measure Theory is not necessarily compact.)

3. Show that a subspace of R^n is bounded $\Leftrightarrow$ it is totally bounded.

4. By considering the sequence of functions in $\mathcal{C}[0, 1]$ defined by

   $$f_n(x) = nx$$

   for $0 \le x \le 1/n$, $f_n(x) = 1$ for $1/n \le x \le 1$, show that $\mathcal{C}[0, 1]$ is not locally compact.

5. Show that a topological space is locally compact $\Leftrightarrow$ there is an open base at each point whose sets all have compact closures.

6. Observe that any discrete space is locally compact. Assuming that there are topological spaces which are not locally compact (we assure the reader that this is true), show that a continuous image of a locally compact space need not be locally compact.

7. Prove the Bolzano-Weierstrass theorem for R^n: if X is a closed and bounded subset of R^n, then every infinite subset of X has a limit point in X.

8. A *countably compact* space is a topological space in which every countable open cover has a finite subcover. Prove that a second countable space is countably compact $\Leftrightarrow$ it is compact.

9. Show that if every open subspace of a space X is compact, then every subspace of X is compact.

10. Let X be an infinite set with the smallest topology in it in which every finite set is closed. Prove that X is compact.
11. Let Y be a subspace of a topological space X. If Z is a non-empty subset of Y, show that Z is compact as a subspace of Y ⇔ it is compact as a subspace of X.
12. Let X be a topological space. If $\{X_i\}$ is a non-empty finite class of compact subspaces of X, show that $\cup_i X_i$, is also a compact subspace of X. If $\{X_i\}$ is a non-empty class of compact subspaces of X each of which is closed, and if $\cap_j X_i$ is non-empty, show that $\cap_{j'} X_j$ is also a compact subspace of X.
13. Prove the converse of the Heine-Borel theorem: every compact subspace of the real line is closed and bounded.
14. All projections, being open mappings, send open sets to open sets. Use the Euclidean plane to show that a projection need not send closed sets to closed sets.
15. Show that the relative topology on a subspace of a product space is the weak topology generated by the restrictions of the projections to that subspace.
16. Let A be a subspace of a Measure Theory X, and show that A is totally bounded ⇔ $\overline{A}$ is totally bounded.
17. Prove in detail that the open rectangles in R^n form an open base.
18. Show that every closed and bounded subspace of the n-dimensional unitary space C^n is compact.
19. Show that a compact Measure Theory is separable.
20. Let f be a mapping of a topological space X into a product space $P_i X_i$, and show that f is continuous ⇔ $p_i f$ is continuous for each projection p_i.
21. Consider the product space defined and discussed in the last paragraph of the text, and show that this space is not second countable. (*Hint:* recall Theorem 18-B, and observe that the index set is uncountably infinite.)
22. Show that a continuous real function f defined on a compact space X attains its infimum and its supremum in the following sense: if a = inf i f(x): x ε X} and b = sup {f(x): x ε X}, then there exist points x_i and x_2 in X such that $f(x_1) = a$ and $f(x_2) = b$.
23. If X is a compact space, and if $\{f_n\}$ is a monotone sequence of continuous real functions defined on X which converges pointwise to a continuous real function f defined on X, show that fn converges

uniformly to f. (The assumption that f_n is a monotone sequence means that either $f_1 \leq f_2 \leq f_3 \leq \ldots$ or $f_1 \geq f_2 \geq f_3 \geq \ldots$)

24. Show that a continuous real or complex function defined on a compact space is bounded. More generally, show that a continuous mapping of a compact space into any Measure Theory is bounded.

25. Show that the union of a finite collection of compact subsets of a space is compact. Is the intersection of two compact subset necessarily compact ?

26. Show that the intersection of any collection of compact subsets of a Hausdorff space is compact : "Hausdorff' is necessary, even for intersection of two compact sects.

27. Prove or disprove : If X is an infinite space with the property that the only compact subsets of X are finite subsets, then X has the discrete topology.

28. Let A be compact subset of a Hausdorff space X and $x \notin A$. Show that there exist disjoint open sets G and H containing x and A respectively.

29. Let R have the left ray topology.

 (i) Prove that this space is not compact.

 (ii) Prove that the subset $A = \{x : x < 0\}$ is not compact.

 (iiii) Prove that the subset $B = \{x : x \leq 0\}$ is compact.

30. Let E be a compaact space, A and B two closed sets without any common point in E, prove that there exists two open sets U, V without any common point such that $A \subset U$ and $B \subset V$.

31. Let A be a compact subset of Measure Theory (X, d) and let be a point of A. Show that there is point $a \in A$ such that d (A, b) = (b, A).

32. Prove that a Measure Theory is compact if and only if every sequence has a cluster point.

33. Prove that a bounded Measure Theory need not be totally bounded.

34. In a locally compact space X, let A be a set such that, for every compact subset K of X, $A \cap K$ is compact. Then prove that A is closed in X.

35. Let X be a locally compact Hausdorff space, and K_1 and K_2 disjoint compact subspaces of X, Show that K_1 and K_2 have disjoint nhds whose closure are compact.

36. (a) Is every open subspace of a locally compact space locally compact ? Give reasons in support of your answer.

    (b) Give an example to show that a subspace of a locally compact space may fail to be locally compact.

37. Show that a Hausdorff space X is locally if and only if each point of X has a neighbourhood base consisting of compact sets.

38. Let A and B be non-empty subsets of a Measure Theory (X, d) such that B is compact. Prove that d (A, B) = 0 if and only if $\bar{A} \cap B$ is non empty

39. Let P be a compact subset of a Measure Theory (X, d) having diameter d (A). Show that there is a pair of point x and y of A such that d (x, y) = d (A).

40. Prove that in a Hausdorff space X, any locally compact subspace A is the intersection of an open set and a closed set and that in a locally compact space X, the intersection of an open set and a closed set is a locally compact subspace of X.

41. Give an example of a Hausdorff space which is not locally compact.

INTRODUCTON OF CONNECTEDNESS

Continuous real functions are often defined on intervals, and functions of this kind have many pleasant properties. For example, such a function assumes as a value every number between any two of its values (the *Weierstrass intermediate value theorem*); furthermore, its graph is a connected subspace of the Euclidean plane. Connectedness is also a basic notion in complex analysis, for the regions on which analytic functions are studied are generally taken to be connected open subspaces of the complex plane. In the portion of topology which deals with continuous curves and their properties, connectedness is of great significance, for whatever else a continuous curve may be, it is certainly a connected topological space. From the intuitive point of view, a *connected*, space is a topological space which consists of a single piece.

This property is perhaps the simplest which a topological space may have, and yet it is one of the most important for the applications of topology to analysis and geometry.' Spaces which are not connected are also interesting. One of the outstanding characteristics of the Cantor set is the extreme degree in which it fails to be connected. Much the same is true of the subspace of the real line which consists of all rational numbers. These spaces are so badly disconnected that they are almost granular in texture. Our purpose in this chapter is to convert these rather vague notions into precise mathematical ideas, and also to establish the fundamental facts in the theory of connectedness which rests upon them.

LOCALLY CONNECTED SPACES

The locally connected space is a topological space with the property that if x is any point in it and G any neighbourhood of x, then G contains a connected neighbourhood of x. This is evidently equivalent to the condition that each point of the space have an open base whose sets are all connected subspaces. Locally connected spaces are quite abundant, every Banach space is locally connected. We know that local compactness is implied by compactness. Local connectedness, however, neither implies, nor is implied by, connectedness. We encountered the concept of a locally compact space, that is, of a space which is compact around each point but need not be compact as a whole. We now study another "local" property which a topological space may have, that of being connected in the vicinity of each of its points. The union of two disjoint open intervals on the real line is a simple example of a space which is locally connected but not connected. A space can also be connected without being locally connected, as the following example shows. Let X be the subspace of the Euclidean plane defined by $X = A \cup B$, where $A = \{(x, y):x = 0 \text{ and } -1 \leq y \leq 1)$ and $B = \{(x, y):0 < x < 1 \text{ and } y = \sin (1/x))$. B is the image of the interval (0,1] under the continuous mapping f defined by

$$f(x) = (x, \sin (1/x)),$$

so B is connected by Theorem 2; and since $X = \overline{B}$, X is connected by Theorem 2 X is not locally connected, however, for it is reasonably easy to see that each point x in A has a neighbourhood which does not contain any connected neighbourhood of x.

We know that the components of an arbitrary topological space X are always closed sets, and from this we see at once that the components of any closed subspace of X are also closed in X. The reader may feel, with some justification, that the components of a well-behaved space ought to be open sets. This is true for locally connected spaces.

Theorem 1:

Let X be a locally connected space. If Y is an open subspace of X, then each component of Y is open in X. In particular, each component of X is open.

Proof:

Let C be a component of Y. We wish to show that C is open in X. Let x be a point in C. Since X is locally connected and Y is open in X, Y contains a connected neighbourhood G of x. It suffices to show that $G \subseteq C$. This will follow at once from the fact that C is a component of Y if we can show that G is connected as a subspace of Y. But this is clear, according to which the topology of G as a subspace of Y is the same as its topology as a subspace of X; for G is connected with respect to the latter topology

The principal applications of local connectedness lie in the theory of continuous curves.

SEPARATED SETS

Definition : *Let (X, $\mathfrak{I}$) be a topological space. Two non-empty subsets A and B of X are said to be $\mathfrak{I}$-separated iff.*

$$A \cap \overline{B} = \varnothing \text{ and } \overline{A} \cap B = \varnothing$$

These two conditions are equivalent to the single condition

$$(A \cap \overline{B}) \cup (\overline{A} \cap B) = \varnothing.$$

Thus, A and are separated if an only if A and B are disjoint and neither of them contains limit point of the other. For we have

$$A \cap \overline{B} = \varnothing \Leftrightarrow [B \cup D(B)] = \varnothing$$

$$\Leftrightarrow (A \cap B) \cup (A \cap D(B)] = \varnothing$$

$$\Leftrightarrow (A \cap B) \Leftrightarrow (A \cap D(B)] = \varnothing$$

$$\Leftrightarrow A \Leftrightarrow D(B) = \varnothing$$

$\Leftrightarrow$ A contains no limit point of B.

Note that any two separated sets are disjoint. But two disjoint sets are not necessarily separated. For example the subsets $A =]-\infty, 0[$ and $B = [0, \infty]$ of **R** are disjoint the not separated with respect to the usual topology since.

$$\overline{A} \cap B =]-\infty, 0] \cap [0, \infty[= \{0\} \neq \varnothing$$

Example:

Consider the following subsets of **R** *(with usual topology).*

$$A =]2, 3[, B =]3, 4] \text{ and } C = [3, 4[.$$

The sets A and B are separated since $\overline{A} = [2, 3]$ and $\overline{B} = [3, 4]$, so that $A \cap \overline{B} = \varnothing$ and $\overline{A} \cap B = \varnothing$. But A and C are not separated since

$$\overline{A} = C = [2, 3] \cap [3, 4[= \{3\} \neq \varnothing.$$

Theorem 1:

Two disjoint sets A and B are separated in a topological space (X, $\mathfrak{I}$) if and only if they are both open and closed in the subspace $A \cup B$.

Proof:

Let the disjoint sets A and B be separated in X so that

$$A \cap Cl_X(B) = \varnothing$$

and $\quad Cl_X(A) \cap B = \varnothing.$

Let $E = A \cup B$. Then $Cl_E(A) = Cl_X(A) \cap E$

$$= Cl_X(A) \cap (A \cup B) = [Cl_X(A) \cap A] \cup [Cl_X(A) \cap B]$$

$$= A \cup \varnothing = A \qquad [\because A \subset Cl_X(A) \text{ and } Cl_X(A) \cap B = \varnothing]$$

Hence, A is closed in the subspace $A \cup B$. Similarly, B is closed in $A \cup B$. Again since $A \cup B = \varnothing$, they are complements of each other in E and hence they are both open in E.

Conversely, let the disjoint sets A and B be both open and closed in $A \cup B$. To show that A and B are separated in A. Since A is closed in E, we have

$$A = Cl_E(A) = [Cl_X(A) \cap E] = Cl_X(A) \cap (A \cup B)$$

$$= [Cl_X(A) \cap A] \cup [Cl_X(A) \cap B] \qquad \text{[Distributive law]}$$

$$= A \cup [Cl_X(A) \cap B] \qquad [\because A \subset Cl_X(A)] \quad ...(1)$$

Since $A \cap B = \varnothing \Rightarrow A \cap [Cl_X(A) \cap B] = \varnothing$, it follows from (1) that $Cl_X(A) \cap B = \varnothing$. Similarly $A \cap [Cl_X(B)] = \varnothing$. Hence A and B are separated in X.

Theorem 2:

Let $(Y, \mathfrak{I}_Y)$ be a subspace of a topological space $(X, \mathfrak{I})$ and let A, B the two subsets of Y. Then A, B are $\mathfrak{I}$-separated iff they are $\mathfrak{I}_Y$-separated.

Proof:

$$Cl_Y(A) = Cl_X(A) \cap Y \text{ and } Cl_Y(B) = Cl_X(B) \cap Y$$

Now $\quad [Cl_Y(A) \cap B] \cup [A \cap Cl_Y(B)]$

$$= [Cl_X(A) \cap Y \cap B] \cup [A \cap Cl_X(B) \cap Y]$$

$$= [Cl_X(A) \cap B] \cup [A \cap Cl_X(B)] \qquad [\because A, B \subset Y]$$

Hence $\quad [Cl_Y(A) \cap B] \cup [A \cap Cl_Y(B)[= \varnothing$

iff $\quad [Cl_X(A) \cap B[\cup [A \cap Cl_X(B)] = \varnothing.$

It follows that A, B are $\mathfrak{I}$-separated.

Theorem 3(a):

If A and B are separated subsets of a space X and $C \subset A$ and $D \subset B$, then C and D are also separated.

Proof:

We are given that

$$A \cap \overline{B} = \varnothing \text{ and } \overline{A} \cap B = \varnothing \quad ...(1)$$

Also $\quad C \subset A \Rightarrow \overline{C} \subset \overline{A}$

and $\quad D \subset B \Rightarrow \overline{D} \subset \overline{B} \quad ...(2)$

It follows from (1) and (2) that

$$C \cap \overline{D} = \varnothing \text{ and } \overline{C} \cap D = \varnothing.$$

Hence C and D are separated.

Theorem 3(b):

Two closed (open) subsets A, B of a topological space are separated if and only if they are disjoint.

Proof:

Since any two separated sets are disjoint, we need only that two disjoint closed (open) sets are separated. It A and B are both disjoint and closed, then

$$A \cap B = \varnothing,\ \overline{A} = A \text{ and } \overline{B} = B$$

so that $\quad \overline{A} \cap B = \varnothing$ and $A \cap \overline{B} = \varnothing$,

showing that A and B are separated. If A and B are both disjoint and open, then A′ and B′ are both closed so that

$$A'^{-} = A' \text{ and } B'^{-} = B'$$

Also $\quad A \cap B = \varnothing$

$\Rightarrow \quad A \subset B'$ and $B \subset A'$

$\Rightarrow \quad \overline{A} \subset \overline{B'} = B'$ and $\overline{B} \subset \overline{A'} = A'$

$\Rightarrow \quad \overline{A} \cap B = \varnothing$ and $\overline{B} \cap A = \varnothing$

$\Rightarrow \quad$ A and B are separated.

Hence the theorem.

CONNECTED AND DISCONNECTED SETS

Intuitively speaking a connected topological space is one which consists of a single piece.

Definition : *Let (X, $\mathfrak{I}$) be a topological space. A subset A of X is said to be* **$\mathfrak{I}$-disconnected** *iff it is the union of two non-empty $\mathfrak{I}$-separated sets, that is, iff there exists two non-empty sets C and D such that $C \cap D = \varnothing$, $\overline{C} \cap D = \varnothing$ and $A = C \cup D$. A is said to be* **connected** *iff it is not disconnected.*

Remark : The empty set is trivially connected. Also every singleton set in a space is connected since it cannot be expressed, as a unit on of two non-empty separated sets.

Definition : *Two points a and b of a topological space X are said to be connected iff they are contained in a connected subset of X.*

Theorem 1:

A topological space X is connected if and only if every non-empty proper subset of X has a non-empty frontier.

Proof:

Let every non-empty proper subset of X have a non-empty frontier. To show that X is connected. Suppose, if possible X is disconnected. Then there exist non-empty disjoint sets G and H both open and closed in X such that $X = G \cup H$. Therefore $G = G^{\circ} = \overline{G}$. But **Fr** $(G) = \overline{G} - G^{\circ}$. Hence **Fr** $(G) = G - G = \varnothing$, which is contrary to our hypothesis. Hence X must be connected.

Conversely, let X be connected and suppose, if possible there exists a non-empty proper subset A of X such that **Fr** $(A) = \varnothing$.

Now $\overline{A} = A^{\circ} \cup$ **Fr** $(A) = A \cup$ **Fr** $(\overline{A})$. Hence $\overline{A} = A^{\circ} = A$ showing that A is both open and closed in X and therefore X is disconnected. But this is a contradiction. Hence every non-empty proper subset of X must have a non-empty frontier.

Theorem 2:

Let $(Y, \mathfrak{I}_Y)$ be a subspace of a topological space $(X, \mathfrak{I})$ and $A \subset Y$. Then A is $\mathfrak{I}$-disconnected if and only if it is $\mathfrak{I}_Y$ disconnected, or equivalently, A is $\mathfrak{I}$-connected iff it is $\mathfrak{I}_Y$-connected.

Proof:

Two non-empty subsets of Y are $\mathfrak{I}$-separated iff they are $\mathfrak{I}_Y$-separated. Therefore A is the union of two $\mathfrak{I}$-separated sets iff it is the union of two $\mathfrak{I}_Y$-separated sets. Hence the result.

Corollary 1 : *Let $(Y, \mathfrak{I}\, Y)$ be a subspace of a topological space $(X, \mathfrak{I})$. Then Y is a subset of X is $\mathfrak{I}$-disconnected if and only if the subsapce $(Y, \mathfrak{I}_Y)$ is disconnected as a subspace in its own right.*

Theorem 3(a):

A topological space X is disconnected if and only if there exists a non-empty proper subset of X which is both open and closed in X.

Proof:

Let A be a non-empty proper subset of X which is both open and closed. We have to show that X is disconnected. Let B = A′. Then B is non-empty since A is a proper subset of X. Moreever, $A \cup B = X$ and $A \cap B = \varnothing$. Since A is both closed and open, B is also both closed and open. Hence $\overline{A} = A$ and $\overline{B} = B$. It follows that $\overline{A} \cap B = \varnothing$ and $A \cup \overline{B} = \varnothing$. Thus X has been expressed as a union of two separated sets and so X is disconnected.

Conversely, let X be disconnected. Then there exist non-empty subsets A and B of X such that $A \cap \overline{B} = \varnothing$, $\overline{A} \cap B = \varnothing$ and

$$A \cup B = X$$

Since $A \subset \overline{A}$, $\overline{A} \cap B = \varnothing \Rightarrow A \cap B = \varnothing$. Hence A = B′. Since B is non-empty, and $B \cup B' = X$, it follows that B = A′ is a proper subset of X. Now $A \cup \overline{B} = X$.

$$[\because A \cup B = X \text{ and } \overline{B}\ B \Rightarrow A \cup \overline{B} \supset A \cup \overline{B}$$

$$\text{always so that } A \cup \overline{B} = X]$$

Also $A \cap \overline{B} = \varnothing \Rightarrow A = (\overline{B})'$ and similarly $B = (\overline{A})'$. Since $\overline{B}$ and $\overline{A}$ are closed sets, it follows that A and B are open sets.

Since A = B′, A is also closed. Thus A is non-empty proper subset of X which is both open and closed.

We have shown incidently that B is also a non-empty proper subset of X which is both open and closed.

Corollary 2 : *A topological space X is connected iff the only non-empty subset of X which is both open and closed in X is X itself.*

Corollary 3 : *Let (X, $\mathfrak{J}$) be a topological space and let $Y \subset X$. Then Y is disconnected iff there exists a non-empty proper subset A of Y which is both open and closed in Y. That is iff there exists a $\mathfrak{J}$-open set G and $\mathfrak{J}$-closed set F such that*

$$A = G \cap Y \text{ and } A = F \cap Y.$$

Theorem 3(b):

A topological space X is disconnected if and only if any one of the following statements holds

(i) *X is the union of two non-empty disjoint open sets.*

(ii) *X is the union of two non-empty disjoint closed sets.*

Proof:

Let X be disconnected. Then there exists a non-empty proper subset A of X which is both open and closed. Then A′ is also both open and closed. Also A ∪ A′ = X. Hence the sets A and A′ satisfy the requirements of (i) and (ii).

Conversely, let X = A ∪ B and A ∩ B = ∅, where A, B are non-empty open sets. It follows that A = B′ so that A is closed as well. Since B is non-empty, A is a proper subset of X. Thus A is a non-empty proper subset of X which is both open and closed. Hence by theorem. X is disconnected. Again, let X = C ∪ D, C ∩ D = ∅ where C, d are non-empty closed sets. Then C = D′, so that C is open as well. Since D is non-empty C is a proper subset of X which is both open and closed. Hence X is disconnected. Thus, it is shown that if any one of the conditions (i) and (ii) hods, then X is disconnected A ∪ B is called a *disconnection* of X.

Corollary 4 : *A subset Y of a topological space X is disconnected if and only if Y is the union of two non-empty disjoint sets both open (closed) in Y.*

Thus, Y is disconnected if and only if there exists non-empty sets G and H both open (closed) in X such that

$$G \cap Y \neq \varnothing,\ H \cap Y \neq \varnothing,\ (G \cap Y) \cap (H \cap Y) = \varnothing$$

and $$(G \cap Y) \cup (H \cap Y) = Y.$$

Corollary 5 : *Let (X, $\mathfrak{I}$) be a topological space and let Y be a subset of X. Then Y is disconnected iff there exists non-empty sets G and H both open (closed) in X such that*

$$G \cap Y \neq \varnothing,\ H \cap Y \neq \varnothing,\ Y \subset G \cup H$$

and $$G \cap H \subset Y - Y.$$

Proof:

Y is disconnected iff there exist non-empty sets G and H both open (closed) in X such that

$$G \cap Y \neq \varnothing,\ H \cap Y \neq \varnothing,\ (G \cap Y) \cap (H \cap Y) = \varnothing$$

and $$(G \cap Y) \cup (H \cap Y) = Y$$

Now $$(G \cap Y) \cap (H \cap Y) = \varnothing$$

$$\Leftrightarrow \quad (G \cap H) \cap Y = \varnothing$$

$$\Leftrightarrow \quad G \cap H \subset X - Y,$$

and $$(G \cap Y) \cup (H \cap Y) = Y$$

$$\Leftrightarrow \quad (G \cup H) \cup (H \cap Y) = Y$$

$\Leftrightarrow \qquad Y \subset G \cup H$

$G \cup H$ is called a *disconnection* of Y.

Theorem 4:

Show that if (X, $\Im$) is disconnected and Ä' is finer than $\Im$, then (X, $\Im'$) is disconnected.

Proof:

Sin (X, $\Im$) *is disconnected*, there is a non-empty proper subset A of X which is both $\Im$-open and $\Im$-closed. Since $\Im'$ is finer than $\Im$, A is $\Im'$ open. Again since A is $\Im$-closed, X – A is $\Im$-open. Since $\Im'$ is finer than $\Im$, X – A is also $\Im'$-open and consequently A is $\Im$-closed. Thus A is a non-empty proper subset of X which is both $\Im'$-open and $\Im$-closed. It follows that (X, $\Im'$) is disconnected.

Theorem 5:

Let (X, $\Im$) be a topological x space and let E be a connected subset of X such that $E \subset A \cup B$ where A and B are separated sets. Then $E \subset A$ or $E \subset B$, that is, E cannot intersect both A and B.

Proof:

Since A B are separated $A \cap \overline{B} = \varnothing$, $\overline{A} \cap B = \varnothing$. Now

$E \subset A \cup B \Rightarrow E = E \cap (A \cup B) = (E \cap A) \cup (E \cap B) \qquad ...(1)$

We claim that at least one of the sets $E \cap A$ and $E \cap B$ is empty. For, it possible, suppose none of thee sets is empty, that is suppose that $E \cap A \neq \varnothing$ and $E \cap B \neq \varnothing$. Then

$$(E \cap A) \cap (E \cap B) \subset (E \cap A) \cap (\overline{E} \cap \overline{B})$$

$$= (E \cap E) \cap (A \cap \overline{B}) = (E \cap \overline{E}) \cap \varnothing = \varnothing.$$

Similarly $(\overline{E \cap A}) \cap (E \cap B) = \varnothing$.

Hence, $E \cap A$ and $E \cap B$ are separated sets. Thus E has been expressed as the union of two non-empty separated sets and consequently E is disconnected. But this is a contradiction. Hence at least one of the sets $E \cap A$ and $E \cap B$ is empty. If $E \cap A = \varnothing$, then (1) gives $E = E \cap B$ which implies that $E \subset B$. Similarly if $E \cap B = \varnothing$, then $E \subset A$. Hence either $E \subset A$ or $E \subset B$.

Corollary 6 : *If E is connected subset of a space X such that $E \subset A \cup B$ where A, B are disjoint open (closed) subsets of X, then A and B are separated.*

Proof:

If A, B are open with $A \cap B = \varnothing$, then

$$A \subset B' = \overline{A} \subset (\overline{B}') = B' \qquad [\because B' \text{ is closed}]$$

$$\Rightarrow \qquad \overline{A} \cap B = \varnothing,$$

Similarly $A \cap \overline{B} = \varnothing$. Hence A, B are separated.

Theorem 6(a):

Let E be a connected subset of a space X. If R is a subset of X such that $E \subset F \subset \overline{E}$, then F is connected. In particular $\overline{E}$ is connected.

Proof:

Suppose F is disconnected. Then there exist non-empty sets A and B such that $A \cap \overline{B} = \varnothing$, $\overline{A} \cap = \varnothing$, and $A \cup B = F$.

Since $E \subset F = A \cup B$, it follows that $E \subset A$ or $E \subset B$. Let $E \subset A$ which implies that

$$\overline{E} \cap \overline{A} = \overline{E} \cap B \subset \overline{A} \cap B = \varnothing.$$

Since $\varnothing$ is a subset of every set, we have $\overline{E} \cap B = \varnothing$. Also

$$A \cup B = F \subset \overline{E} \Rightarrow B \subset F \subset \overline{E} \Rightarrow \overline{E} \in B = B.$$

Hence $B = \varnothing$ which is a contradiction since B is non-empty. Hence F must be connected.

Again since $E \subset \overline{E} \subset \overline{E}$, we see that $\overline{E}$ is connected.

Theorem 6(b):

If every two points of a subset E of a topological space X are contained in some connected subset of E, then E is a connected subset of X.

Proof:

Suppose E is not connected. Then there exists non-empty subsets A and B of X such that $A \cap \overline{B} = \varnothing$, $\overline{A} \cap B = \varnothing$ and $E = A \cup B$. Since A, B are non-empty, there exist a point $A \in A$ and a point $b \in B$. By hypothesis a and b must be contained in some connected subset F of E. Since $F \subset A \cup B$, either $F \subset A$ or $F \subset B$. It follows that either a, b are both in A or both in B. Let $a, b \in A$. Since $B \in B$, we have $A \cap B \neq \varnothing$ which is a contradiction since A, B are disjoint sets. Hence E must be connected.

Theorem 6(c):

Let $\{C_\lambda : l \in \Lambda\}$ be a family of connected subsets of a space X such that

$\cap \{C_\lambda : l \in \Lambda\} \neq \varnothing$. Then $\cup \{C_\lambda : l \in \Lambda\}$ *is a connected set.*

Proof:

Suppose $E = \cup \{C_\lambda : l \in \Lambda\}$ is not connected. Then there exist two non-empty disjoint sets G_1 and G_2 both open in the subspace E of X such that $E = G_1 \cup G_2$. For each λ, $G_1 \cap C_\lambda$ and $G_2 \cap C_\lambda$ are disjoint sets both open in the sub-space C_λ such that

$$(G_1 \cap C_\lambda) \cup (G_2 \cap C_\lambda)$$
$$= (G_1 \cup G_2) \cap C_\lambda = C_\lambda \qquad ...(1)$$

Since C_λ is connected, one of the sets $G_1 \cap C_\lambda$ and $G_2 \cap C_\lambda$ must be empty, say $G_1 \cap C_\lambda = \varnothing$. Then by (1), $G_2 \cap C_\lambda = C_\lambda$ which implies that $C_\lambda \subset G_2 \; \forall_\lambda \in \Lambda$ and hence $\cup \{C_\lambda : l \in \Lambda\} \subset G_2$, that is, $G_1 \cup G_2 \subset G_2$. This gives $G_1 = \varnothing$ which is a contradiction since G_1 is non-empty. Hence E must be connected.

Theorem 7:

Let $\{C_\lambda : l \in \Lambda\}$ be a family of connected subsets of a space such that one of the members of this family intersects every other member. Then $\cup\{C_\lambda : \lambda \in \Lambda)$ is connected.

Proof:

Let C_{λ_0} be a fixed member of the given family such that $C_{\lambda_0} \cap C_\lambda \neq \varnothing$ for every $\lambda \in \Lambda$. Then $D_\lambda = C_{\lambda_0} \cup C_\lambda$ is a connected set for every $\lambda \in \Lambda$.

Now $\quad \cup \{D_\lambda : \lambda \in \Lambda\} = \cup \{C_{\lambda_0} \cup C_\lambda : \lambda \in \Lambda\}$

$$= C_{\lambda_0} \cup [\cup \{C_\lambda : \lambda \in \Lambda]$$
$$= \cup \{C_\lambda : \lambda \in \Lambda\}$$

[$\because C_{\lambda_0}$ is one of the C_λ's]

and $\quad \cap \{D_\lambda : \lambda \in \Lambda\} = \cap \{C_{\lambda_0} \cup C_\lambda : \lambda \in \Lambda\}$

$$= C_{\lambda_0} \cup [\cap \{G_1 : \lambda \in \Lambda\}] \neq \varnothing$$

[$\because C_{\lambda_0}$ is intersects every C_λ, therefore $C_{\lambda_0} \neq \varnothing$]

Hence we have

$$\cup \{D_\lambda : \lambda \in \Lambda\} = \cap \{C_\lambda : \lambda \in \Lambda\} \text{ is connected.}$$

Note: For future discussion, by an interval in **R** we shall understand any of the following sets:

]a, b[,]a, b], [a, b],] $-\infty$, a[,] $-\infty$, a],]a, ∞[, [a, ∞[and **R** with $a \leq b$.

[Note that the condition $a \leq b$ implies that singleton subsets of **R** and empty set $\varnothing$ are also intervals].

Alternately, an interval in **R** is a subsets E of **R** which contains every $p \in \mathbf{R}$ with $x \leq p \leq y$ whenever $x, y \in E$.

Theorem 8(a):

*A subset E of **R** is connected if and only if it is an interval. In particular **R** is connected.*

Proof:

Let E be connected. If E is a singleton set or $\varnothing$, there is nothing to prove. Now let E contain more than one point and suppose if possible, that E is not an interval. Then there exist real numbers a, p, b with $a < p < b$ such that $a, b \in E$ but $p \notin E$. Let G = [p, ∞[$\subset$ E. Then G is non-empty (since $b \in$ G) and proper subset of E (since $a \in E$ but $a \notin G$). Also G is open in E and sine [p, ∞[is closed in R, it follows that G is closed in E. Thus G is a non-empty proper subset of E which is both open and closed in E. Hence E is disconnected. But this is against our hypothesis. Hence E must be an interval.

Conversely, let E be an interval and suppose, if possible, E is disconnected. Then there exist non-empty disjoint sets A and B, both closed in E such that $E = A \cup B$. Choose $a \in A$ and $b \in B$. Since $A \cap B = \varnothing$, a U b. Since E is an interval and $a, b \in E$, we have $[a, b] \subset E = A \cup B$. Let u = sup ([a, b] $\cap$ A). Evidently $a \leq u \leq b$. To each $\varepsilon > 0$, there exists some $v \in [a, b] \cap A$ such that $u - \varepsilon < v \leq u$. This shows that every nhd of u contains a point of $[a, b] \cap A$ and hence a point of A and therefore either $u \in A$ or u is a limit point of A. Since A is closed, in either case, $u \in A$. Then $u \notin B$ [$\because A \cap B = \varnothing$]. Again since $b \in B$, we have $b \neq u$. This gives us the strict inequality $u < b$. Moreover, the definition of u shows that u + e belongs to B for every $\varepsilon > 0$ such that $u + \varepsilon \leq b$. This means that every nhd of u contains a point of B distinct from u and so u is a limit point of B. Since B is closed, $u \in B$. Hence $u \in A \cap B$ which is a contradiction since A and B are disjoint. Hence E must be connected.

Theorem 8(b):

***R** is connected.*

*Proof of this theorem is exactly the same as the proof of the 'if part' of theorem except that E is replaced by **R**.*

CONNECTED SPACES

A *connected space* is a topological space X which cannot be represented as the union of two disjoint non-empty open sets. If $X = A \cup B$, where A and B are disjoint and open, then A and B are also closed, so that X is the union of two disjoint closed sets, and conversely. We see by this that X is connected $\Leftrightarrow$ it cannot be represented as the union of two disjoint non-empty closed sets. It is also clear that the connectedness of X amounts to the condition that ϕ and X are its only subsets which are both open and closed. A *connected subspace* of X is a, subspace Y which is connected as a topological space in its own right. By the definition of the relative topology on Y, this is equivalent to the condition that Y is not contained in the union of two open subsets of X whose intersections with Y are disjoint and non-empty.

Our space X is said to be *disconnected* if it is not connected, that is, if it can be represented in the form $X = A \cup B$, where A and B are disjoint, non-empty, and open; and if X is disconnected, a representation of it in this form (there may be many) is called a disconnection of X.

We begin by proving a theorem which supports a considerable part of the theory of connectedness.

Theorem 1:

The product of any non-empty class of connected spaces is connected.

Proof:

Let $\{X_i\}$ be a non-empty class of connected spaces, and form their product $X = P_i X_i$. We assume that X is disconnected, and we deduce a contradiction from this assumption. By Theorem 4, there exists a continuous mapping f of X onto the discrete two-point space $\{0,1\}$. Let $a = \{a_i\}$ be a fixed point in X, and consider a particular index i_1. We define a mapping f_{i_1} of X_{i_1} into X by means of $f_{i_1}\left(x_{i_1}\right) = \{y_i\}$, where $y_i = a_i$ for $i \neq i_1$ and $y_{i_1} = x_{i_1}$. This is clearly a continuous mapping, so ff_{i_1} is a continuous mapping of X_{i_1} into $\{0,1\}$. Since X_{i_1} is connected, we see by Theorem 4 that ff_{i_1} is constant and that

$$\left(ff_{i_1}\right)\left(x_{i_1}\right) = f(a)$$

for every point x_{i_1} in X_{i_1}. This shows that $f(x) = f(a)$ for all x's in .Y which equal a in all coordinate spaces except X_{i_1}. By repeating this process with another index is, etc., we see that $f(x) = f(a)$ for all x's in X which equal a in all but a finite number of coordinate spaces. The set of all x's of this kind is a dense subset of X, f is a constant mapping. This contradicts the assumption that f maps X onto $\{0, 1\}$, and completes the proof.

As an application of this result, we show that all finite-dimensional Euclidean and unitary spaces are connected.

Theorem 2(a):

The spaces R^n and C^n are connected.

Proof:

We showed in the proof of Theorem 2 that R^n, as a topological space, can be regarded as the product of n replicas of the real line R. We have seen in Theorem 1 that R is connected, so R^n is connected by Theorem E. We next prove that C^n and R^{2n} are essentially the same as topological spaces by exhibiting a homeomorphism f of C^n onto R^{2n}. Let $2 = (z_1, z_2, ..., z_n)$ be an arbitrary element in C^n, and let each coordinate 24 be written out in the form $z_k = a_k + ib_kc$, where a_k and b_k are its real and imaginary parts. We define f by

$$f(z) = (a_1, b_1, a_2, b_2, ..., a_n, b_n).$$

f is clearly a one-to-one mapping of C^n onto R^{2n}, and if we observe that $\|f(2)\| = \|2\|$, it is easy to see that f is a homeomorphism. The fact that R^{2n} is connected now shows that C^n is also connected.

The techniques developed in the next section will make it possible to give an easy proof of a much more general theorem than this, to the effect that any Banach space is connected.

Theorem 2(b):

Any continuous image of a connected space is connected.

Proof:

Let $f:X \to Y$ be a continuous mapping of a connected space X into an arbitrary topological space Y. We must show that f(X) is connected as a subspace of Y. Assume that f(X) is disconnected. As we have seen, this means that there exist two open subsets G and H of Y whose union contains f(X) and whose intersections with f(X) are disjoint and non-empty. This implies, however, that $X = f^{-1}(G) \cup f^{-1}(H)$ is a disconnection of X, which contradicts the connectedness of X.

As a direct consequence of the two theorems just proved, we have the following generalization of the Weierstrass intermediate value theorem.

Theorem 3:

The range of a continuous real function defined on a connected space is an interval.

It is a trivial observation that any two discrete spaces with the same number of points are essentially identical; for any one-to-one mapping of one onto the other (there is at least one) is a homeomorphism, and we may think of them as differing only in the symbols used to designate their points. It is in this sense that there is only one discrete space with any given number of points. The *discrete two-point space*, which is obviously disconnected, is a useful tool in the theory of connectedness. We denote its points by the symbols 0 and 1, and we think of them as real numbers.

Theorem 4:

A topological space X is disconnected <=> there exists a continuous mapping of X onto the discrete two-point space {0,1}.

Proof:

If X is disconnected and $X = A \cup B$ is a disconnection, then we define a continuous mapping f of X onto (0, 1) by the requirement that f(x) = 0 if x is in A and f(x) = 1 if x is in B. This is a valid definition by the fact that A and B are disjoint and their union is X. Since A and B are non-empty and open, f is clearly onto and continuous.

On the other hand, if there exists such a mapping, then X is disconnected; for if X were connected, Theorem 2 would imply that {0, 1} is connected, and this would be a contradiction.

This result is a useful tool for the proof of our next theorem.

THE COMPONENTS OF A SPACE

If a space is not itself connected, then the next best thing is to be able to decompose it into a disjoint class of maximal connected subspaces. Our present objective is to show that this can always be done.

A maximal connected subspace of a topological space, that is, a connected subspace which is not properly contained in any larger connected subspace, is called a *component* of the space. A connected space clearly has only one component, namely, the space itself. In a discrete space, it is easy to see that each point is a component.

The following two theorems will be useful in obtaining the desired decomposition for a general space.

Theorem 1:

Let X be a topological space and A a connected subspace of X. If B is a subspace of X such that $A \subseteq B \subseteq \overline{A}$, *then B is connected; in particular,* $\overline{A}$ *is connected,*

Proof:

Assume that B is disconnected, that is, that there exist two open subsets G and H of X whose union contains B and whose intersections with B are disjoint and non-empty. Since A is connected and contained in $G \cup H$, A is contained in either G or H and is disjoint from the other.

Let us say, just to be specific, that A is disjoint from H. This implies that $\overline{A}$ is also disjoint from H, and since $B \subseteq \overline{A}$, B is disjoint from H. This contradiction shows that B cannot be disconnected, and proves our theorem.

We are now in a position to state and prove the main facts about components.

Theorem 2(a):

If X is an arbitrary topological space, then we have the following: (1) each point in X is contained in exactly one component of X; (2) each connected subspace of X is contained in a component of X, (3) a connected subspace of X which is both open and closed is a component of X, and (4) each component of X is closed.

Proof:

To prove (1), let a: be a point in X. Consider the class $\{C_i\}$ of all connected subspaces of X which contain x. This class is non-empty, since x itself is connected. By Theorem 1, $C = \cup_i C_i$ is a connected subspace of X which contains x. C is clearly maximal, and therefore a component of X, because any connected subspace of X which contains C is one of the C_i's and is thus contained in C. Finally, C is the only component of X which contains x. For if C* is another, it is clearly among the C_i's and is therefore contained in C, and since C* is maximal as a connected subspace of X, we must have C* = C.

Part (2) is a direct consequence of the construction in the above paragraph, for by this construction, a connected subspace of X is contained in the component which contains any one of its points.

We prove (3) as follows. Let A be a connected subspace of X which is both open and closed. By (2), A is contained in some component C. If A is a proper subset of C, then it is easy to see that

$$c = (c \cap A) \cup (G \cap A')$$

is a disconnection of C. This contradicts the fact that C, being a component, is connected, and we conclude that A = C.

Part (4) follows immediately from Theorem 2; for if a component C is not closed, then its closure $\overline{C}$ is a connected subspace of X which properly contains C, and this contradicts the maximality of C as a connected subspace of X.

In view of parts (3) and (4) of this theorem, it is natural to ask if a component of a space is necessarily open. The answer is no, as the following example shows. Let X be the subspace of the real line which consists of all rational numbers. We observe two facts about X. First, if x and z are any two distinct rationals, and if $x < z$, then there exists an irrational y such that $x < y < z$, and therefore

$$x = [X \cap (-\infty, y)] \cup [X \cap (y, +\infty)]$$

is a disconnection of X which separates x and z. It is easy to see from this that any subspace of X with more than one point is disconnected, so the components of X are its points. Second, the points of X are not open, for any open subset of R which contains a given rational number also contains others different from it. Here, then, is a space whose components are its points and whose points are not open. This example also shows that a space need not be discrete in order that each of its points be a component.

Theorem 2(b):

Let X be a topological space. If $\{A_i\}$ is a non-empty class of connected subspaces of X such that $r\backslash A_i$ is non-empty, then $A = \cup_i A_i$ is also a connected subspace of X.

Proof:

Assume that A is disconnected. This means that there exist two open subsets G and H of X whose union contains A and whose intersections with A are disjoint and non-empty. All the A_i's are connected, and each lies in $G \cup H$, so each A_i lies entirely in G or entirely in H and is disjoint from the other. Since $\cap_i A_i$ is non-empty, either all the A_i's lie in G and are disjoint from H, or all lie in H and are disjoint from G. We see by this that A itself is disjoint from either G or H, and this contradiction shows that our assumption that A is disconnected is untenable.

TOTALLY DISCONNECTED SPACES

We have seen that a connected space is one for which no disconnection is possible. We now consider spaces which have a great many disconnections, and which therefore lie, in a manner of speaking, at the opposite end of the connectivity spectrum.

A *totally disconnected space* is a topological space X in which every pair of distinct points can be separated by a disconnection of X. This means that for every pair of points a; and y in X such that $x \neq y$, there exists a disconnection X = A ∪ B with x ε 4 and y ε B. Such a space is evidently a Hausdorff space, and if it has more than one point, it is disconnected. Oddly enough, a one-point space is both connected and totally disconnected.

The discrete spaces are the simplest totally disconnected spaces. A more interesting example is the space discussed at the end of the previous section, that is, the set of all rational numbers considered as a subspace of the real line. The set of all irrational numbers is also a totally disconnected subspace of the real line, and this is proved in much the same way, from the fact that there exists a rational number between any two irrationals. The Cantor set is yet another totally disconnected subspace of the real line, this time one which is compact.

Our first theorem should not come as a surprise to anyone.

Definition : *A topological space (X, $\mathfrak{I}$) is said to be totally disconnected iff for each pair of distinct points x, y ∈ X, there exists a disconnection A ∪ B of X such that x ∈ A and y ∈ B.*

We may remark here that a space consisting of a single point is totally disconnected. Thus, a one point space is both connected and totally disconnected. It follows that every totally disconnected space is Hausdorff.

Theorem 1(a):

Let X be a Hausdorff space. If X has an open base, whose sets are also closed, then X is totally disconnected.

Proof:

Let x and y be distinct points in X. Since X is Hausdorff, x has a neighbourhood G which does not contain y. By our assumption, there exists a basic open set B which is also closed such that x ε B ⊆ G. X = B ∩ B' is clearly a disconnection of X which separates x and y.

If the space X in this theorem is also compact, then the implication can be reversed, and the two conditions are equivalent.

Theorem 1(b):

Let X be a compact Hausdorff space. Then X is totally disconnected ⇔ it has an open base whose sets are also closed.

Proof:

It suffices to assume that X is totally disconnected and to prove that the class of all subsets of X which are both open and closed forms an open base. Let a: be a point and G an open set which contains it. We must produce a set B which is both open and closed such that x ε B ⊆ G. We may assume that G is not the full space, for if G = X, then we can satisfy our requirement by taking B = X. G' is thus a closed subspace of X, and since X is compact, G' is also compact. By the assumption that X is totally disconnected, for each point y in G' there exists a set H_y which is both open and closed and contains y but not x. G' is compact, so there exists some finite class of H_y's, which we denote by $\{H_1, H_2, \ldots, H_n\}$, with the property that its union contains G' but not x. We define H by $H = \bigcup_{i=1}^{n} H_i$, and we observe that since this is a finite union and all the H_i's are closed as well as open, H is both open and closed, it contains G', and it does not contain x. If we now define B to be H', then B clearly has the properties required of it.

Totally disconnected spaces are of considerable significance in several parts of topology, notably in dimension theory.

Theorem 2:

The components of a totally disconnected space are its points.

Proof:

If X is a totally disconnected space, it suffices to show that every subspace Y of X which contains more than one point is disconnected. Let x and y be distinct points in Y, and let $X = A \cup B$ be a disconnection of X with $x \varepsilon A$ and $y \varepsilon B$. It is obvious that

$$Y = (F \cap A) \cap (F \cap B)$$

is a disconnection of Y.

Total disconnectedness is closely related to another interesting property.

Theorem 3:

Let X be a Haudorff space. If X has an open base whose sets are also closed, then X is totally disconnected.

Proof:

Let x and y be two distinct points of X. Since X is Hausdorff, x has an open neighbourhood G such that $y \notin G$. By hypothesis there exists a basic open set B which is also closed such that $x \in B \subset G$. Then B' is also open and closed such that

$$y \in B' \text{ and } x \notin B'$$

Also $B \cap B' = \emptyset$ $B \cup B' = X$. Hence $B \cup B'$ is a disconnection of X such that $x \in B$ and $y \in B'$. It follows that X is totally disconnected.

Theorem 4:

The components of a totally disconnected space $(X, \mathfrak{I})$ are the singleton subsets of X.

Proof:

It is enough to show that every subset of X which contains more than one point is disconnected. Let E be a subset of X consisting of two or more points and let $x, y \in E$ with $x \neq y$. Since X is totally disconnected, there exists a disconnection $A \cup B$ of X with $x \in A$ and $y \in B$.

Let $\quad G = A \cap E$ and $\quad H = B \cap E$

Then $\quad G = \phi$, since $x \in G$ and $y \in H$.

Also $\quad G \cap H = (A \cap E) \cap (B \cap E) = (A \cap B) \cap E$

$= f \cap E = f$

and $\quad G \cup H = (A \cap E) \cup (B \cap E) = (A \cup B) \cap E$

$= X \cap E = E$

This show that E is disconnected. Thus, we have shown that every subset of X containing more than one point is disconnected. It follows that the singleton subsets of X only connected sets and consequently they are the only components of X.

CONTINUITY AND CONNECTEDNESS

Theorem 1:

Continuous image of a connected space is connected.

Proof:

Let $f : X \rightarrow Y$ be a continuous mapping of a connected space X into an arbitrary topological space Y. We wish to show that f [X] is connected as a subspace of Y. Assume that f [X] is disconnected. Then there exists G_1 and G_2 both open in Y such that

$$G_1 \cap f(X] \neq \varnothing, G_2 \cap f[X] \neq \varnothing,$$

$$(G_1 \cap f[X] \cap f(G_2 \cap f[X] = \varnothing$$

and $\quad (G_1 \cap f([X]) \cup (G_2 \cap f[X] = f[X]$

It follows that

$$\begin{aligned} \varnothing &= f^{-1}[\varnothing] = f^{-1}[(G_1 \cap f[X]) \cap (G_2 \cap f[X])) \\ &= f^{-1}[(G_1 \cap G_2] \cap f[X]) \\ &= f^{-1}[G_1] \cap f^{-1}[G_2] \cap f_{-1}[f[X]] \\ &= f^{-1}[G_1] \cap f^{-1}[G_2] \cap X \\ &= f^{-1}[G_1] \cap f^{-1}[G_2]. \end{aligned}$$

and $\quad X = f^{-1}[F[X]] = f^{-1}[(G_1 \cap f[X]] \cup [G_2 \cap f[X])]$

$= f^{-1} [G_1 \cup G_2] \cap f [X]]$

$= f^{-1} [G_1 \cup G_2] \cap f^{-1} [f [X]]$

$= \{f^{-1} [G_1] \cup f^{-1} [G_2]\} \cap X$

$= f^{-1} [G_1] \cup f^{-1} [G_2].$

Since f is continuous and G_1, G_2 are open in Y both intersecting f [X], it follows that f^{-1} $[G_1]$ and f $[G_2]$ are non-empty open subsets of X. Thus X has been expressed as a union of two disjoint non-empty open subsets of X and consequently X is disconnected, which is a contradiction. Hence f [X] must be connected.

Corollary 1 : *If f is a continuous mapping of a connected space onto an arbitrary topological space Y, then Y is connected.*

Proof:

Due to the importance of this corollary, we given an independent proof here. Suppose Y is disconnected. Then there exists a non-empty proper subset G of Y which is both open and closed in Y. Since f is continuous and onto, Y, f^{-1} [G] is a non-empty proper subset of X which is both open and closed in X and therefore X is disconnected which is a contradiction. Hence Y must be connected.

Corollary 2 : *If f is a containing mapping of a connected space X into* **R,** *then f [X] is an interval.*

This corollary leads to the following generalization of the Weierstrass intermediate value theorem.

If $f : X \rightarrow$ **R** *is a continuous map of a connected space X into, then f takes all values between any two of its values.*

Theorem 2:

A topological space X is disconnected if any only if there exists a continuous mapping of X onto the discrete two point space {0, 1}.

Proof:

Let X be disconnected. Then there exist two non-empty disjoint open subsets G_1 and G_2 of X such that $X = G_1 \cup G_2$. Define a mapping f of X onto {0, 1} by setting f (X) = 0 if $x \in G_1$ and f (x) = 1 if $x \in G_2$. Since {0, 1} is discrete, its open sets are $\varnothing$, {0}, {1} and {0, 1}. Since $G_1 \cap G_2 = \varnothing$. The definition of f shows that f^{-1} [{0}] = G1 and f^{-1} [{1}] = G_2. Also f^{-1} [$\varnothing$] = $\varnothing$ and f^{-1} [{0, 1}] = X. Thus, we have shown that the inverse image under f of every open subset of {0, 1} is open in X and therefore f is continuous.

Conversely, if there exists such a mapping then X is disconnected. For if X were connected, the {0, 1} would be connected. But this is impossible since every discrete space is disconnected.

Theorem 3:

A space X is connected if and only if every continuous function f from X into the discrete two points space {0, 1} is constant.

Proof:

First let X be connected and let $y \in f[X] \subset \{0, 1\}$ where $f : X \to \{0, 1\}$ is any continuous function. Then {y} is both open and closed in the discrete space {0, 1} and f is continuous, f–1 [{y}] is closed, open and non-empty. Since X is connected, $f^{-1}[\{y\}] = X$. This shows that f is a constant function.

Conversely, let every continuous function $f : X \to \{0, 1\}$ be constant and suppose, if possible, that X is disconnected. Then there exists a non-empty proper subset A of X which is both open and closed in X. This implies that A′ is also a non-empty proper subset of X which is both open and closed in X. Now consider the characteristic function χ_A of A defined by

$$\chi_A(x) = \begin{cases} 1 \text{ if } x \in A \\ 0 \text{ if } x \in A' \end{cases}$$

Then $$\chi_A^{-1}[\varnothing] = \varnothing,$$

$$\chi_A^{-1}[\{1\}] = A, \ \chi_A^{-1}[\{0\}] = A'$$

and $$\chi_A^{-1}[[0, 1\}] = X.$$

Thus, the inverse image under f of every open subset of {0, 1} is open in X and so χ_A is non-constant, a contradiction. Hence X must be connected.

COMPONENTS

If a topological space is not itself connected then it can be decomposed into a disjoint class of maximal connected subspaces. We shall show that this is always possible.

Definition : *A maximal connected subspace C of a topological space X is called a component of a space.*

Thus C is component of X iff C is connected and C is not properly contained in any larger connected subspace.

A Component of a Subset Y of X is a component of Y with the relative topology, that is, a maximal connected subset of Y.

Since singleton sets of a topological space are connected it is evident that components are non-empty.

Example 1:

Let $(X, \mathfrak{I})$ be connected. Then X has only one component, namely X itself.

Example 2:

Let $(X, \mathfrak{I})$ be a discrete space. Then singleton sets are the only connected subsets of X and consequently they are the maximal connected sets. Hence each singleton set is a component for a discrete space.

Example 3:

Consider the topology

$\mathfrak{I} = \{\varnothing, \{a\}, \{b, c\}, \{a, b, c\}, \{b, c, d, e\}, X\}$

on $X = \{a, b, c, d, e\}$.

The components of X are {a} and {b, c, d, e} since my other connected subset of X is a subset of one of these components as the reader can easily verify. For example, {c, d, e} is a connected subset of X which is a subset of {b, c, d, e}.

Theorem 1:

If X is a topological space, then

(i) *Each point in X is contained in exactly one component of X.*

(ii) *The components of X form a partition of X, that is, any two components are either disjoint or identical and the union of all the components is X.*

(iii) *Each connected subset of X is contained in a component of X.*

(iv) *Each connected subset of X which is both open and closed is a component of X.*

Proof:

(i) Let x be any arbitrary point of X. Let $\{A_\lambda\}$ be the collection of all connected subsets of X which contain x. This collection is non-empty since {x} is connected. Further $\cap A_\lambda \neq \varnothing$ since x is a point of each A_λ. Hence, $A_x = \cup A_\lambda$ is connected. Also A_x contains x. A_x is evidently maximal. For, let B be any connected set such that $A_x \subset B$. Then $x \in B$ so that B is one of the members of the collection $\{A_\lambda\}$ and, hence $B \subset A_x$. Thus $A_x = B$. Hence A_x is a component of X containing x. Let A_x^* be any other component of X containing x. Then

A_x^* is one of the Al's and so $A_x^* \subset A_x$. But A_x^* is maximal as a connected subset of X and hence we must have $A_x^* = A_x$.

(ii) Let $C = \{A_x : x \in X\}$ where A_x is defined as in (i). We claim that C contains all the components of X. By (i) each $A_x \in C$ is a component. And if E is any other component, then E being non-empty, contains some point $x_0 \in X$ and so by (i), we have $E = A_{x_0} \in C$.

We now show that C forms a partition of X. Evidently $Y = \cup \{A_x : x \in X\}$. Again let A_p and A_q be any two distinct components such that $A_p \cap A_q \neq \varnothing$. We shall show that $A_p = A_q$. Let $r \in A_p \cap A_q$. Then $r \in A_p$ and $r \in A_q$. since A_p, A_q are connected subsets containing r, and A_r is a component containing r, we have $A_p \subset A_r$ and $A_p \subset A_r$. But A_p, A_q are components so that we must have $A_p = A_r = A_q$. Hence components of X form partition of X.

(iii) Let C be any connected subset of X. If $C = \varnothing$, then C is contained in every component. If $C \neq \varnothing$, then C contains a point $x_0 \in X$ and so $C \subset A_{x_0}$ by (i).

(iv) Let D be a connected subset of X which is both open and closed. By (iii), $D \subset E$ where E is a component of X and hence E is non-empty. We claim that D = E. For if $D \neq E$, then d is a proper subset of E which is both open and closed in E and therefore E is disconnected, a contradiction. Hence D = E and so D is a component.

Remark : It is natural to ask if a component of a space is necessarily open. The answer is no as the following example shows.

Consider the space **Q** of rational numbers with relativized usual topology on **Q**. We shall show that the only non-empty connected subsets of **Q** are the signleton sets. For let E by any subset of **Q** containing more than one point and let x, y be any two distinct points of E with $x < y$. Then there exists an irrational number z such that $x < z < y$. Since $]-\infty, y[$ and $]y, \infty[$are both open in **R**, if follows that $]-\infty, y[\cap E$ and $]y, \infty[\cap E$ are non-empty disjoint sets both open in E such that

$$E = (] -\infty\ y[\cap E) \cup]y, \infty[\cap E).$$

Hence, E is disconnected. Thus every subset of **Q** containing more than one point is disconnected. Also every singleton set is always connected. Hence the components of **Q** are the singleton sets. But no singleton subset of **Q** is open. For if x is any rational then $\{x\} \neq G \cap \mathbf{Q}$ for any open subset G of **R**. [Note that every open subset of **R** contains infinitely many rationals].

Theorem 2:

Every component of a topological space $(X, \mathfrak{I})$ is closed.

Proof:

Let C be a component of X. Since C is connected, its closure $\overline{C}$ is also connected. Now C, being a component, is a maximal connected set. Hence $\overline{C} \subset C$. But $C \subset \overline{C}$. Hence $\overline{C} = C$ and C is closed.

LOCALLY CONNECTED SPACES

Definition : *A topological space X is said to be locally connected at a point x iff every open neighbourhood of x contains a connected open neighbourhood of x, that is iff the collection of all connected open neighbourhoods of x forms a local base at x. The space X is said to be locally connected iff it is locally connected at each of its points.*

Remark : *Examples 3 and 4 show that local connectedness neither implies nor implied by connectedness.*

Theorem 1:

A topological space X is locally connected if and only if the components of every open subspace of X are open in X.

Proof:

First not that since the subspaces considered are open sets in X, a set is open in the subspace iff it is open in X. Now suppose that the components of every open subspace of X are open in X. Let a point x and an open subspace Y containing x be given. The x belongs to some component C_x of Y. By hypothesis, Cx is open and connected in Y. Hence C_x is open and connected in X. It follows that X is totally connected.

Conversely, let X be locally connected. Let Y be an open set in X and C a component of Y. Then by definition of local connectedness, for every $x \in C$, there exists a connected open set G_x in Y containing x, $G_x \cup C = C$ so that $G_x \subset C$. Thus C contains a nhd of each of its points in Y and so C is open in Y. Since Y is open in X, it follows that C is open in X.

Theorem 2(a):

A topological space X is locally connected if and only if components of every open subset of X are open in X.

Proof:

Suppose X is locally connected. Let Y be any open subspace of X and C any component of Y. We must show that C is open X. Let x be any point of C. By definition of local connectedness, there exists an open connected nhd G of x in X such that $x \in G \subset Y$, G is open and connected in Y. Since C is a maximal connected set in Y, we must have $x \in G \subset C$. Thus C is a nhd in Y of each of its points and so C is open in Y and since Y is open in X, it follows that C is open in X.

Conversely, suppose the components of every open subspace of X are open in X. Let U be any open nhd of x ∈ X. There exists a component C of U such that x ∈ C (i). By hypothesis, C is open in X. Also C being a component, is connected. Thus, we have shown that every open nhd of x contains a connected open nhd of x and therefore X is locally connected.

Theorem 2(b):

The image of a locally connected space under a mapping which is both continuous and open is locally connected.

Proof:

Let f be a continuous and open mapping of a locally connected space X into an arbitrary space Y. To show that the image f [X] is locally connected. Let y be an point of f [X} and V any open nhd of y in Y. Then There exists x ∈ X such that y = f (x). Since f is continuous, f–1 [V] is open in X and contains x since V contains f (x). Thus f–1 [V] is a open nhd of x. Since X is locally connected, there exists a connected open nhd U of x such that x ∈ U ⊂ f–1 [V] which implies that

$$y = f(x) \in f[U] \subset V$$

since f is an open mapping, f [U] is an open subset of f [X]. Again since f is continuous and U is q connected in X. f [U] is connected in f [X]. Thus we have shown that to each open nhd V of y ⊂ f [X], there exists a connected open nhd f [U] of y such that y ∈ f [U] ⊂ V. Hence f [X] is locally connected.

ARCWISE CONNECTIVITY

Definition : *Let E be a subset of a space X and let I = [0, 1]. Then a path in E joining two points a and b of E is a continuous mapping f = I → E such that f (0) = a and f (1) = b. We call a the initial point and b the terminal point of the path.*

Definition : *A subset E of space X is said to be arcwise connected iff every two points in E can be joined by some path in E.*

Theorem 1:

An arcwise connected subset of a space is connected.

Proof:

Let E be an arcwise connected subset of a space X. If E = ∅, then E is connected. Since I = [0,1] is connected , f [I] is connected for any continuous map f : I → E. Let a, be any tow points of E. By arc connectivity, there exists a continuous map f : I → E such that f (0) = a and f (I) = b that is, a, b ∈ f (1). Thus every two points of E are contained in a connected subspace f [I] of E and therefore E is connected.

SOLVED EXAMPLES

Example 1:

Give an example of a space which is connected but not locally connected.

Solution:

Consider the subspace Y of the Euclidean plane **R2** defined by Y = A ∪ B where

$$A = \{(0, y)\} : y \in [-1, 1]\} \text{ and}$$
$$B = \{(x, y) : x \in]0, 1] \text{ and } y = 1/x\}$$

Now B is the image of the interval]0, 1[under the continuous mapping f defined by f (x) = x sin (1/x). Since]0, 1] is connected, B is also connected. It is easy to see that every nhd of a point x ∈ A contains an infinite number of points of B. Hence every point of A is a limit point of B. Also no point outside A ∪ B can be a limit point of B. Hence $\overline{B} = A \cup B = Y$.

It then follows that $\overline{B}$ i.e., Y is connected. However, X is not locally connected; since it can be easily shown that each point x ∈ A has nhd which does not contain any connected nhd of x.

Example 2:

Consider the topology

$$\mathfrak{I} = \{\varnothing, \{a\}, \{c, d\}, \{a, c, d\}, \{b, c, d, e\}\ X\}$$

on the set X = {a, b, c, d, e} and show that

(i) (X, $\mathfrak{I}$) is disconnected.

(ii) Y = {b, d, e} is a connected subset of X.

Solution:

(i) Here {a} is open. Also {a} is closed since it is the complement of the open set {b, c, d, e}. Thus, {a} is a non-empty proper subset of X which is both $\mathfrak{I}$-open and $\mathfrak{I}$-closed. Hence (X, $\mathfrak{I}$) is disconnected.

(ii) The relative topology $\mathfrak{I}$Y on Y is given by

$\mathfrak{I}$Y = {∅, {d}, Y}.

Since ∅ and Y are the only subsets of Y which are both open and closed in Y, it follows that the subspace (Y, $\mathfrak{I}$Y) is connected. Accordingly, Y is a connected subset of X.

Example 3:

Consider the topology

$$\mathfrak{I} = \{\varnothing, \{a\}, \{b, c\}, X\} \text{ on } X = \{a, b, c\}$$

Then the space (X, ∅) is disconnected since {a} is a non-empty proper subset of X which is both $\mathfrak{I}$-open and $\mathfrak{I}$-closed.

Example 4:

Let X = {a, b, c, d} and let $\mathfrak{I}$ {∅, {a}, {a, b} {a, b, c}, X}.

Solution:

Here $\mathfrak{I}$-closed subsets of X are:

X, {b, c, d{, {c, d}, {c, d}, {d} and ∅.

We set that there exists no non-empty proper subset of X which is both $\mathfrak{I}$-open and $\mathfrak{I}$-closed. Hence (X, $\mathfrak{I}$) is connected.

Example 5:

Show that

*(i) Every indiscrete topological space (X, **I**) where X is non-empty is connected.*

*(ii) Every discrete space (X, **D**) where X contains more than one point is disconnected.*

Solution:

Since there is no non-empty proper subset of X which is both I-open and I-closed, it follows that (X, **I**) is connected.

(ii) Since X contains more than one point, every singleton subset of X is a non-empty proper subset of X which is both **D**-open and **D**-closed. Hence (X, **D**) is disconnected.

Example 6:

Consider the lower limit topological **S** *and* **R** *and show that the space (**R**, **S**) is disconnected.*

Solution:

Let A = [0, 1] Then A is a non-empty proper subset of **R** which is S-open. Now

R – [0, 1[=] – ∞, 0 [∪ [1, ∞[.

Since **S** is finer than **U** and]–∞, 0[is U-open, it follows that]–∞, 0[is S-open also. Also [1, ∞[is **S**-open. Hence **R** – [0, 1[is **S**-open, being the union of two **S**-open subsets of **R**. Therefore [0, 1[is **S**-closed. Thus [0, 1[is a non-empty proper subset of **R** which is the both **S**-open and **S**-closed. It follows that the space (**R**, **S**) is disconnected.

Example 7:

Show that if (X, $\mathfrak{I}$) is connected and $\mathfrak{I}'$ is coarser than $\mathfrak{I}$, then (X, $\mathfrak{I}'$) is connected.

Hint: Assume that (X, $\mathfrak{I}'$) is disconnected that (X, $\mathfrak{I}$) is disconnected.

Example 8:

Show that every discrete space is totally disconnected.

Solution:

Let (X, D) be any discrete space and let x, y be any two distinct points of X. Then A = {x} and B = X – {x} are both non-empty open disjoint sets whose union is X such that x ∈ A and y ∈ B. It follows that (X, D) is totally disconnected.

Example 9:

Show that the space **(R, S)** *is totally disconnected.*

Solution:

Let x, y be any two distinct real numbers and let x < y. Let A =]–∞, y[and B = [y, ∞[. Then A, being **U**-open is also **S**-open since **S** is finer than **U**. Also B is **S**-open. Since x < y, we have x ∈ A. Also y ∈ B. Further A ∪ B = **R** and A ∩ B = ∅. Hence A ∪ B is a disconnection of **R** with x ∈ A and y ∈ B. It follows that **(R, S)** is totally disconnected.

Example 10:

Give and example of a topological space (X ,$\mathfrak{I}$) having a non-empty proper $\mathfrak{I}$-open subset Y such that Y is compact.

Solution:

Consider any co-finite topological space (X, $\mathfrak{I}$) on an infinite set X. Let p be any point of X. Then the set Y = X – {p} is a non-empty proper, $\mathfrak{I}$-open subset of X, such that Y is compact. The set Y is open since its complement is {p} which is finite.

We now show that Y is compact.

Let $$\mathbf{C} = \{G_\lambda : \lambda \in \Lambda\}$$

be an open cover of Y. By definition of $\mathfrak{I}$, each contains all except at most a finite number of points of Y and finite number of points can be contained in the union of a finite subcollection of **C**.

Thus it is shown that every open cover of Y has a finite subcover and therefore Y is compact.

Example 11:

Show by means of an example that locally compact space need not be compact.

Solution:

Consider any discrete space X where X is infinite. Then X is not compact since the collection of all singleton sets in an infinite open cover of X which has no finite subcover. But X is locally compact. For let x be any point of X. Then {x} is a nhd of x whose closure. For let x compact subset of X, being finite. Hence, every point of x has a nhd whose closure is compact

Example 12:

Show that if (X, $\mathfrak{I}$) is compact and $\mathfrak{I}'$ is coarser than $\mathfrak{I}$, then (X, $\mathfrak{I}'$) is compact.

Solution:

Let the collection

$$\mathbf{C} = \{G_\lambda : \lambda \in \Lambda\}$$

a $\mathfrak{I}'$-open cover of X. Since $\mathfrak{I}$ is finer than $\mathfrak{I}'$ each G_λ is also $\mathfrak{I}$-open. Hence **C** is also a $\mathfrak{I}$-open cover of X. Since (X, $\mathfrak{I}$) is compact **C** has a finite subcover. It follows that (X, $\mathfrak{I}'$) is compact.

Example 13:

Is every indiscrete space compact ? Show that no infinite discrete space is compact.

Solution:

Yes. For (X, I) is any indiscrete space, then the only open cover of Xis {X} which is finite since it consists of single member of X.

Let (X, **D**) be any discrete space where X is infinite. Since each singleton subset of X is **D**-open, if follows that the collection

$$\mathbf{C} = \{\{x\} : x \in X\}$$

is **D**-open infinite cover of X which cannot have a finite subcover since if we remove a single member of **C**, then it will not cover X. Hence (X, **D**) is not compact.

Example 14:

Give and example of a topological space $(X, \Im)$ having a non-empty proper $\Im$-open subset Y such that Y is compact.

Solution:

Consider any co-finite topological space $(X, \Im)$ on an infinite set X. Let p be any point of X. Then the set $Y = X - \{p\}$ is a non-empty proper, $\Im$-open subset of X, such that Y is compact. The set Y is open since its complement is $\{p\}$ which is finite.

We now show that Y is compact.

Let $$\mathbf{C} = \{G_\lambda : \lambda \in \Lambda\}$$

be an open cover of Y. By definition of $\Im$, each contains all except at most a finite number of points of Y and finite number of points can be contained in the union of a finite subcollection of **C**.

Thus it is shown that every open cover of Y has a finite subcover and therefore Y is compact.

Example 15:

Show that every co-finite topological space $(X, \Im)$ is compact.

Solution:

Let **C** be any $\Im$-open cover of X. By definition of $\Im$, every member of **C** contains all except a finite number of points and this finite number of points can be contained in the union of a finite sub-collection of **C**. This means that **C** has a finite subcover. Hence $(X, \Im)$ is compact.

Example 16:

*Show that the set **Q** of all rational numbers with the relativized usual topology is totally disconnected.*

Solution:

Let x, y be any two distinct rational numbers and let $x < y$. Then there exists an irrational number p such that $x < p < y$. Let $A = \mathbf{Q} \cap]-\infty, p[$ and $B = \mathbf{Q} \cap], \infty[$.

Then $A \cup B = (Q \cap) -\infty,$

$$p [) \cup (\mathbf{Q} \cap]p, \infty[) = \mathbf{Q}$$

Also $A \cap B = \varnothing$. Hence $A \cup B$ is a disconnection of **Q** with $x \in A$ and $y \in B$. It follows that **Q** is totally disconnected.

Example 17:

Show that the set of all irrational numbers with the relativized usual topology is totally disconnected.

Solution:

Solution of this problem is on the same lines as that of the preceding example since between any two irrational numbers there is a rational number.

Example 18:

Given an example of a locally connected space which is totally disconnected.

Solution:

Every discrete space is locally connected but totally disconnected.

Example 19:

Show that every discrete space is locally connected.

Solution:

Let x be an arbitrary point of a discrete space X. We know that every subset of a discrete space is open and that every singleton set is connected. Hence {x} is a connected open nhd of x. Also every open hand of x must contain {x}. Hence X is locally connected.

Example 20:

Given two examples of locally connected spaces which are not connected.

Solution:

(i) let X be a discrete space containing more than one point. Then X is disconnected. For it $x \in X$, then {x} is a non-empty proper subset of X which is both open and closed in X. X is locally connected.

(ii) Let Y be the union of two disjoint open intervals on the real line, that is, let $Y =]a, b[\cup] c, d[$ where $a < b \leq c < d$. Then Y as a subspace of **(R, U)** is locally connected but not connected as shown below. Since [a, b[and [c, d[are open in **R**, they are also open in Y.

Thus Y is the union of two non-empty disjoint sets both open in Y and therefore Y is disconnected. To show that Y is locally connected, let x be any point of Y and G_x any open nhd of x in Y. Then we can always find $\varepsilon > 0$ such that $]x - \varepsilon, x + \varepsilon[\subset G_x$. But $]x - \varepsilon, x + \varepsilon[$ is connected in **R** and hence connected in Y.

Thus every open nhd of x in Y contains an open connected nhd of x in Y and therefore Y is locally connected.

Example 21:

Let A be a compact subset of a Measure Theory (X, d). Show that foe every $B \subset X$, there exists a point $x_0 \in A$ such that

$$d\,(x_{0,}\ B) = d\,(A,\ B).$$

Solution:

Let d (A, b) = η

Now d (A, B) = inf f {d (x, y) : $x \in A$, $y \in B$}

It follows that for every $n \in N$, there exists

$$x_n \in A \text{ and } y_n \in B$$

such that $n \leq d\,(x_n, y_n) < n + \frac{1}{n}$.

Now A is compact $\Rightarrow$ A is sequentially compact. Then by definition of sequential compactness, the sequence $<x_n>$ has a subsequence which converges to a point $x_0 \in A$. We assert that

$$d\,(x_0, B) = d\,(A, B) = \eta.$$

Suppose, if possible, d $(x_0, B) > \eta$, that is, suppose

$$d\,(x_0, B) > \eta + \varepsilon \text{ where } \varepsilon > 0.$$

Since a subsequence of $< x_n >$ converges to x, there exists $n_0 \in N$ *i.e.*,

$$d\,(x_0, x_{n_0}) < \frac{1}{2}\,\varepsilon$$

and $\quad d\,(x_{n_0}, y_{n_0}) < \eta + \frac{1}{\eta_0} < \eta + \frac{1}{2}\varepsilon$

But then, $d\,(x_0, x_{n_0}) + d\,(x_{n_0}, x_{n_0})$

$$< \frac{1}{2}\varepsilon + \eta + \frac{1}{2}\varepsilon = \eta + \varepsilon = d\,(x_0, B) \leq d\,(x_0, y_{n_0}).$$

This contradicts triangle inequality. It follows that

$$d\,(x_0, B) = d\,(A, B).$$

Example 22:

Let a be a compact subset of a Measure Theory (X, d) and let B be a closed subset of X, such that

$$A \cap B = \varnothing$$

Prove that d (A, B) > 0

Solution:

If possible, let d (A, B) = 0 Then by the previous example, there exists $x_0 \in A$, such that

$$d(x_0, B) = d(A, B) = 0$$

Since B is a closed subset of X, it must contain all those points whose distance from B is zero. It follows that $x_0 \in B$. Thus.

$$x_0 \in A \text{ and } x_0 \in B.$$

Hence $x_0 \in A \cap B$.

But this is a contradiction since $A \cap B = \varnothing$.

Hence d (A, B) > 0.

EXERCISES

1. Show that a connected subspace of a locally connected space is locally connected if X is the real line. Is this true if X is an arbitrary locally connected space?
2. Let X be a locally connected space and C a component of a subspace Y of X. Show that

   $$Fr(C) \subset Fr(Y).$$

3. Prove that components of a totally disconnected space are its points.
4. If f is a continuous map of a space X into another space Y show that f maps every arcwise connected subset of X onto an arcwise connected subset of Y.
5. Show that Cantor's set is totally disconnected.
6. Give an example to show that components need not be open.
7. Consider the Cantor's set Γ with the relativized usual topology. Show that components of Γ are the points.

   [Hints : If $x \in \Gamma$, then among the intervals removed from [0, 1] in the process of constructing Γ, there are intervals arbitrarily close to x on process of constructing Γ, there are intervals close to x on either side, and each such interval induces a disconnection of Γ].
8. Give an example of a topological space all of whose components are open.
9. If f is continuous map of a topological space X into another space Y, show that f maps every connected subset of X onto a connected subset of Y.

10. Show that a topological space X is connected if and only if there exists no continuous function from X onto the discrete two point space {0, 1}.
11. If two topological spaces X and Y are homeomorphic, there exists a homeomorphism $f: X \to Y$, that is, f is one-one onto and continuous and $f^{-1} : Y \to X$ is also one-one onto and continuous.
12. Consider the space (R, U) and let

$$A =]0, 1[, B =]1, 2[\text{ and } C = [1, 2]$$

Show that A and B are separated but A and C are not separated.
13. If A and B are subsets of a space X and both A and B are closed or both open, show that A – B is separated from B – A.
14. Let A and B be subsets of a space X with $A \cup B = X$. If their complements A' and B' are separated sets, prove that for every subset S of X, $\overline{S} = (\overline{S \cap A} \cap A) \cup (\overline{A \cap B} \cap B)$.
15. Prove that a topological space X is locally connected if the components of every open subspace of X are open in X.
16. A connected subspace of a locally connected space X is locally connected if X is the real line. Why? Is this true if X is an arbitrary locally connected space?
17. Show that a compact locally connected space has a finite number of components.
18. Show that the graph of a continuous real function defined on an interval is a connected subspace of the Euclidean plane.
19. Show that if a connected space has a non-constant continuous real function defined on it, then its et of points is unaccountably infinite.
20. Show that connectedness is a topological invariant.
21. Show that the image of a locally connected space under a mapping which is both continuous and open is locally connected.
22. Prove that the product of any non-empty class of connected locally connected spaces is locally connected.
23. If $A_1, A_2, ..., A_n, ...$ is a sequence of connected subspaces of a topological space each of which intersects its successor, show that $\bigcup_{n=1}^{\infty} A_n$ An is connected.
24. Show that a subset of a topological space X is connected if and only if A is not the union of two non-empty separated sets.

25. Show that the union of any non-empty class of connected subspaces of a topological space each pair of which intersects is connected.
26. In Theorem 5 we proved that a product space is connected if its coordinate spaces are. Devise a different proof of this fact, based on Theorem 1, for the case in which there are only two coordinate spaces.
27. Prove that the product of any non-empty finite class of locally connected spaces is locally connected.
28. Show that the product of an arbitrary non-empty class of locally connected spaces can fail to be locally connected.
29. Use Theorem 1 to prove that an arbitrary Banach space B is connected.
30. Let B be an arbitrary Banach space. A *convex set* in B is a non-empty subset S with the property that if x and y are in S, then

$$z = x + t(y - x) = (1 - t)x + ty$$

is also in S for every real number t such that $0 \leq t \leq 1$. Intuitively, a convex set is a non-empty set which contains the segment joining any pair of its points. Prove that every convex subspace of B is connected. Prove also that every sphere (open or closed) in B is convex, and is therefore connected.

31. Show that an open subspace of the complex plane is connected $\Leftrightarrow$ every two points in it can be joined by a polygonal line:
32. Consider the union of two open discs in the complex plane which are externally tangent to each other. State whether this subspace of the plane is connected or disconnected, and justify your answer. Do the same when one disc is open and the other closed, and when both are closed.
33. Consider the following subspace of the Euclidean plane: $\{\{x, y\}: x \neq 0 \text{ and } y = \sin(1/x)\}$. Is this connected or disconnected? Why? Answer the same questions for the subspace $\{(x, y): x \neq 0,$ and $y = \sin(1/z)\} \cup \{(x, y): x = 0 \text{ and } -1 \leq y \leq 1\}$.
34. Prove that the product of any non-empty class of totally disconnected spaces is totally disconnected.
35. Prove that a totally disconnected compact Hausdorff space is homeomorphic to a closed subspace of a product of discrete two-point spaces.
36. Consider the topology

$\mathfrak{I} = \{\varnothing, \{a\}, \{a, b\}, \{a, b, c\}, \{a, b, c, d\}, X\}$ on $X = \{a, b, c, d, e\}$ and show that $\{X, \mathfrak{I}\}$ is disconnected.

37. Show that a confinite topological space X is connected if X is infinite and is connected if X is finite.

38. Let C be a connected subset of a space X. If A, B are separated sets such that $C' = A \cup B$, show that the set, $A \cup C$ and $B \cup C$ are connected.

39. Let E be a subset of a topological space. Prove that $\overline{E}$ is connected if and only if E is not the union of two non-empty sets A and B such that $\overline{A} \cap \overline{B} = \phi$.

40. Let A, B be non-empty subsets of a space X, each of which is closed in $A \cup B$. If $A \cup B$ and $A \cap B$ are connected, show that A is connected (and so is B).

41. Show that each of the following subspaces of the space of real numbers with the usual topology is disconnected.

    (i) Any finite set.

    (ii) $]0, 1[\cup] 2, 3 [\cup] 7, 16[$.

    (iii) Q i.e., the st of all rational numbers.

    [Hint : If x and y are any two distinct rationals with $x > y$, then there exists an irrational z such that $x < z < y$ and therefore $Q = (Q \cap]-\infty, z]) \cup (Q \cap], z, \infty[)$ is disconnection of Q].

    (iv) $R - Q$ i.e., the set of all irrational numbers.

    (v) $\left\{x: x = \frac{1}{n}, n \in N \text{ or } x = 0\right\}$.

42. Let X be a space with the property that given any $x \in X$ and any neighbourhood N of x, there exists a neighbourhood M of x such that $\overline{M}$ is properly contained in N. Show that X is connected.

43. Prove that a subspace of a topological space X is disconnected $\Leftrightarrow$ it can be represented as the union of two non-empty sets each of which is disjoint from the closure (in X) of the other.

3

BANACH AND HILBERT SPACE

INTRODUCTION

The *Hahn-Banach theorem* guarantees that a Banach space is richly supplied with continuous linear transformations, and makes possible an adequate theory of conjugate spaces. The *open mapping theorem* enables us to give a satisfactory description of the projections on a Banach space, and has the important closed graph theorem as one of its consequences. We use the *uniform boundedness theorem* in our discussion of the conjugate of an operator on a Banach space. Banach space is a linear space which is also, in a special way, a complete metric space. This combination of algebraic and metric structures opens up the possibility of studying linear transformations of one Banach space into another which have the additional property of being continuous.

The objects of greatest interest in connection with any linear space are the linear transformations on that space. In our treatment of Banach spaces, we took advantage of the metric structure of such a space by focusing our attention on its operators. A Banach space, however, is still a bit too general to yield a really rich theory of operators. One fact that did emerge, which is of great significance for our present work, is that with each operator T on a Banach space B there is associated an operator T* (its conjugate) on the conjugate space B*. One of the central properties of a Hilbert space H is that there is a natural correspondence between H and its conjugate space H*. If T is an operator on H, this correspondence makes it possible to regard the conjugate T* as acting on H itself (instead of H*), where it can be compared with T. These ideas lead to the concept of the adjoint of an operator on a Hilbert space, and they make it easy to understand the importance of operators.

THE DEFINITION AND SOME SIMPLE PROPERTIES

The Banach spaces studied in the previous chapter are little more than linear spaces provided with a reasonable notion of the length of a vector. The main geometric concept missing in an abstract space of this type is that of the angle between two vectors. The theory of Hilbert spaces does not hinge on angles in general, but rather on some means of telling when two vectors are orthogonal.

In order to see how to introduce this concept, we begin by considering the three-dimensional Euclidean space R^3. A vector in R^3 is of course an ordered triple $x = (x_1, x_2, x_3)$ of real numbers, and its norm is denned by

$$\| x \| = (| x_1 |^2 + | x_2 |^2 + | x_3 |^2)^{1/2}.$$

In elementary vector algebra, the inner product of x and another vector $y = (y_1, y_2, y_3)$ is denned by

$$(x, y) = x_1y_1 + x_2y_2 + x_3y_3,$$

and this inner product is related to the norm by

$$(x, x) = \| x \|^2.$$

We assume that the reader is familiar with the equation

$$(x, y) = \| x \| \, \| y \| \cos \theta,$$

where θ is the angle between z and *y,* and also with the fact that z and *y* are orthogonal precisely when (x, y) = 0.

Most of these ideas can readily be adapted to the three-dimensional unitary space C^3. For any two vectors $x = (x_1, x_2, x_3)$ and $y = (y_1, y_2, y_3)$ in this space, we define their inner product by

$$(x, y) = x_1\bar{y}_1 + x_2\bar{y}_2 + x_3\bar{y}_3. \quad ...(1)$$

Complex conjugates are introduced here to guarantee that the relation

$$(x, x) = \| x \|^2$$

remains true. It is clear that the inner product defined by (1) is linear as a function of z for each fixed y, and is also conjugate-symmetric, in the sense that $\overline{(x, y)} = (y, x)$. In this case, it is no longer possible to think of (x, y) as representing the product of the norms of *x* and *y* and the cosine of the angle between them, for (x, y) is in general a complex number. Nevertheless, if the condition (x, y) = 0 is taken as the definition of orthogonality, then this concept is just as useful here as it is in the real case.

With these ideas as a background, we are now in a position to give our basic definition. A *Hilbert space* is a complex Banach space whose norm arises from an *inner product,* that is, in which there is defined a complex function (x, y) of vectors x and y with the following properties:

(1) $(\alpha x + \beta y, z) = \alpha (x, z) + \alpha(y, z)$;

(2) $\overline{(x, y)} = (y, x)$;

(3) $(x, x) = \| x \|^2$.

It is evident that the further relation

$$(x, \alpha y + \beta z) = \bar{\alpha}(x, y) + \bar{\beta}(x, z)$$

is a direct consequence of properties (1) and (2).

The reader may wonder why we restrict our attention to complex spaces. Why not consider real spaces as well ? As a matter of fact, we could easily do so, and many writers adopt this approach. There are a few places in this chapter where complex scalars are necessary, but the theorems involved are not crucial, and we could get along with real scalars without too much difficulty. It is only in the complex case, however, that the theory of operators on a Hilbert space assumes a really satisfactory form. This will appear with particular clarity in the next chapter, where we make essential use of the fact that every polynomial equation of the nth degree with complex coefficients has exactly n complex roots (some of which, of course, may be repeated). For this and other reasons, we limit ourselves to the complex case throughout the rest of this book.

The following are the main examples of Hilbert spaces. In accordance with the above remarks, the scalars in each example are understood to be the complex numbers.

Example 1:

The space l_2^n, with the inner product of two vectors

$$x = (x_1, x_2, \ldots, x_n) \text{ and}$$
$$y = (y_1, y_2, \ldots, y_n) \text{ defined by}$$
$$(x, y) = \sum_{i=1}^{n} x_i \bar{y}.$$

It is obvious that conditions (1) to (3) are satisfied.

Example 2:

The space l_2, with the inner product of the vectors $x = \{x_1, x_2, \ldots, x_n \ldots\}$ and $y = y_1 i\, y_2, \ldots, y_n, \ldots\}$ defined by

$$(x. y) = \sum_{n=1}^{\infty} x_n, \bar{y}_n$$

The fact that this series converges—and thus defines a complex number —for each x and y in l_2 is an easy consequence of Cauchy's inequality.

Example 3:

The space L_2 associated with a measure space X with measure m, with the inner product of two functions f and g denned by

$$(f, g) = \int f(x)\overline{g(x)}\, dm(x).$$

This Hilbert space is of course not part of the official content of this book, but we mention it anyway in case the reader has some knowledge of these matters.

As our first theorem, we prove a fundamental relation known as the Schwarz inequality.

Theorem:

If B is a complex Banach space whose norm obeys the parallelogram law, and if an inner product is defined on B by (2), *then B is a Hilbert space.*

Proof:

All that is necessary is to make sure that the inner product defined by (2) has the three properties required by the definition of a Hilbert space. This is easy in the case is best treated by splitting it into two parts:

$$(x + y, e) = (z, z) + (y, z),$$

and $(\alpha x, y) = \alpha\,(\xi, \psi)$. The first requires the parallelogram law, and the second follows from the first. We ask the reader to work out the details.

This result has no implications at all for our future work. However, it does provide a satisfying geometric insight into the place Hilbert spaces occupy among all complex Banach spaces: they are precisely those in which the parallelogram law is true.

THE DEFINITION AND SOME EXAMPLES

The normed linear space is a linear space N in which to each vector x there corresponds a real number, denoted by $\| x \|$ and called the norm of x, in such a manner that

(1) $\| x \| \geq 0$, and $\| x \| = 0 \Leftrightarrow x = 0$;

(2) $\| x + y \| \leq \| x \| + \| y \|$

(3) $\| ax \| = | a |\, \| x \|$.

The non-negative real number $\| x \|$ is to be thought of as the length of the vector x. If we regard $\| x \|$ as a real function defined on N, this function is called the *norm* on N. It is easy to verify that the normed linear space N is a metric space with respect to the metric d defined by $d(x, y) = \| x - y \|$. A *Banach space* is a complete normed linear space. Our main interest in this chapter is in Banach spaces, but there are several points in the body of the theory at which it is convenient to have the basic definitions and some of the simpler facts formulated in terr s of normed linear spaces. For this reason, and also to emphasize the role of completeness in theorems which require this assumption, we work in the more general context whenever possible. The reader will find that the deeper theorems, in which completeness hypotheses are necessary, often make essential use of Baire's theorem.

Several simple but important facts about a normed linear space are based on the following inequality:

$$\| x \| - \| y \| I \leq \| x - y \| \qquad ...(1)$$

To prove this, it suffices to prove that

$$\| x \| - \| y \| \leq \| x - y \| \qquad ...(2)$$

follows from (2) that we also have

$$-(\| x \| - \| y \|) = \| y \| - \| x \| \leq \| y - xx \| = \| -(x - y) \| = \| x - y \|,$$

which together with (2) yields (1). We now prove (2) by observing that $\| x \| = \| (x - y) + y \| \leq \| x - y \| + \| y \|$. The main conclusion we draw from (1) is that the norm is a continuous function:

$$x_n \rightarrow x \Rightarrow \| x_n \| \rightarrow \| x \|.$$

This is clear from the fact that $| \| x_n \| | - \| x \| \leq \| x_n - x \|$, since $x_n \rightarrow x$ means that $\| x_n - x \| \rightarrow 0$. In the same vein, we can prove that addition and scalar multiplication are jointly continuous.

$$x_n \rightarrow x \text{ and } y_n \rightarrow y$$

$$\Rightarrow \quad x_n + y_n \rightarrow x + y$$

$$\text{and} \quad \alpha_n \rightarrow \alpha \text{ and } x_n \rightarrow x$$

$$\Rightarrow \quad \alpha_n x_n \rightarrow \alpha x.$$

These assertions follow from

$$\| (x_n + y_n) - (x + y) \| = \| (x_n - x) + (y_n - y) \|$$

$$\| x_n - x \| + \| y_n - y \|$$

and

$$\| \alpha_n x_n - \alpha x \| = \| \alpha_n (x_n - x) + (\alpha_n - a) x \|.$$

Our first theorem exhibits one of the most useful ways of forming new normed linear spaces out of old ones.

Theorem:

Let M be a closed linear subspace of a normed linear space N. If the norm of a coset x + M in the quotient space N/M is defined by

$$\| x + M \| = \inf\{\| x + m \| : m \varepsilon M\}, \qquad ...(3)$$

then N/M is a normed linear space. Further, if N is a Branch space, then so is N/M.

Proof:

We first verify that (3) defines a norm in the required sense. It is obvious that $\| x + M \| \geq 0$, and since M is closed, it is easy to see that $\| x + M \| = 0 \Leftrightarrow$ there exists a sequence $\{m_k\}$ in M such that $\| x + m_k \| \to 0 \Leftrightarrow$ x is in M $\Leftrightarrow$ x + M = M = the zero element of N/M. Next, we have $\| x + M) + (y + M) \| = \| (x + y) + M \| = \inf \{\| x + y + m \| : m \varepsilon M\} = \inf \{ \| x + y + m + m' \| : m \text{ and } m' \leq M\} = \inf \{\| (x + m) + (y + m') \| : m \text{ and } m' \varepsilon M\} \leq \inf \{ \|x + m \| + \| y + m' \| : m \text{ and } m' \varepsilon M] = \inf \{\| y + m \| : m \varepsilon M\} + \inf \| y + m' \| : m' \varepsilon M\} = \| x + M \| + \| y + M \|$. The proof of $\| \alpha (x + M) \| = | a |x + M \|$ is similar.

Finally, we assume that N is complete, and we show that N/M is also complete. If we start with a Cauchy sequence in N/M, then it suffices to show that this sequence has a convergent subsequence. It is clearly possible to find a subsequence $\{x_n + M \}$ of the original Cauchy sequence such that $\| (x_1 + M) - (x_1 + M\} \| \leq 1/2$, $\| (x_2 + M) - (x_3 + M) \| < 1/4$ and, in general, $\| (x_1 + M) - (x_{n+1} + M) \| < 1/2$.

We prove that this sequence is convergent in N/M. We begin by choosing any vector y_1 in $x_1 + M$, and we select y_2 in $x_2 + M$ such that $\| y_1 \; y_2 \| < 1/4$. We next select a vector y_3 in $x_3 + M$ such that $\| y_2 - y_3 \| < 1/4$. Continuing in this way, we obtain a sequence {yn} in N such that $\| y_n - y_{n+1} \| < 1/2^n$. If m < n, then

$$\| y_m - y_n \| = \| y_m - y_{m+1} (+ (y_{m+1} - y_{m+2}) + \ldots$$

so $\{y_n\}$ is a Cauchy sequence in N. Since W is complete, there exists a vector y in N such that $y_n \to y$. It now follows from $\| (x_n + M) - (y + M) \| \leq \| y_n - y \|$ that $x_n, + M \to y + M$, so N/M is complete. In the following sections and chapters, we shall often have occasion to consider the quotient space of a normed linear space with respect to a closed linear subspace. In accordance with our theorem, a quotient space of this kind can always be regarded as a normed linear space in its own right.

We now describe some of the main examples of Banach spaces. In each of these, the linear operations are understood to be defined either coordinate wise or point wise, whichever is appropriate in the circumstances.

Example 1:

The Banach space of primary interest to us is the space $\mathcal{C}(X)$ of all bounded continuous scalar-valued functions defined on a topological space X, with the norm given by

$$\| f \| = \sup | f(x) |.$$

This norm is sometimes called the uniform norm, because the statement that f_n converges to f with respect to this norm means that f_n converges to f uniformly on X. The fact that this space is complete amounts to the fact that if f is the uniform limit of a sequence of bounded continuous functions, then f itself is bounded and continuous. If, as above, we consider n-tuples and sequences as functions defined on {1, 2, . . . , n} and (1, 2, . . . , n, . . .}, then the spaces l_∞^n and l_∞ are the special cases of $\mathscr{C}$(X) which correspond to choosing X to be the sets just mentioned, each with the discrete topology.

Many important properties of a Banach space are closely linked to the shape of its *closed unit sphere*, that is, the set $S = \{x : \| x \| \leq 1\}$. One basic property of S is that it is always convex, in the sense that if x and y are any two vectors in S, then the vector $z = \alpha x + \beta y$ is also in S, where α and β are non-negative real numbers such that $\alpha + \beta = 1$; for $\| z \|| = \| \alpha x + \beta y \| \leq \alpha \| x \| + \beta \| y \| \leq \alpha + \beta = 1$. In this connection, it is illuminating to consider the shape of S for certain simple examples. Let our underlying linear space be the real linear space R^2 of all ordered pairs $x = (x_1, x_2)$ of real numbers. As we have seen, there are many different norms which can be defined on R^2, among which are the following: $\| x \|_1 = | x_1 + | x_2 |$; $\| x \|_2 = (| x_1 |^2 + | x_2 |^2)^{1/2}$ and $\| x \|_\infty = \max \{| x_1 |, | x_2 |\}$. In the first case, S is the square with vertices (1, 0), (0, 1), (–1, 0), (0, – 1); in the second, it is the circular disc of radius 1; and in the third, it is the square with vertices (1, 1), (– 1, 1), (–1, –1), (1, –1). If we consider the norm defined by

$$\| x \|_p = (|x_1 |^p + | x_2 |p)^{1/p}$$

where $1 \leq p \leq \infty$, and if we allow p to increase from 1 to ∞, then the corresponding S's swell continuously from the first square mentioned to the second. We note that S is truly "spherical" $\Leftrightarrow p = 2$. These considerations also show quite clearly why we always assume that $p \geq 1$; for if we were to define $\|| x \|_p$ with $p < 1$, then $S = \{x : \| x \|_p \leq 1 \}$ would not be convex. For $p < 1$, therefore, formula (12) does not yield a norm. In the above examples, we have exhibited several different types of Banach spaces, and there are yet others which we have not mentioned. Amid this diversity of possibilities, it is well to realize that any Banach space can be regarded–from the point of view of its linear and norm structures alone–as a closed linear subspace of $\mathscr{C}$*(X)* for a suitable compact Hausdorff space *X*. We prove this below, in our discussion of the natural embedding of a Banach space in its second conjugate space.

Example 2:

Consider the linear space of all bounded sequences $x = | x |, x_1, x_2 . . . , x_n, . . .\}$ *of scalars. By analogy with Example 5, we define the norm by*

$$|| x || = sup \mid x_n |,$$

and we denote the resulting Banach space by l_∞. The set *c* of all convergent sequences is easily seen to be a closed linear subspace of l_∞ and is therefore itself a Banach space. Another Banach space in this family is the subset c_0 of c which consists of all convergent sequences with limit 0.

Example 3:

The linear spaces R^n and C^n of all n-tuples $x = (x_1, x_2, \ldots, x_n)$ of real and complex numbers can be made into normed linear spaces in an infinite variety of ways, as we shall see below. If the norm is defined by

$$\| x \| = \left(\sum_{i=1}^{n} |x_i|^2 \right)^{1/2} \qquad \text{...(4)}$$

then we get the n-dimensional Euclidean and unitary spaces familiar to us from our earlier work. We denoted these spaces by R^n and C^n in Part 1 of this book, and we know by the theorems that both are Banach spaces.

Each of the following examples consists of n-tuples of scalars, sequences of scalars, or scalar-valued functions defined on some none empty set, where the scalars are the real numbers or the complex numbers.

We do not normally specify which system of scalars is to be used, and it should be emphasized that both possibilities are allowed unless the contrary is clearly stated.

Also, we make no distinction in notation between the real case and the complex case. When it turns out to be necessary to distinguish these two cases, we do so verbally, by referring, for instance, to "the complex space –." These conventions are in accord with the standard usage preferred by most mathematicians, and they enable us to avoid a good deal of cumbersome notation and many unnecessary case distinctions.

Example 4:

Let p be a real number such that $1 \le p < \infty$. We denote by l_p^n the space of all n-tuples $x = (x_1, x_2, \ldots, x_n)$ of scalars, with the norm defined by

$$\| x \|p = \left(\sum_{i=1}^{n} |x_i| p \right)^{1/p} \qquad \text{...(5)}$$

Obviously the special case of (5) which corresponds to $p = 2$, so the real and complex spaces l_2^n are the n-dimensional Euclidean and unitary spaces R^n and C^n. It is easy to see that (5) satisfies conditions (1) and (3) required by the definition of a norm. In Problem 4 we outline a proof of the fact that (5) also satisfies condition (2), that is, that $\| x + y \|_p \le \| x \|_p + \| y \|_p$. The completeness of Inf follows from substantially the same reasoning as that used in the proof of so l_p^n is a Banach space.

Example 5:

We again consider a real number p *with the property that* $1 < p < \infty$ *(and we denote by* lp *the space of all sequences*

$$x = \{x_1, x_2, \ldots, x_n, \ldots\}$$

of scalars such that $\Sigma_{n=1}^{\infty} |x_n|^p < \infty$, with the norm defined by

$$\| x \|_p = \left(\sum_{i=1}^{n} |x_n| p \right)^{1/p} \qquad \ldots(6)$$

The reader will observe that the real and complex spaces It are precisely the infinite-dimensional Euclidean and unitary spaces R^∞ and C^∞ defined. The proof of the fact that I_p actually is a Banach space requires arguments similar to those used.

The Banach spaces discussed in these examples are all special cases of the important L_p spaces studied in the theory of measure and integration. A detailed treatment of these spaces is outside the scope of this book, but we can describe them loosely as follows. An L_p space essentially consists of all measurable functions defined on a measure space X with measure m which are such that $| f(x) |^p$ is integrable, with

$$\| f \|^p = (\int |f(x)|^p \, dm(x))^{1/p} \qquad \ldots(7)$$

taken as the norm. In order to include the l_p^n spaces and l_p within the theory of L_p spaces, we have only to consider the sets $\{1, 2, \ldots, n\}$ and $\{1, 2, \ldots, n, \ldots\}$ as measure spaces in which each point has measure 1, and to regard n-tuples and sequences of scalars as functions defined on these sets. Since integration is a generalized type of summation.

Example 6:

Just we start with the linear space of all n-tuples $x = (x_1, x_2, \ldots, x_n)$ *of scalars, but this time we define the norm by*

$$\| x \| = \max \{| X_1, | | x_2 |, \ldots, | x_n |\}. \qquad \ldots (1)$$

This Banach space is commonly denoted by l_∞^n and the symbol $\| x \|_\infty$ is occasionally used for the norm given by (1). The reason for this practice lies in the interesting fact that

$$\| x \|_\infty = \lim \| x \|_p \text{ as } p \to \infty$$

that is, that

$$\max \{| x_i |\} = \lim \left(\sum_{i=1}^{n} |x_i| p \right)^{1/p} \text{ as } p \to \infty \qquad \ldots(2)$$

We briefly inspect the case n = 2 to see why this is true. Let $x = (x_1, x_2)$ be an ordered pair of real numbers with x_1 and $x_2 \geq 0$. It is clear that $\| x \|_\infty = \max \{x_1, x_2\} \leq (x_1{}^p + x_2{}^p)^{1/p} = \| x \|_p$. If $x_1 = x_2$, then $\lim \| x \|_p = \lim (x_1{}^p + x_2{}^p)^{1/p} = \lim (2x_2p^p)^{1/p} = x_2 = \| x |_\infty$. And if $x_1 < x_2$, then

$$\begin{aligned}\lim \| x \|_p &= \lim (x_1{}^p + x_2{}^p)^{1/p} = \lim ([(x_1/x_2)^{1/p} + 1]x_2{}^p)^{1/p} \\ &= \lim [(x_1/x_2)^p + 1]^{1/p} x_2 = x_2 = \| x \|_\infty.\end{aligned}$$

ORTHONORMAL SETS

An *orthonormal set* in a Hilbert space *H* is a non-empty subset of *H* which consists of mutually orthogonal unit vectors; that is, it is a non-empty subset $\{e_i\}$ of *H* with the following properties:

(1) $i \neq j \Rightarrow e_i \perp e_{ji}$,

(2) $\| e_i \| = 1$ for every i.

If H contains only the zero vector, then it has no orthonormal sets. If H contains a non-zero vector x, and if we normalize x by considering $e = x/\| z \|$, then the single-element set {e} *is* clearly an orthonormal set. More generally, if $\{x_i\}$ is a non-empty set of mutually orthogonal non-zero vectors in H, and if the x_i's are normalized by replacing each of them by $e_i = x_i/\| x_i \|$, then the resulting set $\{e_i\}$ is an orthonormal set.

Example:

Similarly, if en is the sequence with 1 in the nth place and 0's elsewhere, then $\{e_1, e_2, \ldots, e_n, \ldots\}$ *is an orthonormal set in* l_2.

At the end of this section, we give some additional examples taken from the field of analysis.

Every aspect of the theory of orthonormal sets depends in one way or another on our first theorem.

Theorem 1:

Let $[e_1, e_2, \ldots, e_n]$ *be a finite orthonormal set in a Hilbert space H. If x is any vector in H, then*

$$\sum_{i=1}^{n} |(x, e_i)|^2 \leq \| x \|^2 \qquad \text{...(1)}$$

$$\textit{further,} x - \sum_{i=1}^{n} (x, e_i) e_i \perp e_j \ \textit{for each } j. \qquad \text{...(2)}$$

Proof:

The follows from a computation similar to that used in proving Schwarz's inequality:

$$0 \le \| x - \sum_{i=1}^{n} (x, e_i) e_i \|^2$$

$$= (x - \sum_{i=1}^{n} (x, e_i) e_i, x - \sum_{j=1}^{n} (x, e_j) e_j)$$

$$= (x, x) - (x - \sum_{i=1}^{n} (x, e_i) \overline{(x, e_i)}, - \sum_{j=1}^{n} (x, e_j) \overline{(x, e_i)}$$

$$+ \sum_{i=1}^{n} \sum_{j=1}^{n} (x, e_i) \overline{(x, e_j)} (e_i, e_j)$$

$$= \| x \|^2 - \sum_{i=1}^{n} |(x, e_i)|^2$$

To conclude the proof, we observe that

$$(x - \sum_{i=1}^{n} (x, e_i) e_i, e_j) = (x, e_j) - \sum_{i=1}^{n} (x, e_i)(e_i, e_j) = (x, e_j) - (x, e_j) = 0$$

From which statement follows at once.

The reader should note that the inequality can be given the following loose but illuminating geometric interpretation: the sum of the squares of the components of a vector in various perpendicular directions does not exceed the square of the length of the vector itself. This is usually called *Bessel's inequality,* though, as we shall see below, it is only a special case of a more general inequality with the same name. In a similar vein, relation (2) says that if we subtract from a vector its components in several perpendicular directions, then the result has no component left in any of these directions.

Our next task is to prove that both parts of Theorem 1 generalize to the case of an arbitrary orthonormal set. The main problem here is to show that the sums in (1) and (2) can be defined in a reasonable way when no restriction is placed on the number of e_i's under consideration. The key to this problem lies in the following theorem.

Theorem 2:

If $\{e_i\}$ is an orthonormal set in a Hilbert space H, and if x is any vector in H, then the set $S = \{e_i : (x, e_i) \neq 0\}$ is either empty or countable.

Proof:

For each positive integer *n,* consider the set

$$S_n = \{e_i : |(x, e_i)|^2 > \| x \|^2/n\}.$$

By Bessel's inequality, *Sn* contains at most *n —I* vectors. The conclusion now follows from the fact that $S = \cup_{n=1}^{\infty} s_n$. As our first application of this result, we prove the general form of Bessel's inequality.

Theorem 3:

(Bessel's Inequality) : If *$\{e_i\}$ is an orthonormal set in a Hilbert space H, then*

$$\Sigma \mid (x_i, e_i)^2 \mid \leq \| x \|^2 \qquad ...(1)$$

for every vector x in H.

Proof:

Our basic obligation here is to explain what is meant by the sum on the left of (3). Once this is clearly understood, the proof is easy. As in the preceding theorem, we write S = $\{e_i:(x,e_i) \neq 0\}$. If S is empty, we define S $|(x, e_i)|^2$ to be the number 0. We now assume that *S* is non-empty, and we see from Theorem B that it must be finite or countably infinite. If S is finite, it can be written in the form $S = [e_1, e_2, \ldots, e_n\}$ for some positive integer *n*. In this case, we define $\Sigma\, |(x,ei)\,|^2$ to be $\Sigma_{i=1}^{n} |(x, e_i)|^2$, which is clearly independent of the order in which the elements of *S* are arranged. All that remains is to consider the case in which *S* is countably infinite. Let the vectors in *S* be arranged in a definite order:

$$\Sigma = \{e_1, e_2, \ldots, e_n, \ldots\}.$$

By the theory of absolutely convergent series, if $\Sigma_{i=1}^{\infty} |(x, e_n)|^2$ converges, then every series obtained from this by rearranging its terms also converges, and all such series have the same sum. We therefore define $\Sigma|(x, e_i)|^2$ to be $\Sigma_{n=1}^{\infty} |(x, e_n)|^2$, and it follows from the above remark that $\Sigma|(x, e_i)|^2$ is a non-negative extended real number which depends only m *S,* and not on the arrangement of its vectors.

$$\sum_{n=1}^{\infty} |(x, e_n)|^2 \leq \|x\|^2$$

and since it follows from (1) that no partial sum of the series on the left of (4) can exceed $\| x \|^2$, it is clear that (4) itself is true. The second part of Theorem A is generalized in essentially the same way.

Theorem 4:

If $\{e_i\}$ is an orthonormal set in a Hilbert space H, and if x is an arbitrary vector in H, then

$$x - \Sigma\, (x,e_i)\, e_i \perp e_j \qquad ...(1)$$

for each j.

Proof:

As in the above proof, we define S $(x,e_i)e_i$, for each of the various cases, and we prove (5) as we go along. We again write

$$\Sigma = \{e_i : (x,e_i) \neq 0\}.$$

When *S* is empty, we define $\Sigma\,(x,e_i)e_i$ to be the vector 0, and we observe that (5) reduces to the statement that $x - 0 = x$ is orthogonal to each e_j, which is precisely what is meant by saying that *S* is empty. When S is non-empty and finite, and can be written in the form

$$\Sigma = \{e_1, e_2, \ldots, e_n\},$$

We define $\Sigma\,(x,e_i)\,e_i$ to be $\Sigma_{n=1}^{n} (x,e_i)\,e_i$; and in this case, (5) reduces to (2), which has already been proved.

By Theorem B, we may assume for the remainder of the proof that *S* is countably infinite. Let the vectors in S be listed in a definite order: $S = \{e_1, e_2, \ldots, e_n, \ldots\}$. We put $s_n = \Sigma_{n=1}^{n} (x,e_i)\,e_i$, and we note that for $m > n$ we have

$$\| x_m - s_n \|^2 = \sum_{i=n+1}^{m} (x,e_i)\,e_i \,\|^2 = \sum_{i=n+1}^{m} |(x,e_i)|^2$$

Bessel's inequality shows that the series $\Sigma_{n=1}^{\infty} |(x, e_n)|^2$ converges, so $\{s_n\}$ is a Cauchy sequence in H; and since H is complete, this sequence converges to a vector s, which we write in the form $s = \Sigma_{n=1}^{\infty} (x, e_n)e_n$. We now define $\Sigma\,(x,e_i)\,e_i$ to be $\Sigma_{n=1}^{\infty} (x, e_n)e_n$, and—deferring for a moment the question of what happens when the vectors in S are rearranged—we observe that (5) follows from (2) and the continuity of the inner product :

$$\begin{aligned}(x - \Sigma(x,e_i)e_i, e_j) &= (x - s_1, e_j) = (x, e_j) - (x,e_j,) - (\lim s_n\, e_j)\\ &= (x, e_j) - \lim (s_n, e_j) = (x,e_j) - (x,e_j) = 0.\end{aligned}$$

All that remains is to show that this definition of $\Sigma(x,e_i)\,e_i$ is valid, in the sense that it does not depend on the arrangement of the vectors in *S*. Let the vectors in *S* be rearranged in any manner:

$$S = \{f_1, f_2, \ldots,, f_n, \ldots\}.$$

We put $s'_n = \Sigma_{i=1}^{\infty} (x, f_i)f_i$, and we see — as above — that the sequence *[$',,}* converges to a limit s', which we write in the form $s' = \Sigma_{n=1}^{\infty} (x, f_n)f_n$. We conclude the proof by showing that *s'* equals s. Let $\varepsilon > 0$ be given, and let n_0 be a positive integer so large that if $n \geq n_0$, then $\| s_n - s \| < \varepsilon$, $\| s'_n - s' \| < \varepsilon$, and $\Sigma_{i=n_0+1}^{\infty} |(x, e_i)|^2 < \varepsilon^2$. For some positive integer $m_0 > n_0$, all terms of s_{n_0}, occur among those of s'_{m_0}, so s'_{m_0} sum of terms of the form $(x,e_i)\,e_i$ for $i = n_0 + 1, n_o + 2,$ yields $\| s'_{m_0} - s_{n_0} \|^2 \leq \Sigma_{i=n_0+1}^{\infty} |x, e_i)|^2 < \varepsilon^2$, so $\| s'_{m_0} - s_{n_0} \|$ < and

$$\| s' - s \| \geq \| s' - s'_{m_0} \| + \| s'_{m_0} - s \| < \varepsilon + \varepsilon = 3e.$$

Since is arbitrary, this shows that s ' = s.

Let H be a non-zero Hilbert space, so that the class of all its orthonormal sets is non-empty. This class is clearly a partially ordered set with respect to set inclusion.

An orthonormal set $\{e_i\}$ in H is said to be *complete* if it is maximal in this partially ordered set, that is, if it is impossible to adjoin a vector e to $\{e_i\}$ in such a way that $\{e_i, e)$ is an orthonormal set which properly contains $\{e_i\}$.

Theorem 5:

Every non-zero Hilbert space contains a complete orthonormal mal set.

Proof:

The statement follows at once from Zorn's lemma, since the union of any chain of orthonormal sets is clearly an upper bound for the chain in the partially ordered set of all orthonormal sets.

Orthonormal sets are truly interesting only when they are complete. The reasons for this are presented in our next theorem.

Theorem 6:

Let H be a Hilbert space, and let $\{e_i\}$ be an orthonormal set in H. Then the following conditions are all equivalent to one another:

(1) $\{e_i\}$ *is complete;*

(2) $x \perp \{e_i\}$ =>* = 0;

(3) *if x is an arbitrary vector in H, then* $x — \Sigma\,(x, e_i)\,e_i$;

(4) *if x is an arbitrary vector in H, then* $\|\,x\,\|^2 = \Sigma|(x,e_i)\,|^2$.

Proof:

We prove that each of the conditions (1), (2), and (3) implies the one following it and that (4) implies (1).

(1) ⇔ (2). If (2) is not true, there exists a vector $x \neq 0$ such that $x \perp \{e_i\}$. We now define e by $e = x/\,\|\,x\,\|$, and we observe that $\{e_i\; e\}$ is an orthonormal set which properly contains $\{e_i\}$. This contradicts the completeness of $\{e_i\}$.

(2) ⇒ (3). By Theorem D, $x — \Sigma\,(x, e_i)\,e_{i\,=}\,0$ is orthogonal to $\{e_i\}$, so (2) implies that $x - S\,(x,e_i)\,e_i = 0$, or equivalently, that $y = 2\,(x, e_i)\,e_i$.

(3) ⇒ (4). By the joint continuity of the inner product, the expression in (3) yields

$$\|\,x\,\|^2 = (x, x) = (\Sigma(x, e_i)\,e_i\,\Sigma\,(x, e_j)\,e_j) = \Sigma\,(x, e_i)\,\overline{(x, e_i)} = \Sigma\,|\,(x, e_i)|^2.$$

(4) $\Rightarrow$ (1). If $\{e_i\}$ is not complete, it is a proper subset of an orthonormal set $\{e_i, e\}$. Since *e* is orthogonal to all the e_i's, (4) yields $\| e \|^2 = \Sigma|(e, e_i)|^2 = 0$, and this contradicts the fact that e is a unit vector.

There is some standard terminology which is often used in connection with this theorem. Let $\{e_i\}$ be a complete orthonormal set in a Hilbert space *H*, and let x be an arbitrary vector in *H*. The numbers (x, e_i) are called the *Fourier coefficients* of x, the expression $x = \Sigma (x, e_i) e_i$ is called the *Fourier expansion* of x, and the equation $\| x, e_i |\} = \Sigma |(x,e_i)|^2$ is called *Parseval's equation*—all with respect to the particular complete orthonormal set $(e_i\}$ under consideration.

Example:

Many orthonormal sets of great interest and importance in analysis can be obtained conveniently by applying the Gram-Schmidt process to sequences of simple functions.

(a) In the space L_2 associated with the interval [—1,1], the functions x^n (n = 0, 1, 2, . . .) are linearly independent. If we take these functions to be the x_n's in the Gram-Schmidt process, then the e_n's are the normalized Legendre polynomials.

(b) Consider the space L_2 over the entire real line. If the x_n's here are taken to be the functions $x_n e^{-x^2/2}$ (n = 0, 1, 2, . . .), then the corresponding e_n's are the normalized Hermite functions.

(c) Consider the space L_2 associated with the interval $[0,+\infty)$. If the x_n's are the functions $x^n e^{-x}$ (n = 0, 1, 2, . . .), then the e_n's are the normalized Laguerre functions.

Each of the orthonormal sets described in the above example can be shown to be complete in its corresponding Hilbert space. The analysis involved in a detailed study of these matters is quite complicated and has no proper place.

THE CONJUGATE SPACE H*

We pointed out in the introduction to this one of the fundamental properties of a Hilbert space H is the fact that there is a natural correspondence between the vectors in H and the functionals in H*. Our purpose in this section is to develop the features of this correspondence which are relevant to our work with operators in the rest of t chapter.

Let y be a fixed vector in H, and consider the function f_y defined on H by $f_y(x) = (x, y)$. It is easy to see that f_y is linear, for.

$$\begin{aligned} f_y(x_1 + x_2) &= (x_1 + x_2, y) \\ &= (x_1, y) + (x_2, y) \end{aligned}$$

$$= f_y(x_1) + f_y(x_2)$$

and
$$f_y(\alpha x) = (\alpha x, y)$$
$$= \alpha (x, y)$$
$$= \alpha f_y(x).$$

Further, f_y is continuous and is therefore a functional, for Schwarz's inequality gives

$$| f_y(x) | = | x, y) | \leq \| x \| \| y \|,$$

which shows that $\| fy \| \leq \| y \|$. Even more, equality is attained here, that is, $\| f_y \| = \| y \|$. This is clear if $y = 0$; and if $y \neq 0$, it follows from

$$\| fy \| = \sup | \{f_y(x) | : \| x \| = 1\}$$
$$\geq \left| f_y \left(\frac{y}{\|y\|} \right) \right|$$
$$= \left| \left(\frac{y}{\|y\|}, y \right) \right| = \| y \|.$$

To summarize, we have seen that $y \to f_y$ is a norm-preserving mapping of H into H*. This observation would be of no more than passing interest if it were not for the fact that every functional in H* arises in just this way.

Theorem:

Let H be a Hilbert space, and let f be an arbitrary functional in H. Then there exists a unique vector y in H such that*

$$f(x) = (x, y) \qquad ...(1)$$

for every x in H.

Proof:

It is easy to see that if such a y exists, then it is necessarily unique. For if we also have $f(x) = (x, y')$ for all x, then $(x, y') = (x, y)$ and $(x, y' - y) = 0$ for all x; and since 0 is the only vector orthogonal to every vector, this implies that $y' - y - 0$ or $y' = y$.

We now turn to the problem of showing that y does exist. If $f = 0$, then it clearly suffices to choose $y = 0$. We may therefore assume that $f \neq 0$. The null space M of f is thus a proper closed linear subspace of H, and by theorem there exists a non-zero vector y_0 which is orthogonal to M . We show that if a is a suitably chosen scalar, then the vector $y = \alpha y_0$ meets our requirements. We first observe that no matter what a may be, (1) is true for every x in M; for $f(x) = 0$ for such an x, and since x is orthogonal to y_0, we also have $(x, y) = 0$. This allows us to focus our attention on choosing a in such a way that (1) is true for $x = y_0$. The condition this imposes on a is that

$$f(y_0) = (y_0, \alpha y_0) = \bar{\alpha} \; \| y0 \|^2.$$

We therefore choose α to be $\overline{f(y_0)} / \|y_0\|^2$, and it follows that (1) is true for every x in M and for $x = y_0$. It is easily seen that each x in H can be written in the form $x = m + \beta y_0$ with m in M: all that is necessary is to choose β in such a way that $f(x - \beta y_0) = f(x) - f(x) - \beta f(y_0) = 0$, and this is accomplished by putting $\beta = f(x)/f(y_0)$. Our conclusion that (1) is true for every x in H now follows at once from

$f(x) = f(m + \beta y_0) = f(m) + \beta f(y_0) = (m, y) + \beta (y_0, y) = (m + \beta y_0, y) = (x, y)$.

This result tells us that the norm-preserving mapping of H into H* defined by

$$y \to f_y \text{ where } f_y(x) = (x, y), \qquad ...(2)$$

is actually a mapping of H onto H*. It would be pleasant if (2) were also a linear mapping. This is not quite true, however, for

$$f_{y_1} + y_2 = f_{y_1} + f_{y_2} \text{ and } f_{\alpha y} = \bar{\alpha} f_y. \qquad ...(3)$$

It is an easy consequence of (3) that the mapping (2) is an isometry, for $\| f_x - fy \| = \| f_{x-y} \| = \| x - y \|$. We state several interesting additional facts about this mapping (and what it enables us to do) in the problems, and we leave their verification to the reader. It should be remembered, however, that the real significance of this entire circle of ideas lies in its influence on the theory of the operators on H. We begin the treatment of these matters in the next section.

ORTHOGONAL COMPLEMENTS

Two vectors *x* and *y* in a Hilbert space *H* are said to be *orthogonal* (written $x \perp. y$) if $(x, y) = 0$. The symbol ± is often pronounced "perp." Since $\overline{(x, y)} = (y, x)$, we have $x \perp y \Leftrightarrow y \perp. x$. It is also clear that $x \perp 0$ for every *x*, and $(x, x) = \| x \|^2$ shows that 0 is the only vector orthogonal to itself. One of the simplest geometric facts about orthogonal vectors is the *Pythagorean theorem:*

$$x \perp y \Rightarrow \| x + y \|^2 = \| x - y \|^2 = \| x \|^2 + \| y \|^2.$$

A vector *x* is said to be orthogonal to a non-empty set *S* (written $x \perp S$) if $x \perp y$ for every *y* in *S*, and the *orthogonal complement* of S – denoted by $S^{\perp}$ — is the set of all vectors orthogonal to *S*. The following statements are easy consequences of the definition:

$$\{0\}^{\perp} = H; H^{\perp} = \{0\};$$
$$S \cap S^{\perp} \subseteq \{0\};$$
$$S_1 \subseteq S_2 \Rightarrow S_1^{\perp} \supseteq S_2^{\perp};$$

($S^{\perp}$ is a closed linear subspace of H.

It is customary to write $(S^{\perp})^{\perp}$ in the form $S^{\perp\perp}$. Clearly, $S \subseteq S^{\perp\perp}$. Let M be a closed linear subspace of *H*. We know that *M-1* is also a closed linear subspace, and that *M* and Mx are disjoint in the sense that they have only the zero vector in common. Our aim in this section is to prove that $H = M \rightarrow M^{\perp}$, and each of our theorems is a step in this direction.

Theorem 1:

If M and N are closed linear subspaces *of a Hilbert space H such that M $\perp$ N, then the linear subspace M + N is also closed.*

Proof:

Let z be a limit point of M + N. It suffices to show that z is in M + *N*. There certainly exists a sequence $\{z_n\}$ in M + *N* such that $z_n \rightarrow z$. By the assumption that M $\perp$ *N*, we see that M and *N* are disjoint, so each z_n can be written uniquely in the form $z_n = x_n + y$, where x_n is in *M* and *yn* is in *N*. The Pythagorean theorem shows that $\| z_m - z_n \|^2 = \| x_m = \| x_m - x_n \|^2 + \| y_m - y_n \|^2$, so $\{x_n\}$ and $\{y_n\}$ are C..uchy sequences in M and N. M and N are closed, and therefore complete, so there exist vectors x and y in M and N such that $x_n \rightarrow x$ and $y_n \rightarrow y$. Since $x + y$ is in $M + N$, our conclusion follows from the fact that $z = \lim z_n = \lim (x_n + y_n) = \lim x_n + \lim y_n = x + y$.

The way is now clear for the proof of our principal theorem.

Theorem 2:

If *M is a closed linear subspace of a Hilbert space H, then* $H = M \oplus M^{\perp}$.

Proof:

Since M and $M^{\perp}$ are orthogonal closed linear subspaces of H, Theorem C shows that $M + M^{\perp}$ is also a closed linear subspace of *H*. We prove that $M + M^{\perp}$ equals *H*. If this is not so, then by Theorem 2 there exists a vector $z_0 \perp 0$ such that $z_0 \perp (M + M^{\perp})$. This non-zero vector must evidently lie in $M^{\perp} \cap M^{\perp\perp}$; and since this is impossible, we infer that $H = M + M^{\perp}$. To conclude the proof, it suffices to observe that since M and $M^{\perp}$ are disjoint, the statement that $H = M + M^{\perp}$ can be strengthened to $H = M \oplus M^{\perp}$.

The main effect of this theorem is to guarantee that a Hilbert space is always rich in projections. In fact, if M is an arbitrary closed linear subspace of a Hilbert space H, then Theorem 4 shows that there exists a projection defined on H whose range is M and whose null space is $M^{\perp}$.

Theorem 3:

Let M be a closed linear subspace of a Hilbert space H, let x be a vector not in M, and let d be the distance from x to M. Then there exists a unique vector y_0 in M such that $\| x - y_o \| = d$.

Proof:

The set $C = x + M$ is a closed convex set, and d is the distance from the origin to C. By Theorem 2, there exists a unique vector z_0 in C such that $\| z_0 \| = d$. The vector $y_0 = x - z_0$ is easily seen to be in M, and $\| x - y_o \| = \| z_o \| =$ d. The uniqueness of j/n follows from the fact that if y_i is a vector in M such that $y_1 \neq y_0$ and $\| x - y_1 \| = d$, then $z_1 = x - y_1$ is a vector in C such that $z_1 \neq z_0$ and $\| z_1 \| = d$, which contradicts the uniqueness of z_o.

We use this result to prove

AN OPERATOR OF THE ADJOINT

Throughout the rest of this chapter, we focus our attention on a fixed but arbitrary Hilbert space H, and unless we specifically state otherwise, it is to be understood that H is the context for all our discussions and theorems.

Let T be an operator on H. That T gives rise to an operator T* (its conjugate) on H*, where T* is denned by

$$(T^*f)\,x = f(T_x)\ 1.$$

We also saw that the mapping $T \to T^*$ is an isometric isomorphism of $\mathcal{B}$(H) into $\mathcal{B}$(H*) which reverses products and preserves the identity transformation. In the same way, T* gives rise to an operator T** on H**; and since H is reflexive, it follows that T** = T when H** is identified with H by means of the natural imbedding. These statements depend only on the fact that H is a reflexive Banach space. We now bring its Hilbert space character into the picture, and we use the natural correspondence between H and H* discussed in the previous section to pull T* down to H. The details of this procedure are as follows. Let y be a vector in H, and f_y, its corresponding functional in H*; operate with T* on f_y, to obtain a functional $f_y = T^*f_y$, and return to its corresponding vector z in H. There are three mappings under consideration here, and we are forming their product:

$$y \to f_y \to T^*f_y = f_x \to z. \qquad ...(1)$$

We write z = T*y, and we call this new mapping T* of H into itself the adjoint of T. The same symbol is used for the adjoint of T as for its conjugate because these two mappings are actually the same if H and H * are identified by means of the natural correspondence. It is easy to keep track of whether T* signifies the conjugate or the adjoint of T by noticing whether it operates on functionals or on vectors. The action of the adjoint can be linked more closely to the structure of H by observing that for every vector x we have $(T^*f_y)\,x = f_y$ $(Tx) = (Tx, y)$ and $(T^*f_y)\,x = f_x\,(x) = (x, z) = (x, T^*y)$, so that

$$(Tx, y) = (x, T^*y) \qquad ...(2)$$

for all x and y. Eqn. (16) is much more than merely a property of the adjoint of T, for it uniquely determines this adjoint. The proof is simple, if T′ is any mapping of H into itself such that (Tx, y) = (x,T′ y) or all x and y, then (x,T′y) = (x,T*y) for all z, so T′y = T*y;[1] and since the latter is true for all y, T′ = T*.

Our remarks in the above paragraph have shown that to each operator T on H there corresponds a unique mapping T* of H into itself (called the adjoint of T) which satisfies relation for all x and y. There is a more direct but less natural approach to these ideas, one which avoids any reference to the conjugate of T. If y is fixed, it is clear that the expression (Tx, y) is a scalar-valued continuous linear function of x. By Theorem 1, there exists a unique vector z such that (Tx, y) = (x, z) for all x. We now write z = T*y, and since y is arbitrary, we again have relation (2) for all x and y. The fact that T* is uniquely determined by (2) follows just as before.

The principal value of our approach to the definition of the adjoint (as opposed to that just mentioned) lies in the motivation it provides for considering adjoints at all. We can express this by emphasizing that an operator on a Banach space always has a conjugate which operates on the conjugate space; and when the Banach space happens to be a Hilbert space, then, as we have seen, the natural correspondence discussed in the previous section makes it almost inevitable that we regard the conjugate as an operator on the space itself. Once the definition of the adjoint is j fully understood, however, there is no further need to mention conjugates. All our future work with adjoints will be based on Eq. (2), I and from this point on, the symbol T* will always signify the adjoint of T (and never its conjugate).

As our first step in exploring the properties of adjoints, we verify that T* actually is an operator on H (all we know so far is that it maps H into itself). For any y and 2, and for all x, we have

$$\begin{aligned}(x, T^*(y + z)) &= (Tx, y + z) = (Tx, y) + (Tz, z)\\ &= (x, T^*y) + (x,T^*z) = (x, T^*y + T^*z),\end{aligned}$$

so
$$T^*(y + z) = T^*y + T^*z.$$

The relation $T^*(\alpha y) = \alpha T^*y$

is proved similarly, so T* is linear. It remains to be seen that T* is continuous; and to prove this, we note that

$$\| T^*y \|^2 = (T^*y, T^*y)$$
$$= (TT^*y,y) \leq \| TT^*y \| \| y \| \leq \| T \| \| T^*y \|$$

implies that $\| T^*y \| \leq \| T \| \, |y \|$ for all y, so

$$\| T^* \| \leq \| \ \|.$$

These facts tell us that $T \rightarrow T^*$ is a mapping of $\mathcal{B}$(H) into itself. This mapping is called the *adjoint operation* on *$\mathcal{B}$(H).*

Theorem:

The adjoint operation $T \rightarrow T^*$ on (B(H) has the following properties:

(1) $(T_1 + T_2)^* = T_1,^* + T_2,^*,$

(2) $(\alpha T)^* = \bar{\alpha}\, T^*,$

(3) $(T_1 T_2)^* = T_2{}^* T_1 V^*,$

(4) $T^{**} = T,$

(5) $\| T^* \| = \| T \|$

(6) $\| T^* \| = \| T \|^2.$

Proof:

The arguments used in proving (1) to (4) are all essentially the same. As an illustration of the method, we observe that (3) follows from the fact that for all x and y we have

$$(x, (T_1T_2)^*y) = (T_1T_2x, y) = (T_2x, T_1{}^*y) = (x, T_2{}^*T_1{}^*y)$$

To prove (5), we note that we already have $\| T^* \| < \| T \|$, and if we apply this to T* instead of T and use (4), we obtain $\| T \| = \| T^{**} \| \leq \| T^* \|$. Half of (6) follows from (5) and the inequality (5), for

$$\| T^*T \| \leq \| T^* \| \| T \| \| T \| = \| T \|^2,$$

and the fact that $\| T \|^2 \leq \| T^*T \|$ is an immediate consequence of

$$\| Tx \|^2 = (Tx, Tx) = (T^*Tx, x) \leq \| T^*Tx \| \| x \| \leq \| T^*T \| \| x \|^2$$

The presence of the adjoint operation is what distinguishes the theory of the operators on H from the more general theory of the operators on a reflexive Banach space. I

n the next three sections, we use this operation as a tool by means of which we single out for special study certain types of operators on *H* whose theory is particularly complete and satisfying.

OPERATORS OF THE SELF-ADJOINT

There is an interesting analogy between the set $\mathcal{B}(H)$ of all operators on our Hilbert space H and the set C of all complex numbers. This can be summarized by observing that each is a complex algebra together with a mapping of the algebra onto itself ($T \to T^*$ and $z \to \bar{z}$) and that these mappings have similar properties.

We shall see that this analogy is quite useful as an intuitive guide to the study of the operators on H. The most significant difference between these systems is that multiplication in the algebra $\mathcal{B}(H)$ is in general non-commutative, and it will become clear as we proceed that this is the primary source of the much greater structural complexity of $\mathcal{B}(H)$.

The most important subsystem of the complex plane is the real line, which is characterized by the relation $z = \bar{z}$. By analogy, we consider those operators A on H which equal their adjoints, that is, which satisfy the condition $A = A^*$. Such an operator is said to be self-adjoint. The self-adjoint operators on H are evidently those which are related in the simplest possible way to their adjoints.

We know that $0^* = 0$ and $I^* = I$, so 0 and I are self-adjoint. If A_1 and A_2 are self-adjoint, and if α and β are real numbers, then

$$(\alpha A_1 + \beta A_2)^* = \bar{\alpha} A_1^* + \bar{\beta} A_2^* = \alpha A_1 + \beta A_2$$

shows that $\alpha A_1 + \beta A_2$ is also self-adjoint. Further, if $\{A_n\}$ is a sequence of self -adjoint operators which converges to an operator *A*, then it is easy to see that *A* is also self -ad joint; for

$$\| A - A^* \| \le \| A - A_n \| + \| A_n - A_n^* \| + \| (A_n^* - A)^* \| = \| A - A_n \|$$
$$+ \| (A_n - A)^* \| = \| A - A_n \| + \| A_n - A \| = 2\| A_n - A \| \to 0$$

shows that $A - A^* = 0$, so $A = A^*$. These remarks yield our first theorem.

Theorem 1:

The self-adjoint operators in $\mathcal{B}(H)$ form a closed real linear subspace of $\mathcal{B}(H)$–and therefore a real Banach space–which contains the identity transformation.

The reader will notice that we have said nothing here about the product of two self-adjoint operators. Very little is known about such products, and the following simple result represents almost the extent of our information.

Theorem 2:

If A_1 AND A_2 are self-adjoint operators on H, then their product A_1A_2 is self-adjoint $\Leftrightarrow A_1A_2 = A_2A_1$.

Proof:

This is an obvious consequence of

$(A_1A_2)^* = A_2^*A_1^* = A_2A_1$.

The order properties of self-adjoint operators are more interesting, and we devote the remainder of the section to establishing some of the simpler facts in this direction.

If *T* is an arbitrary operator on *H,* it is easy to see that

$$T = 0 \Leftrightarrow (Tx, y) = 0$$

for all x and y. It is also clear that $T = 0 \Rightarrow (Tx, x) = 0$ for all x. We shall need the converse of this implication.

Theorem 3:

If T is an operator on H for which (Tx, x) = 0 for all x, then T = 0.

Proof:

It suffices to show that (Tx, y) = 0 for any x and y, and the proof of this depends on the following easily verified identity:

$$(T(\alpha x + \beta y), \alpha x + \beta y) - |\alpha|^2 (Tx, x) - |\beta|^2 (Ty, y)$$

$$= \alpha\bar{\beta}\,(Tx, y) + \bar{\alpha}\beta\,(Ty, x). \qquad ...(1)$$

We first observe that by our hypothesis, the left side of (1)—and therefore the right side as well–equals 0 for all α and β. If we put $\alpha = 1$ and $\beta = 1$, then (1) becomes

$$(Tx, y) + (Ty, x) = 0, \qquad ...(2)$$

and if we put $\alpha = i$ and $\beta = 1$, we get

$$i\,(Tx, y) - i\,(Ty, x) = 0. \qquad ...(3)$$

Dividing (3) by i and adding the result to (2) yields 2(Tx, y) = 0, so (Tx, y) = 0 and the proof is complete.

It is worth emphasizing that this proof makes essential use of the fact that the scalars are the complex numbers (and not merely the real numbers).

We now apply this result to proving our next theorem, which indicates that self-adjoint operators are linked to real numbers by stronger ties than might be suspected from the loose analogy that led to their definition.

Theorem 4:

An operator T on H is self-adjoint $\Leftrightarrow$ (Tx, x) is real far all x.

Proof:

If T is self-adjoint, then

$$\overline{(Tx, x)} = (x, Tx) = (x, T^*x) = (Tx, x)$$

shows that (Tx, x) is real for all x. On the other hand, if (Tx, x) is real for all z, then (Tx, x) = (Tx, x) = (x, T*x) = (T*x, x) or

$$([T - T^*]x, x) = 0$$

for all x. By Theorem C, this implies that T – T* = 0, so T = T*.

This theorem enables us to define a respectable and useful order relation on the set of all self-adjoint operators. If A_1 and A_2 are self-adjoint, we write $A_1 \le A_2$ if $(A_1x, x) \le (A_2x, x)$ for all x. The main elementary facts about this relation are summarized in

Theorem 5:

The real Banach space of all self-adjoint operators on H is a partially ordered set whose linear structure and order structure are related by the following properties:

(1) *if $A_1 \le A_2$, then $A_1 + A \le A_2 + A$ for every A;*

(2) *if $A_1 \le A_2$ and $\alpha \ge 0$, then $\alpha A_1 \le \alpha A_2$.*

Proof:

The relation in question is obviously reflexive and transitive. To show that it is also antisymmetric, we assume that $A_1 \le A_2$ and $_2 \le A_1$. This implies at once that $([A_1 - A_2,]x, x) = 0$ for all x, so by Theorem C, $A_1 - A_2 = 0$ and $A_1 = A_2$. The proofs of properties (1) and (2) are easy. For instance, if $A_1 \le A_2$, so that $(A_1x, x) \le (A_2x, x)$ for all *x*, then $(A_1x, x) + (Ax, x) \le (A_2x, x) + (Ax, x)$ or $([A_1 + A]x, x) \le ([A_2 + A]x, x)$ for all x, so $A_1 + A \le A_2 + A$. The proof of (2) is similar.

A self-adjoint operator A is said to be *positive* if A ≥ 0, that is, if (Ax, x) ≥ 0 for all x. It is clear that 0 and I are positive, as are T*T and TT* for an arbitrary operator T.

Theorem 6:

*If A is a positive operator on H, then I + A is non-singular. In particular, I + T*T and I + TT* are non-singular for an arbitrary operator T on H.*

Proof:

We must show that I + A is one-to-one onto as a mapping of H into itself. First, it is one-to-one, for

$$(I + A)x = 0 \Rightarrow Ax = -x$$

$$\Rightarrow \quad (Ax, x) = (-x, x) = -\|x\|^2 \geq 0 \Rightarrow x = 0.$$

We next show that the range M of I + A is closed. It follows from $\|(I+A)x\|^2 = \|x\|^2 + \|Ax\|^2 + 2(Ax, x)$–and the assumption that A is positive–that $\|z\| \leq \|(I + A)x\|$. By this inequality and the completeness of H, M is complete and therefore closed. We conclude the proof by observing that M = H, for otherwise there would exist a nonzero vector x_0 orthogonal to M, and this would contradict the fact that $(x_0, [I + A]x_0) = 0 \Rightarrow \|x_0\|^2 = -(Ax_0, x_0) \leq 0 \Rightarrow x_0 = 0$.

If the reader wonders why we fail to show that the partially ordered set of all self-adjoint operators is a lattice, the reason is simple: it isn't true. As a matter of fact, this system is about as far from being a lattice as a partially ordered set can be, for it can be shown that two operators in the set have a greatest lower bound $\Leftrightarrow$ they are comparable. This whole situation is intimately related to questions of commutativity for algebras of operators and is too complicated for us to explore here. For further details.

NORMAL AND UNITARY OPERATORS

An operator N on H is said to be normal if it commutes with its adjoint, that is, if NN* = N*N. The reason for the importance of normal operators will not become clear until the next chapter. We shall see that they are the most general operators on H for which a simple and revealing structure theory is possible. Our purpose in this section is to resent a few of their more elementary properties which are necessary or our later work.

It is obvious that every self-adjoint operator is normal, and that if y is normal and a is any scalar, then aN is also normal. Further, the limit N of any convergent sequence {Nk} of normal operators is normal; 'or we know that $N^{**} \to N^*$, so

$$\|NN^* - N^*N\| \leq \|NN^* - N_kN_k^*\| + \|N^*N_k^* - N_k^*N_k\| + \|N_k^*N_k - N^*N\| = \|NN^* - N_kN_k^*\| + \|N_k^*N_k - N^*N\| \to 0,$$

which implies that $NN^* - N^*N = 0$.

These remarks prove

Theorem 1:

The set of all normal operators on H is a closed subset of (&(H) which contains the set of all self-adjoint operators and is closed under scalar multiplication.

It is natural to wonder whether the sum and product of two normal operators are necessarily normal. They are not, but nevertheless, we can say a little in this direction.

Theorem 2:

If N_1 and N_2 are normal operators on H with the property that either commutes with the adjoint of the other, then $N_1 + N_2$ and $N_1 + N_2$ are normal.

Proof:

It is clear by taking adjoints that

$$N_1N_2{*}= N_2{*}\, N_1 \Leftrightarrow N_2\, N_1{*} = N_1{*}N_2$$

so the assumption implies that each commutes with the adjoint of the other. To show that $N_1 + N_2$ is normal under the stated conditions, we have only to compare the results of the following computations:

$$(N_1 + N_{2,}) (N_1 + N_1)^* = (N_1 + N_2) (N_1{*} + N_1{*})$$
$$= N_1N_1{*} + N_1N_1{*} + N_2N_2{*}$$

and $$(N_1 + N_2)^*(N_1 + N = (N_1{*} + N_2{*}) (N_1 + N_2)$$
$$= N_1{*}N_1 + N_1{*}N_2{*} + N_2{*}\, N_1 {}_+ N_2{*}\, N_2$$

The fact that N_1N_2 is normal follows similarly from

By definition, a self-ad joint operator A is one which satisfies the identity $A^*x = Ax$. Many properties of self -adjoint operators do not depend on this, but only on the weaker identity $\| A^*x \| = \| Ax \|$. Our next theorem shows that all such properties are shared by normal operators.

Theorem 3:

An operator T on H is normal $\Leftrightarrow \| T^*x \| = \| Tx \|$ for every x.

Proof:

In view of Theorem 3, this is implied by the fact that

$\| T^*x \| = \| Tx \| \Leftrightarrow \| T^*x \|^2 = \| Tx \|^2 \Leftrightarrow (T^*x, T^*x)$

$= (Tx, Tx) \Leftrightarrow (TT^*x, x) = (T^*Tx, x) \Leftrightarrow ([TT^* - T^*T]\, x, x) = 0.$

The following consequence of this result will be useful in our later work.

Theorem 4:

If N is a normal operator on H, then $\| N^2 \| = \| N \|^2$.

Proof:

The preceding theorem shows that

$$\| N^2x \| = \| NNx \| = \| N^*Nx \|$$

for every x, and this implies that $\| N^2 \| = \|N^*N \|$. By theorem A, we have $\| N^*N \| = \| N \|^2$, so the proof is complete.

We know that any complex number z can be expressed uniquely in the form $z = a + ib$ where a and b are real numbers, and that these real numbers are called the real and imaginary parts of z and are given by $a = (z + \bar{z})/2$ and $b = (z - \bar{z})/2i$. The analogy between general operators and complex numbers, and between self-adjoint operators and real numbers, suggests that for an arbitrary operator T on H we form $A_1 = (T + T^*)/2$ and $A_2 = (T - T^*)/2i$. A_1 and A_2 are clearly self-adjoint, and they have the property that $T - A_1 + i A_2$. The uniqueness of this expression for T follows at once from the fact that

$$T^* = A_1 - i A_2.$$

The self-adjoint operators A_1 and A_2 are called the real part and the imaginary part of T.

We emphasized earlier that the complicated structure of $\mathcal{B}(H)$ is due in large part to the fact that operator multiplication is in general non-commutative. Since our future work will be focused mainly on normal operators, it is of interest to see–as the following theorem shows–that the existence of non-normal operators can be traced directly to the non-commutativity of self-adjoint operators.

Theorem 5:

If T is an operator on H, then T is normal ⇔ its real and imaginary parts commute.

Proof:

If A_1 and A_2 are the real and imaginary parts of T, so that $T = A_1 + iA_2$ and $T^* = A_1 - iA_2$, then

$$TT^* = (A_1 + iA_2)(All - iA_2) = A_1{}^2 + A_2{}^2 + i (A_2A_1 - A_1A_2)$$

and $T^*T = (A_1 - iA_2)(A_1 + iA_2) = A_1{}^2 + A_2{}^2 + i (A_1A_2 - A_2A_1).) =$

It is clear that if $A_1A_2 = A_2A_1$, then $TT^* = T^*T$, Conversely, if $TT^* = T^*T$, then $A_1A_2 - A_2A_1 - A_1A_2$, so $2A_1A_2 = 2A_2A_1$ and $A_1A_2 = A_2A_1$.

Perhaps the most important subsystem of the complex plane after the real line is the unit circle, which is characterized by either of the equivalent identities $| z | - 1$ or $z\bar{z} = \bar{z}z = 1$. An operator U on H which satisfies the equation $UU^* = U^*U = I$ is said to be unitary. Unitary operators–which are obviously normal–are thus the natural analogues of complex numbers of absolute value 1. It is clear from the definition that the unitary operators on H are precisely the non-singular operators whose inverses equal their adjoints. The geometric significance of these operators is best understood in the light of our next theorem.

Theorem 6:

If T is an operator on H, then the following conditions are all equivalent to one another:

(1) $T^*T = I$,

(2) $(Tx, Ty) = (x, y)$ *fo, all x and y,*

(3) $\| Tx \| = \| x \|$| *for all x.*

Proof:

If (1) is true, then $(T^*Tx, y) = (x, y)$ or $(Tx, Ty) = (x, y)$ for all x and y, so (2) is true; and if (2) is true, then by taking $y = x$ we obtain $(Tx, Tx) = (x, x)$ or $\| Tx \|^2 = \| x \|^2$ for all x, so (3) is true. The fact that (3) implies (1) is a consequence of Theorem 3 and the following chain of implications:

$$\| Tx \| = \| x \| \Rightarrow \| Tx \|^2 = \| x \|^2$$

$$\Rightarrow \quad (Tx, Tx) = (x, x) \Rightarrow \Rightarrow (T^*Tx, x)$$

$$= (x, x) \Rightarrow ([T^*T - I]\, x, x) = 0.$$

An operator on H with this theorem is simply an isometric isomorphism of H into itself. That an operator of this kind need not be unitary is easily seen by considering the operator on h defined by

$$T\,\{x_1, x_2, \ldots\} = \{0, x_1, x_2, \ldots\},$$

which preserves norms but has no inverse. These ideas lead at once to

COMPLETENESS METRIC SPACE THEOREMS

Theorem 1:

Let (X, d) be a metric space. Show that $F \subset X$, F is closed $\Leftrightarrow$ F' is open.

Proof:

Let (X, d) be a metric space .

Let C be a closed subset of X, so that $D(F) \subset F$.

To prove that F' is open in X.

Let $X \in F'$ be arbitrary. Then $X \notin F$.

$$D(F) \subset F.\ x \notin F \Rightarrow x \notin D(F)$$

$$\Rightarrow (S_{r(x)} - \{x\}) \cap F = \varnothing \text{ for some } r > 0.$$

$$\Rightarrow S_{r(x)} \cap F = \varnothing \qquad (\because x \notin F)$$

$\Rightarrow S_{r}(x) \subset X - F$

$\Rightarrow S_{r}(x) \subset F'$.

$\therefore$ Given $x \in F'$, $\exists$ any an open sphere $S_{r(x)}$ s.t.

$S_{r}(x) \subset F'$.

By definition, this proves that F' is open.

Conversely suppose that F' is open in X.

To prove that F is closed in X.

Let $x \in F'$ be arbitrary, then $x \notin F$.

$\because$ F' is open, $\exists$ $r \subset R^+$ s.t., $S_{r}(x) \subset F'$

This $\Rightarrow S_{r}(x) \cap F = \varnothing$

$\Rightarrow (S_{r}(x) - \{x\}) \cap F = \varnothing$

$\Rightarrow x \notin D(F)$.

Thus, any $x \in F' \Rightarrow x \notin D(F)$.

i.e., any $x \in X - F$

$\Rightarrow x \in X - D(F)$

This $\Rightarrow$ $X - F \subset X - D(F)$

or $D(F) \subset F$.

This $\Rightarrow$ F is closed.

Theorem 2:

In any metric space (X, d) each open sphere is an open set.

Proof:

Let (X, d) be a metric space. Let $S_{r0}(x0)$ be an open sphere in X.

To prove that $S_{r0}(x0)$ is an open set.

Let $x \in S_{r0}(x_0)$ be arbitrary, then $d(x, x_0) < r_0$

Write $r = r_0 - d(x, x_0)$, ...(1)

By definition, $S_{r0}(x0) = \{y \in X : d(y, x_0) < r_0\}$

$S_{r}(x) = \{y \in X : d(y, x) < r\}$.

We claim $S_{r}(x) \subset S_{r0}(x0)$

Let $y \in S_{r}(x)$ be arbitrary.

Then $d(x, y) < r$

$d(y, x_0) \leq d(y, x) + d(x, x_0)$

$$< r + d(x, x_0) = r_0.$$

$\therefore$ $\qquad d(y, x_0) < r_0$ $\qquad$ on using (1)

$\Rightarrow$ $\qquad y \in S_{r0\,(x0)}$

and $\qquad y \in S_{r\,(x)}$

$\Rightarrow$ $\qquad y \in S_{r0\,(x0)}$

$\Rightarrow$ $\qquad S_{r\,(x)} \subset S_{r0\,(x0)}$

Thus we have shown that

given any $x \in S_{r0\,(x0)}$, $\exists\, r > 0$ s.t. $S_{r\,(x)} \subset S_{r0\,(x0)}$.

By definition, this proves that $S_{r0\,(x0)}$ is an open set.

Theorem 3(a):

Frechet Space: *Let F be the set of infinite sequences of real numbers. x, y, z ∈ F, then*

$$x = < x_n > = < x_1, x_2 \ldots >,$$

$$y = < y_n >, \; z = < z_n >$$

where $\qquad x_n, y_n, z_n \in R$

Proof:

We define a map

$$d : F \times F : \rightarrow R \text{ s.t.}$$

$$d(x,y) = \sum_{n=1}^{\infty} \frac{1}{2^n} \frac{(x_n - y_n)}{\left[1 + \left|x_n - y_n\right|\right]}$$

To show that d is metric on F.

(i) $d(x, y) \geq 0$ For $|x_n - y_n| \geq 0 \; \forall_n$

(ii) $d(x, y) = 0 \Leftrightarrow x = y$

For $d(x,y) = 0 \Leftrightarrow \sum_{n=1}^{\infty} \frac{1}{2^n} \frac{(x_n - y_n)}{\left[1 + \left|x_n - y_n\right|\right]} = 0$

$$\Leftrightarrow \frac{1}{2^n} \frac{(x_n - y_n)}{\left[1 + \left|x_n - y_n\right|\right]} = 0 \; \forall \; n$$

$$\Leftrightarrow |x_n + y_n| = 0 \; \forall \; n$$

$$\Leftrightarrow |x_n + y_n| = \forall \; n$$

$$\Leftrightarrow x = y$$

(iii) $d(x, y) = d(y, x)$

For $|x_n + y_n| = |y_n + x_n|$

(iv) $d(x, y) \geq d(x, z) + d(z, y)$

Here we use the fact that

$$\frac{|\alpha + \beta|}{1 + |\alpha + \beta|} \leq \frac{|\alpha|}{1 + |\alpha|} + \frac{|\beta|}{1 + |\beta|}$$

In view of this we have

$$d(x, y) = \sum_{n=1}^{\infty} \frac{|x_n - y_n|}{1 + |x_n - y_n|} \cdot \frac{1}{2^n}$$

$$= \sum_{n=1}^{\infty} \frac{1}{2^n} \cdot \frac{|(x_n - z_n) + (z_n - y_n)|}{1 + |(x_n - z_n) + (z_n - y_n)|}$$

$$\geq \sum_{n=1}^{\infty} \frac{1}{2^n} \cdot \frac{|x_n - z_n|}{1 + |x_n - z_n|} + \sum_{n=1}^{\infty} \frac{1}{2^n} \cdot \frac{|z_n - y_n|}{1 + |z_n - y_n|}$$

$= d(x, z) + d(z, y)$.

Thus d is metric on F. The pair (F, d) is a metric space and this metric space is called *Frechet space.*

Theorem 3(b):

Let (X, d) be a metric space. Suppose A is a closed subset of X and x is a point of X s.t. $x \notin A$. Then $\exists$ disjoint open subsets U and V of X s.t. $x \in U$, $A \subset V$.

Proof:

$x \notin A \Rightarrow \exists\ \delta > 0$ s.t. $S_\delta(x) \cap A = \phi$.

Choose ε, η with $0 < \varepsilon < \eta < \delta$.

write $U = S_\varepsilon(x)$, $V = X - S_\eta[x]$.

Since each open sphere is an open set and each closed sphere is a closed set. Also complement of closed set is open set. These prove that U and V are open sets s.t. $U \cap V = \phi$.

For $U = S_\varepsilon(x) \subset S_\eta(x) \subset S_\delta(x)$.

Also $A \subset V$, by construction of V.

Theorem 4(a):

Let $<f_n(x)>$ be a sequence of continuous functions defined on a metric space (X,d). Let this sequence converge uniformly to f on X. Then f(x) is continuous on X.

Or

Uniform limit of a sequence of continuous function is continuous.

Proof:

Since $<f_n(x)>$ converges uniformly to f on (X, d). Hence given $\varepsilon > 0$, $\exists\, n_0 \in N$ independent of $x \in X$ s.t. $n \geq n_0$.

$$\Rightarrow d(f_n(x), f(x) < \varepsilon/3 \quad ...(1)$$

Let $a \in X$ be arbitrary. To prove that f is continuous on X, we have to prove that f is continuous at x = a. For this we have to show that given $\varepsilon > 0$, $\exists\, \delta > 0$, 0 s.t. $d(x, a) < \delta$.

$$\Rightarrow d(f(x), f(a) < \varepsilon \quad ...(2)$$

Continuity of f_n at $a \in X$

$$\Rightarrow d(f_n(x), f(x) < \varepsilon/3 \text{ for } d(x, a)\, ,\, \delta \quad ...(3)$$

By (1), $d(f_n(a), f(a)) < \varepsilon/3 \ \forall\, n \geq n_0$ ...(4)

If $d(x, a) < \delta$, then

$d(f(x), f(a)) \leq d[f(x), f_n(x)] + d[f_n(x), f_n(a) + d[f_n(a), f(a)]$

$$< \frac{\varepsilon}{3} + \frac{3}{3} = \frac{\varepsilon}{3} o = \varepsilon \text{ by (1), (3), \& (4)}$$

or, $d(f(x), f(a)) < \varepsilon$ for $d(x, a) < \delta$. Hence the result (2).

Theorem 4(b):

In a metric space (X, d), prove that

F is closed $\Leftrightarrow$ D (F) $\subset$ F.

Or

Prove that a subset F of a metric space X contains all its limit points iff X – F is open.

Proof :

Let (X, d) be a metric space and $F \subset X$.

We know that F is closed $\Leftrightarrow$ X – F is open.

Step I : Let X – F be open so that F is closed.

Aim. D (F) $\subset$ F.

Let $x \in X - F$ be arbitrary. The X – F is an open set containing x s.t. $(X - F) \cap F = \phi$.

This $\Rightarrow$ is not a limit point of F

$$\Rightarrow x \notin D(F) \Rightarrow x \in X - D(F)$$

Thus, " $x \in X - F \Rightarrow x \in X - D(F)$

$\therefore \quad X - F \subset X' - D(F)$

or, $\quad D(F) \subset F$

Step II : Given $\quad D(F) \subset F \quad$ ...(1)

To prove F is closed.

Let $y \in X - F$, then $y \notin F$

$$y \notin F,\ D(F) \subset F \Rightarrow y \notin D(F)$$

$$\Rightarrow \exists \text{ open set } G \text{ with } y \in G \text{ s.t.}$$

$$(G - \{y\}) \cap F = \phi$$

$$\Rightarrow G \cap F = \phi \text{ as } y \notin F$$

$$\Rightarrow G \subset X - F$$

Thus we have shown that

$$\text{any } y \in X - F \Rightarrow \exists \text{ open set } G \text{ with } y \in G \text{ s.t. } G \subset X - F$$

$$\text{This} \Rightarrow X - F \text{ is open} \Rightarrow \text{is closed.}$$

Theorem 5:

If a_i, b_i ($i = 1, 2 \ldots n$) are non- negative real numbers and $p > 1$, then

$$\sum_{i=1}^{n}(a_i + b_i)^p \le \sum_{i=1}^{n} a_i^p + \sum_{i=1}^{n} b_i^p$$

Proof :

Let $f(x) = 1 + x^p - (1 + x)^p$.

Then taking derivative ; it can be shown that

$$f(x) \ge 0 \text{ for } x \ge 0.$$

or $\quad 1 + x^p - (1 + x)^p \ge 0.$

or $\quad (1 + x)^p \le 1 + x^p$

put $x = a_i/b_i$, we get

$$\left(1 + \frac{a_i}{b_i}\right)^p \le 1 + \left(\frac{a_i}{b_i}\right)^p$$

or, $(a_i + b_i)^p \le (a_i)^p + (b_i)^p$

or $$\sum_{i=1}^{n}(a_i + b_i)^p \le \sum_{i=1}^{n} a_i^p + \sum_{i=1}^{n} b_i^p$$

Theorem 6:

(or l_2-space) : *Let H be the set of infinite recl sequences*

$< x_n >$ s.t. $\sum_{i=1}^{n} x_i^{\ 2} < \infty$.

Let x , y ∈ H s. t $x = < x_n >$, $y = < y_n >$. Define a map

$$d : H \times H \to R \text{ s.t. } d(x, y) = \left[\sum_{i=1}^{\infty} (x_i - y_i)^2 \right]^{1/2}$$

Aim : d is a metric in H.

(1) To prove that d is well defined map

$$d(x, y) = [\Sigma \{x_i + (-y_i)\}^2]^{1/2}$$
$$\leq (\Sigma x_i^2)^{1/2} + (\Sigma (-y_i)^2)^{1/2} \text{ by Minkowski inequality}$$
$$= \Sigma x_i^2 + \Sigma y_i^2 < \infty \text{ , by given condition,}$$

or, $d(x, y) < \infty$,

meaning thereby d s a finite real number. Hence d is well defined map.

(2) $d(x, y) \geq 0$

For x_i , y_i, are real

⇒ $(x_i - y_i)^2 \geq 0$

⇒ $d(x , y) \geq 0$

(3) $d(x , y) = d(y , x)$

For $(x_1 - y_i)^2 = (y_i - x_i)^2$

⇒ $\{ \Sigma (x_1 - y_i)^2\}^{1/2} = \{\Sigma (y_i - x_i)^2\}^{1/2}$

⇒ $d(x , y) = d(y , x)$

(4) $d(x , z) \leq d(x, y) + d(y, z)$.

where $z = < z_n > \in H$.

$$d(x,z) = \left[\sum_{i=1}^{\infty} (x_i - z_i)^2 \right]^{1/2}$$

$$= [\Sigma \{(x_i - y_i) + (y_i - z_i)\}^2]^{1/2}$$
$$\leq [\Sigma (x_i - y_i)^2]^{1/2} + [\Sigma (y_i - z_i)^2]^{1/2},$$

by Minkowski inequality

$$= d(x, y) + d(y, z).$$

The above facts prove that d is a metric on H. The space (H, d) is called Hilbert space or l_2-space.

Theorem 7(a):

Let (x, ρ) be a metric space, Show that ∅ and X are open sets in X.

Proof :

Let (X, ρ) be a metric space.

To prove that ∅ and X are open sets .

To prove that ∅ is open in X, it suffices to show that

any $x \in \Phi \Rightarrow \exists\ \varepsilon > 0$, s.t. $S_{\varepsilon\,(x)} \subset \varnothing$.

Since ∅ does not contain any element and hence this condition is automatically fulfilled.

Corresponding to every point x ∈ X, ∃ an open sphere with its centre at x which is contained in X, showing there by X is an open .

Theorem 7(b):

In a metric space (X, ρ), ∅ and X are closed sets.

Proof :

Let (X, ρ) be a metric space.

To prove that ∅ and X are closed sets.

∵ $D(\varnothing) = \varnothing \subset \varnothing$

∴ $D(\varnothing) \subset \varnothing$

∵ This ⇒ ∅ is a closed set.

All the limiting points of X belong to X. For X is the universal set.

i.e., any $x \in D(X) \Rightarrow x \in X$

∵ $D(X) \subset X$.

This ⇒ X is a closed set.

Theorem 8:

Let (X, d) be a metric space. Let $\langle x_n \rangle$ be a Cauchy sequence in X and $\langle y_n \rangle$ be a sequence in X s.t.

$$d(x_n, y_n) < \frac{1}{n}\ \forall\ n \in N.$$

Then show that

(i) $\langle y_n \rangle$ is a Cauchy sequence,

(ii) $\langle y_n \rangle$ converges to x iff $\langle x_n \rangle$ converges to x.

Proof :

Let $\langle x_n \rangle$ be a Cauchy sequence. Then given $\varepsilon > 0$.

$$\exists\ n_0 \in N \text{ s.t. } m, n \geq n_0$$

$$\Rightarrow \quad d(x_n, x_m) < \varepsilon/3 \qquad ...(1)$$

Choose $m_0 \in N$ s.t. $\dfrac{1}{m_0} < \dfrac{\varepsilon}{3}$.

Then $\quad n \geq m_0 \Rightarrow \dfrac{1}{n} \leq \dfrac{1}{m_0} < \dfrac{\varepsilon}{3}$.

or, $\quad d(x_n, y_n) < \dfrac{1}{n} \leq \dfrac{1}{m_0} < \dfrac{\varepsilon}{3}$.

or, $\quad d(x_n, y_n) \leq \dfrac{\varepsilon}{3}\ \forall\ n \geq m_0 \qquad ...(2)$

Choose $k_0 = \max\{n_0, m_0\}$. Then $\forall$ n, m $\geq k_0$, we have

$$d(y_n, y_m) \leq d(y_n, x_n) + d(x_n, x_m) + d(x_m, y_m)$$

$$< \frac{\varepsilon}{3} + \frac{\varepsilon}{3} + \frac{\varepsilon}{3} \text{ by (1) \& (2)}$$

or, $\quad d(y_n, y_m) < \varepsilon$,

Showing that $\langle y_n \rangle$ is a Cauchy sequence.

(ii) Let $\lim x_n = x$, then $\exists\ m_1 \in N$ s.t.

$$d(x_n, x) < \varepsilon/3\ \forall\ n \geq m_1 \qquad ...(3)$$

But $\quad d(y_n, x) \leq d(y_n, x_n) + d(x_n, x) \qquad ...(4)$

Choose $\quad m_2 = \max\{m_1, m_0\}$

Using (2) and (3), in (4).

$$d(y_n, x) < \frac{\varepsilon}{3} + \frac{\varepsilon}{3} = \frac{2\varepsilon}{3} < \varepsilon\ \forall\ n \geq m_2$$

This $\Rightarrow \lim y_n = x$

Similarly, $y_n \to x \Rightarrow x_n \to x$.

Theorem 9:

Any sequence in a metric space converges to the unique point.

Or, A sequence in a metric space can converge to almost one point.

Or, If $\langle a_n \rangle$ is a sequence in a metric space (X, ρ) s.t. $a_n \to a_0$, $a_n \to a_0'$, then $a_0 = a_0'$.

Proof :

Let $\langle a_n : n \in N \rangle$ be a sequence in metric space (X, ρ) s.t.

$a_n \to a_0, a_n \to a_0'$,

To prove that $a_0 = a_0'$.

Suppose the contrary.

Then $a_0 \neq a_0'$.

Let $\rho(a_0, a_0') = 2r$.

Then $S_r(a_0) \not\in S_r(a_0') = \phi$.

$\because$ $a_n \to a_0, a_n \to a_0'$

$\therefore$ Given any $r > 0$, $\exists\ n_0, n_0' \in N$ s.t.

$$\forall\ n \geq n_0 \Rightarrow a_n \in S_r(a_0)$$

$$\forall\ n \geq n_0' \Rightarrow a_n \in S_r(a_0')$$

Taking $k_0 \Rightarrow \max\{n_0, n_0'\}$, we obtain

$$\forall\ n \geq k_0 \Rightarrow a_n \in S_r(a_0)\ a_n \in S_r(a_0')$$

$$\Rightarrow a_n \in S_r(a_0) \cap S_r(a_0')$$

$$\Rightarrow S_r(a_0) \cap S_r(a_0') \neq \phi.$$ A contradiction.

Hence our initial assumption $a_0 \neq a_0'$ is wrong. Consequently

$$a_0 = a_0'.$$

Theorem 10(a):

Let $\langle x_n \rangle$ be a sequence in a metric space (X, d) and the sets A_n be defined by

$$A_1 = \{x_1, x_2, x_3, \ldots\},$$

$$A_2 = \{x_2, x_3, \ldots\},$$

$$A_n = \{x_n, x_{n+1}, \ldots\},\ \forall n \in N.$$

Then the sequence $< x_n >$ is a Cauchy sequence iff $d(A_n) = 0$ as $n \to \infty$.

Proof :

Let $\langle x_n \rangle$ be a sequence in a metric space (X, d) and let

$$A_n = \{x_n, x_{n+1}, \ldots\},\ \forall\ n \in N$$ so that

and $$\left.\begin{aligned} n < m &\Rightarrow A_m \subset A_n \\ r \geq n &\Rightarrow x_r \in A_n \end{aligned}\right\} \qquad \ldots(1)$$

Step (i) : Let $\langle x_n \rangle$ be a Cauchy sequence so that given any

$\varepsilon > 0$, $\exists\ n_0 \in N$ s.t.

$$n, m \geq n_0 \Rightarrow d(x_n, x_m) < \varepsilon \qquad \ldots(2)$$

To prove that $d(A_n) = 0$ as $n \to \infty$.

Let $n \geq n_0$, then

$$d(A_n) = \sup \{d(x_r, x_s) : x_r, x_s \in A_n\}$$
$$= \sup \{d(x_r, x_s) : r, s \geq n\} < \varepsilon, \text{ by (1), (2)}$$

$\therefore \quad d(A_n) < \varepsilon \ \forall\ n \geq n_0$

This $\Rightarrow \lim_{n \to \infty} d(A_n) = 0$

Step (ii) : Let us suppose $\lim_{n \to \infty} d(A_n) = 0$, so that given any

$\varepsilon > 0, \exists\ n_0 \in N$ s.t.

$n \geq n_0 \Rightarrow d(A_n) < \varepsilon$

$\Rightarrow \quad d(A_{n0}) < \varepsilon.$

To prove that $< x_n >$ is a Cauchy sequence, it is enough to prove that

$d(x_r, x_s) < \varepsilon \ \forall\ r, s \geq n_0$.

Now, $r, s \geq n_0$

$\Rightarrow \quad x_r, x_s \in A_{n0}$, by (1)

$\Rightarrow \quad d(x_r, x_s) < \varepsilon$

$\because \quad d(A_{n0}) < \varepsilon$

Theorem 10(b):

In any metric space, any closed sphere is a closed set.

Proof:

Let $Sr_{0\ [x0]}$ denote a closed sphere in a metric space (X, d).

To prove that $Sr_{0\ [x0]}$ is a closed set.

For this we must show that $S'\, r_0\, [x_0]$ is open in X.

Let $x \in S'\, r_{0\ [x0]}$ be arbitrary,

$x \in S'\, r_{0\ [x0]} \Rightarrow x \notin S'\, r_{0\ [x0]}$

$\Rightarrow d(x, x_0), <, \neq r_0,$

$\because Sr_{0\ [x0]} = \{ y \in X : d(y, x_0) \leq r_0\}$

$\Rightarrow d(x, x_0) > r_0$

$\Rightarrow d(x, x_0) - r_0 > 0$

$\Rightarrow r > 0$, on taking $r = d(x, x_0) - r_0$ ...(1)

We claim $S_{r\,(x)} \subset S'\, r_{0\ [x0]}$.

Let $y \in S_{r\,(x)}$ be arbitrary,

so that, $d(y, x) < r$.

$\therefore \quad d(x, x_0) \le d(x, y) + d(y, x_0)$.

$\therefore \quad d(y, x_0) \ge d(x, x_0) - d(x, y)$.

$> d(x, x_0) - r = r_0$ on using (1)

$\therefore \quad d(y, x_0) > r_0$

$\Rightarrow y \notin Sr_{0\,[x0]}$

Thus, any $y \in S_{r\,(x)} \Rightarrow y \in S'r_{0\,[x0]}$

This$\Rightarrow$ $S_{r\,(x)} \subset S'r_{0\,[x0]}$

$\therefore$ Given any $x \in S'r_{0\,[x0]}$, $\exists\, r > 0$ s.t. $S_{r\,(x)} \subset S'r_{0\,[x0]}$

This proves that $S'r_{0\,[x0]}$ is open in x.

Theorem 11:

In any metric space X, the empty set ϕ and the full space X are open sets.

Proof:

To show that ϕ is open, we must show that each point in ϕ is the center of an open sphere contained in ϕ; but since there are no points in ϕ, this requirement is automatically satisfied. X is clearly open, since every open sphere centered on each of its points is contained in X.

We have seen that [0, 1) is not open as a subset of the real line. However, if we consider [0, 1) as a metric space X in its own right, as a subspace of the real line, then [0, 1) is open as a subset of X, since from this point of view it is the full space.

This apparent paradox disappears when we realize that points outside of a given metric space have no relevance to any discussion taking place within the context of that space. A set is open or not open only with respect to a specific metric space containing it, never on its own.

Our next theorem justifies the adjective in the expression "open sphere."

Theorem 12(a):

Let X be a metric space. A subset G of X is open $\Leftrightarrow$ it is a union of open spheres.

Proof:

We assume first that G is open, and we show that it is a union of open spheres. If G is empty, it is the union of the empty class of open spheres. If G is non-empty, then since it is open, each of its points is the center of an open sphere contained in it, and it is the union of all the open spheres contained in it.

We now assume that G is the union of a class S of open spheres. We must show that G is open. If S is empty, then G is also empty, and by Theorem A, G is open. Suppose that S is non-empty. G is also non-empty.

Let a; be a point in G. Since G is the union of the open spheres in S, x belongs to an open sphere $S_r(x_0)$ in S. By Theorem B, x is the center of an open sphere $S_{r_1}(x) \subseteq S_r(x_0)$. Since $S_r(x_0) \subseteq G$, $S_{r_1}(x) \subseteq G$ and we have an open sphere centered on x and contained in G. G is therefore open.

The fundamental properties of the open sets in a metric space are those stated in

Theorem 12(b):

Every non empty open set $G \subset R$ is the union of a countable collection of pairwise disjoint open intervals.

Proof:

Let G be a non-empty open subset of R and $x \in G$ be arbitrary. Then $\exists$ an interval I_x with centre x s.t. $x \in G$.

Let C_x denote the union of all such open intervals then $C_x = \cup \{I_x : x \in I_x \subset G\}$. Evidently C_x is the largest open interval containing x s.t. $x \in C_x \subset G$.

Moreover this also suggests that $G = \cup C_x$.

$\forall$ x, y $\in C_y$ we have either

$C_x \cap C_y = \phi$ or $C_x \cap C_y \neq \phi$

Consider the case $C_x \cap C_y \neq \phi$.

Evidently $C_x \cap C_y$ is an open interval containing both x and y

s.t. $C_x \subset C_x \cup C_y \subset G$,

$$C_y \subset C_x \cup C_y \subset G.$$

By def. of C_x, and C_y, $C_x \cup C_y \subset C_x$,

$$C_x \cup C_y \subset C_y$$

Hence $\quad C_x = C_x \cup C_y = C_y$

or $\quad C_x = C_y$

$\therefore \forall$ x, y $\in$ G

we have either $\quad C_x \cap C_y = \phi$

or $\quad C_x = C_y$

s.t. $G = \cup C_x$.

Second Part: It remains to show that the intervals C_x form a countable collection. Let $G = \{C_x\}$ be a collection of open intervals s.t. $C_n \neq \Phi \ \forall \ n$ and $C_n \cap C_m = \phi$ for $n \neq m$.

To prove that G is countable.

We know that each C_n contains an infinite number of rationals. Let K be the set of all rationals belonging to $\cup C_n$. Define a map $f : K \to G$ s.t.

$f(x) = C_x$ where $C_x \in G$ s.t. $x \in C_x$.

For each $x \in K$, there is unique open interval C_x containing x, the rational number. Hence f is a well defined map. For each $C_a \in G$, has at least one rational $a \in K$ s.t. $f(a) = C_a$, showing thereby f is onto.

Also $f(x) = f(y) \Rightarrow C_x = C_y$

where $x \in C_x, y \in C_y \Rightarrow x = y$

Now f is one-one onto. Hence $K \sim G$. But K is countable. Hence G is countable.

For $K \subset Q$, Q is countable $\Rightarrow$ K is countable.

Theorem 13:

Let (X, d) be a metric space and let $G \subset X$ be an arbitrary set. Show that G is open $\Leftrightarrow$ it is a union of open spheres.

Proof:

Let (X, d) be a metric space. Let $G \subset X$ be an open set.

To prove that G is a union of open spheres.

If $G = \varnothing$, the proof is trivial,

If $G \neq \varnothing$, then by definition of an open set

any $x \in G \Rightarrow \exists \ r \in R^+$ s.t. $S_r(x) \subset G$ ...(1)

In view of this statement we are justified to write.

$$G = \bigcup_i \{S_{ri}(x_i) : x_i \in G\}$$

where each $r_i \in R^+$ is s.t. $S_{ri}(x_i)$ satisfies the condition (1).

Conversely suppose that (X, d) is a metric space. Let $G \subset X$ be expressible as a union of open spheres, *i.e.*, $G = \bigcup_i S_{ri}(x_i)$.

To prove that G is an open set.

Let $x \in G$ be arbitrary.

Then x is a centre of some open sphere, say.

$$S_{(x,\, r)} \in \{S_{ri(xi)}\}$$

Then $S_{r\,(x)} \subset \bigcup_i S_{ri\,(xi)} = G,$

or $S_{r\,(x)} \subset G,$

$\therefore$ any $x \in G \Rightarrow \exists\, r \in R^+$ s.t. $S_{r\,(x)} \subset G.$

By definition this proves that G is an open set.

Theorem 14:

Let (X, d) be a metric space. Prove that

(i) an arbitrary union of open sets is open in X.

(ii) any finite intersection of open sets is open.

Proof:

Let (X, d) be a metric space.

Let $G_i \subset X$ be open X $\forall\, i \in N$

Let $G = \bigcap_{i=1}^{\infty} G_i,\ G = \bigcap_{i=1}^{n} G_i$

To prove that

(i) G is open in X.

(ii) G is open in X.

(i) By Theorem 7, each G_i is a union of open spheres and hence G is the union of all open spheres arising in this way.

Hence g is open in X.

(ii) If G $= \Phi$, then G is open in X. For Φ is open in X.

If $G \neq \Phi$, then $\exists\, x \in \bigcap_{i=1}^{n} G_i$ so that

$x \in G_i$ for $1 \le i \le n.$

$x \in G_i$, G_i is open

$\Rightarrow$ $\exists\, r_i \in R^+$ s.t. $S_{r\,(x)} \subset G_i$ ...(1)

This is true for each i (= 1, 2, ..., n).

Take $r = \min \{r_1, r_2 ..., r_n\}$

Then $r \le r_i$ for $1 \le i \le n.$

This $\Rightarrow S_{r\,(x)} \subset S_{ri\,(x)}$ for $1 \le i \le n$

$\Rightarrow S_{r\,(x)} \subset S_{ri\,(x)} \subset G_i$ by (1)

$\Rightarrow S_{r\,(x)} \subset G_i$

$\Rightarrow S_{r\,(x)} \subset G = \bigcap_{i=1}^{n} G_i = G \Rightarrow S_{r\,(x)} \subset G$

$\therefore$ any $x \in G \Rightarrow \exists\, r \in R^+$ s.t. $S_{r\,(x)} \subset G$

By definition, this $\Rightarrow$ G is open in X.

Remark: The intersection of an infinite number of open sets may or may not be open. If we take an open set,

$A_n = (-1/n\ ,\ 1/n)$, then

$\bigcap_{n=1}^{\infty} A_n = \{0\} =$ closed set]

Theorem 15:

In any metric space (X, d) , show that

(i) an arbitrary intersection of closed sets is closed.

(ii) any; finite union on closed sets is closed.

Proof:

Let (X, d) be a metric space

Let $F_i \subset X$ be a closed set $\forall\ i \in N$

Let $F = \bigcap_{i=1}^{\infty} F_i,\ F = \bigcup_{i=1}^{n} F_i$

To Prove that F and F are closed sets in X.

By theorem 4, F_i is closed $\Leftrightarrow F_i'$ is open.

By theorem 8, $\bigcup_{i=1}^{\infty} F_i', \bigcap_{i=1}^{n} F_i'$ are open in X. ...(1)

By, De Morgan's law, $\bigcup_{i=1}^{\infty} F_i', = (\bigcap_{i=1}^{\infty} F_i) = F'$

and $\bigcap_{i=1}^{\infty} F_i' = (\bigcup_{i=1}^{\infty} F_i) = F'$

From (1), F' , F' are open in X

i.e., F, F are closed in X. [by Theorem 4]

The union of an infinite number of closed sets may or may not be closed.

Let $F_n = \left[0, \frac{n}{n+1}\right]$

Then $\bigcup_{i=1}^{\infty} F_n = [0,\ 1)$ =semi open set

$\therefore \quad \bigcup_{n=1}^{\infty} F_n$ is not closed, though each F_n is closed.

Examples:

To find the boundary of set of integers Z and set of rational Q.

$Z^o = \cup \{ G \subset R : G \text{ is open and } G \subset Z \} = \phi$

(1) For every subset of R contains fractions also.

Similarly $\quad (R - Z)^o = \phi$

$$b(Z) = R - Z^o \cup (R - Z)^o = R - \phi \cup \phi = R$$

or, $\quad b(Z) = R$

Similarly $\quad b(Q) = R$

(2) Give two examples of limit points.

(i) If $A = \left\{1 + \frac{1}{n} : n \in N\right\},$

$$A = \left\{2, \frac{3}{2}, \frac{4}{3}, \frac{5}{4}, \frac{6}{5}, \ldots\right\}, \text{then}$$

1 is limit point of A for $\lim_{n \to \infty} \left(1 + \frac{1}{n}\right) = 1.$

(ii) If $A = \left\{1, \frac{1}{2}, \frac{1}{3}, \frac{1}{4}, \ldots\right\} = \left\{\frac{1}{n} : n \in N\right\},$

then 0 is the limit point of A.

For $\lim_{n \to \infty} \frac{1}{n} = 0.$

(3) Since N has no limit points and so D (N) = ϕ*. For every open sphere centred at any point of N other than that point.*

$$\overline{N} = N \cup D(N) = N \cup \phi = N$$

or $\quad \overline{N} = N \quad$ *($\because$ N is closed.)*

so that N' is open and so $(N')^o = N'.$

$$ext(N) = (N')^o = N'$$

$$b(N) = [N^o \cup ext(N)]' = [\phi \cup ext(N)']$$

$$= (N')' = N$$

Note that N is non-dense set.

For $\quad (\overline{N})^o = N^o = \phi$

Theorem 16:

A subset in a metric space is open iff it is a nbd of each of its point.

Proof:

Let A be a subset of a metric space (X, d).

Step I : Given A is and of each of its points.

Aim : A is an open set. Recall that a set N is called nbd of a point $x \in X$ if $\exists$ open set

$G \subset X$ s.t. $x \in G \subset N$.

X s.t. $p \in G_p \subset A$.

It is true $\forall$ $p \in A$.

Take $A = \cup \; \{G_p : p \in G_p, G_p$ is open set, $G_p \subset A\}$

= An arbitrary union of open sets

= open set.

$\therefore$ A is an open set.

Step II : Let A be an open subset of X.

Aim : A is a nbd of each of its points. By assumption, we can write

$$p \in A \subset A \;\forall\; p \in A$$

This $\Rightarrow$ A is a nbd of each of its points.

Theorem 17:

In any metric space X, each closed sphere is a closed set.

Proof:

Let $S_r[x_0]$ be a closed sphere in X. By Theorem B, it suffices to show that its complement $S_r[x_0]'$ is open. $S_r[x_0]'$ is open if it is empty, so we may assume that it is non-empty. Let a; be a point in $S_r[x_0]'$. Since $d(x, x_0) > r$, $r_1 = d(x, x_0) - r$ is a positive real number. We take r_1 as the radius of an open sphere Sr_1 (x) centered on x, and we show that $S_r[x_0]'$ is open by showing that $S_{r_1}(x) \subseteq S_r[x_0]'$. Let y be a point in $S_{r_1}(x)$, so that $d(y, x) < r_1$. On the basis of this and the fact that $d(x_0, x) < d(x + y) + d(y, x)$, we see that $d(y, xa_0) > d(x, x_0) - d(y, x) > d(x, x_0) - n = d(x, x_0) - [d(x, x_0) - r] = r$, so that y is in $S_r[x_0]'$.

The main general facts about closed sets are those given in our next theorem.

Theorem 18(a):

A subset A of metric subspace (Y, d) of a metric space (X, d) is open in Y iff $\exists$ an open set G in (X, d) s.t. $A = Y \cap G$.

Proof:

Let $A \cap Y$ and let (Y, d) be a subspace of a metric space (X, d).

Then $A \subset Y \subset X$ ...(1)

Step I : Let A be open in Y. ...(2)

Aim : $A = Y \cap G$ for some open set G in X. By def. of open set,

$\forall\ y \in A \Rightarrow \exists$ open sphere $S_{1\,(Y,\,r)}$ in s.t.

$$S_{1\,(Y,\,r)} \subset A \quad ...(3)$$

Then A is expressible as

$$A = \cup\ \{S_{1\,(Y,\,r)} : y \in A\} \quad ...(4)$$

Suppose S(Y, r) is an open sphere in X.

Then $S_{1\,(Y,\,r)} = Y \cap S(Y, r)$

Now (4) is expressible as

$$A = \cup\ \{S_{(Y,r)} \cap Y : y \in A\}$$
$$= Y \cup [S_{(Y,r)} : y \in A\}$$

or, $A = Y \cap G,$

where $G = \cup\ \{S_{(Y,r)} : y \in A\}$

= arbitrary union of open spheres in X

= open set in X.

$\therefore$ $A = Y \cap G$, where G is open in X.

For each open sphere is an open set.

Step II : Let $A = Y \cap G$ ...(5)

where G is an open set in X.

Aim : A is open in Y.

By def. of open set in X,

$\forall\ x \in G \subset\ \Rightarrow \exists$ open sphere $S_{(x,\in)}$ in X. s.t.

$$S_{(x,\in)} \subset G \quad ...(6)$$

Write $S_{1(x,\hat{I})} = Y \cap S_{(x,\hat{I})}$ ...(7)

Then $S_{1(x,\in)}$ is an open sphere in Y.

By (6), $S_{(x,\in)} \cap Y \subset G \cap Y$

Using (7), we get $S_{1(x,\in)} \subset G \cap Y = A$, by (5)

or, $x \in S_{1(x,\in)} \subset A$

This $\Rightarrow$ A is open in Y, be def. of open set.

Theorem 18(b):

The necessary and sufficient condition that a subset A of a metric subspace (Y, d) of a metric space (X, d) be closed in (Y, d) is that ∃ a set F closed in (X, d) s.t.

$$A = Y \cap F.$$

Proof:

Firstly we shall prove a lemma :

A is open in Y ⇔ A Y ∪ G, G is open in X.

Now we come to the proof of the actual theorem.

F is closed in X ⇒ X – F is open in X.

A is closed in Y ⇔ Y – A is open in Y.

Be the lemma,

$$\therefore \qquad Y - A = Y \cap (X - F)$$

or, $Y - Y \cap (X - F) = A$

or, $Y - (Y \cap X - Y \cap F) = A$

or, $Y - (Y - Y \cap F) = A$

or, $Y \cap F = A$

Theorem 19(a):

If A_1 is an open set in (X_1, d) and A_2 is an open set in (X_2, d_2), then $A_1 \times A_2$ is open in the product metric space (X,d), where $X = X \times X_2$.

Proof:

Let $x = (x_1, x_2) \in A_1 \times A_2$ be arbitrary, then $x_1 \in A_1$, $x_2 \in A_2$. By def. of open set,

∃ open spheres $S(x_1, r_1)$, $S(x_2, r_2)$ s.t.

$$\left.\begin{array}{l} S(x_1, r_1) \subset A_1 \\ S(x_2, r_2) \subset A_2 \end{array}\right\} \qquad \text{...(1)}$$

Let $r = \min. \{r_1, r_2\}$, then $r \le r_1$, $r \le r_2$.

$$\therefore \qquad S(x_1, r) \subset S(x_1, r_1) \subset A_1$$

$$S(x_2, r) \subset S(x_2, r_2) \subset A_2$$

This $\Rightarrow S(x_1, r) \times S(x_2, r_1) \subset A_1 \times A_2$

$$\Rightarrow x = (x_1, x_2) \in S(x, r)$$

$$\Rightarrow S(x_1, r) \times S(x_2, r_1) \subset A_1 \times A_2$$

$$\Rightarrow x \in S(x, r) \subset A_1 \times A_2.$$

It is true $\forall\ x \in A_1 \times A_2$

$\therefore$ $A_1 \times A_2$ is open subset of (X, d).

Here S (x, r) is an open sphere in (X, d).

Theorem 19(b):

Let $a = (a_1, a_2, \ldots, a_n)$, $b = b_1, b_2, \ldots, b_n)$ be two n tuples or real or complex numbers, then

$$\|a+b\| \le \|a\| + \|b\|$$

$$\left[\sum_{i=1}^{n}|a_i + b_i|^2\right]^{1/2} \le \left(\sum_{i=1}^{n}|a_i|^2\right)^{1/2} + \left(\sum_{i=1}^{n}|b_i|^2\right)^{1/2}$$

Proof:

We have $\|a + b\|^2 = \sum_i |a_i + b_i|^2$,

or $\|a\|^2 = \sum_i |a_i|^2$

$\|b\|^2 = \sum_i |b_i|^2$

where $\sum_i$ *means* $\sum_{i=1}^{n}$

$$\|a + b\|^2 = \sum_i |a_i + b_i|^2 = \sum_i |a_i + b_i|.|a_i + b_i|$$

$$\le \sum_i |a_i + b_i|.(|a_i + b_i|)$$

$$\le \sum_i |a_i + b_i|.\ |a_i| + \sum_i |a_i + b_i|.\ |b_i|$$

$$\le \sum_i |(a_i + b_i)a_i| + \sum_i |(a_i + b_i)b_i|$$

Applying Cauchy Schwarz inequality to each sigma or R.H.S., we get

$$\|a + b\|^2 \le \left(\sum_i |a_i + b_i|^2\right)^{1/2} \left(\sum_i |a_i^2|\right)^{1/2}$$

$$+ \left(\sum_i |a_i + b_i|^2\right)^{1/2} . \left(\sum_i |b_i^2|\right)^{1/2}$$

$$= \|a+b\| . \|a\| + \|a+b\| . \|b\|$$

Dividing by $\|a+b\|$, we get

$$\|a+b\| \leq \|a\|+\|b\|$$

Theorem 20:

Let (X, d) be metric space and $A \subset X$. Show that

$\overline{A} = \{x \in X : d(x, A) = 0\}$.

Proof:

Let (X, d) be metric space and $A \subset X$.

We claim $\overline{A} = \{x \in X : d(x, A) = 0\}$

Suppose $d(x, A) = \in > 0$

Then (i) $x \notin A$ and

(ii) $(S \in/2_{(x)} - \{x\}) \cap A = \Phi$

Evidently (ii) $\Rightarrow x \notin D(A)$.

Thus, $d(x, A) = \notin$

$\Rightarrow x \notin A, x \notin D(A)$

$\Rightarrow x \notin A \cup D(A) = \overline{A}$

$\Rightarrow x \in \overline{A}$

Finally, $d(x, A) = \varepsilon \Rightarrow x \notin \overline{A}$

Again $d(x, A) = 0 \Rightarrow$

(iii) $x \in A$,

(iv) $(S_{r(x)} - \{x\}) \cap A \neq \Phi$

(iv) $\Rightarrow x \in D(A)$

Finally, $d(x, A) = 0 \Rightarrow x \in A$ or $x \in D(A)$

$\Rightarrow x \in A \cup D(A) = \overline{A} \Rightarrow x \in \overline{A}$

Thus, we have shown that

$d(x, A) = 0 \Rightarrow x \in \overline{A}$

and $d(x, A) = \varepsilon > 0 \Rightarrow x \notin \overline{A}$

Furthermore, $d(x, A) = \geq 0$.

Consequently the required result follows :

Theorem 21:

Let (X, d) be a metric space. A subset A of X is closed if given any $x \in X - A$, $d(x, A) \neq 0$.

Proof :

Let (X, d) be a metric space and $A \subseteq X$ be an arbitrary closed set.

To prove that

given any $x \in X - A$, $d(x, A) \neq 0$

A is closed $\Rightarrow$ $X - A$ is open.

By definition of open set,

any $x \in X - A \Rightarrow \exists\, r \in R^+$ s.t. $S_{r(x)} \subset X - A \Rightarrow S_{r(x)} \cap A = \Phi$

$\Rightarrow d(x, A) \geq r \Rightarrow d(x, A) \neq 0$.

Conversely let A be any subset of a metric space (X, d). Let

any $x \in X - A \Rightarrow d(x, A) \neq 0$

To prove that A is closed.

Let $x \in X - A$ be arbitrary so that, by assumption,

$$d(x, A) = r \neq 0 \Rightarrow S_{r(x)} \cap A = \phi$$

$\Rightarrow$ $S_{r(x)} \subset X - A$

$\therefore$ $x \in X - A \Rightarrow \exists\, r \in R^+$ s.t. $S_{r(x)} \in X - A$.

By definition, this $\Rightarrow \lambda - A$ is open

$\Rightarrow$ A is closed.

Theorem 22(a):

Let (X, d) be a metric space and let

$$d'(x, y) = \min\{1, d(x, y)\}.$$

Then d' is a metric for X.

Proof:

Let (X, d) be a metric space so that

(i) $d(x, y) \geq 0$

(ii) $d(x, y) = d(y, x)$

(iii) $d(x, y) = 0$ iff $x = y$

(iv) $d(x, y) + d(y, x) \geq d(x, z)$, where $x, y, z \in X$.

Let $d'(x, y) = \min, \{1, d(x, y)\}$.

Then $$d'(x,y) = \begin{cases} 1 \\ or\ d(x,y) \end{cases} \quad ...(1)$$

(2) $d'(x, y) \geq 0$, according to (1) and (i),

(3) $d'(x, y) = d'(y, x)$.

For if $d'(x, y) = 1$, then $d'(y, x) = 1$ so that $d'(x, y) = d'(y, x)$.

Again if $d'(x, y) = d(x, y)$, then $d'(y, x) = d(y, x)$ using (ii) and combining the last two, we get

$$d'(x, y) = d'(y, x)$$

$\therefore$ In either case, $d'(x, y) = d'(y, x)$

(4) $d'(x, y) = 0$ iff $x = y$.

For $d'(x, y) = 0 \Leftrightarrow d(x, y) = 0$ according to (1)

$\Leftrightarrow x = y$ according to (iii).

(5) To prove $d'(x, y) + d'(y, z) \geq d'(x, z)$

If $d'(x, y) = 1$, then $d'(x, z) = 1$, $d'(y, z) = 1$ and

so $d'(x, y) + d'(y, z) > d'(x, z)$

If we take $d'(x, y) = d(x, y)$, then

$d'(y, z) = d(y, z)$,

$d'(x, z) = d(x, z)$

and so $d'(x, y) + d'(y, z) = d(x, y) + d(y, z)$

$\geq d(x, z) = d', (x, z)$ d by (iv)

or, $d'(x, y) + d'(y, z) \geq d'(x, z)$

$\therefore$ (5) is true in both cases.

From what has been it follows that d' is a metric on X.

Theorem 22(b):

If $1 < p < \infty$ *and* $\frac{1}{p} + \frac{1}{q} = 1$, *then*

$ab \leq \frac{a^p}{p} + \frac{b^q}{q}$ *where a, b are two non-negative reals.*

Proof :

if $a = 0$ or $b = 0$ the result is obvious.

Now consider the case in which $a \neq 0$, $b \neq 0$.

Let $f(t) = 1 - \lambda + \lambda t - t^\lambda$ for $0 < \lambda < 1$.

Then $f'(t) = \lambda - \lambda t^{\lambda - 1}$

$f'(t) = 0 \Rightarrow \lambda(1 - t^{\lambda-1})$,

where $\lambda, \lambda - 1 \neq 0$

$\Rightarrow t = 1$

$f''(t) = -\lambda(\lambda - 1)t^{\lambda - 2}$.

At $t = 1$, $f''(t) = -\lambda(\lambda - 1) = +$ ive as $0 < \lambda < 1$.

Thus, $\frac{df}{dt} = 0, \frac{d^2f}{dt^2} >$ at $t = 1$.

This proves that f is minimum at t = 1.

$\therefore$ $f(1) \le f(t)$

This $\Rightarrow 1 - \lambda + \lambda - 1 \le 1 - \lambda + \lambda t - t^\lambda$

or $1 - \lambda + \lambda t - t^\lambda \ge 0$.

or $(1 - \lambda) + \lambda t \ge t^\lambda$ (Put $\lambda = 1/p$)

or $\left(1 - \frac{1}{p}\right) + \frac{t}{p} \ge \frac{t^1}{p}$

But $\frac{1}{p} + \frac{1}{q} = 1$

Hence $\frac{1}{q} + \frac{t}{p} \ge \frac{t^1}{p}$. Put $t = \frac{a^p}{b^q}$.

Then $\frac{1}{q} + \frac{a^p}{pb^q} \ge \frac{a}{b^{q/p}}$

or $\frac{b^q}{q} + \frac{a^p}{p} \ge ab^{q-q/p} = ab$.

or $\frac{a^p}{p} + \frac{b^q}{q} \ge ab$

as $\frac{1}{p} + \frac{1}{q} = 1 \Rightarrow q - \frac{q}{p} = 1$

Theorem 22(c):

Let A be a subset of a metric space (X, d).

Then A is non-dense in X $\Leftrightarrow (\overline{A})' = X - \overline{A}$ *is dense in X.*

Proof :

Recall that a is dense in X if $\overline{A} = x$ and A is nowhere dense in X if $\text{int}(\overline{A}) = \phi$.

A is non-dense (or no where dense) in X

$\Leftrightarrow \text{int}(\overline{A}) = \phi \Leftrightarrow \exists$ no open nbd N of any point of $\overline{A}$ s.t. $N \subset \overline{A}$

$\Leftrightarrow \overline{A}$ contains no nbd. $\Leftrightarrow S_r(x) \not\subset \overline{A}\ \forall\ r > 0\ \&\ \forall\ x \in X$

$\Leftrightarrow S_r(x) \cap (\overline{A})' \ne \phi$

$\Leftrightarrow$ x is adherent point of $(\overline{A})'\ \forall\ x \in X$

$\Leftrightarrow [(\overline{A})'] = X = \Leftrightarrow (\overline{A})'$ is dense in X.

Theorem 22(d):

In any metric space, prove that difference between the two sides of a triangle is less than the third side.

Or

Let (X, d) be a metric space and let x, y, z be any three points of X. Then $|d(x, z) - d(y, z)| \le d(x, y)$

Proof :

Let x, y, z denote the co-ordinates of vertices of a triangle.

If we show that $|d(x, z) - d(y, z)| \le d(x, y)$, the result will follow.

We know that

$$d(x, y) + d(y, z) \ge d(x, z)$$

Similarly $d(y, x) + d(x, z) \ge d(y, z)$

From Which $d(x, y) \ge d(x, z) - d(y, z)$ ...(1)

and $d(y, x) \ge -[d(x, z) - d(y, z)]$

The last gives $-d(x, y) \le d(x, z) - d(y, z)$ ...(2)

For $d(x, y) = d(y, x)$.

Combining (1) and (2),

$$-d(x, y) \le d(x, z) - d(y, z) \le d(x, y)$$

This $\Rightarrow |d(x, z) - d(y, z)| \ge d(x, y)$.

Theorem 23(a):

Let (X, d) be a metric space, then $\forall\, x, y, x', y'$, *in X,* $|d(x, y) - d'(x' y')| \le d(x, x') + d(y, y')$.

Proof:

For proving the required result, we make repeated use of "triangle inequality."

$$d(x, y) \le d(x, x') + d(x', y)$$

$$\le d(x, + d(x', y') + d(y',$$

or, $d(x, y) - d(x', y') \le d(x, + d(y',$ ...(1)

as $d(y, = d(y',$

Again $d(x', y') \le d(x', + d(x, y')$

$$\le d(x', + d(x, y) + d(y,$$

Using the fact that

d (a, b) = d (b, a),

we get $\quad$ d (x', y') – d (x, y) ≤ d (x, + d (y,

or, $\;$ d (x, y) – d (x', y') ≥ – [d (x, + d (y, $\quad$ ...(2)

Combining (1) and (2), we get,

– [d (x, + d (y, ≤ d (x, y) – d (x', y') ≤ d (x, + d (y,

or, $\;$ |d (x, y) – d (x', y')| ≤ d (x, + d (y,

Theorem 23(b):

Let (X, d) be a metric space. If $\langle x_n \rangle$ and $\langle y_n \rangle$ are sequences in X, such $x_n \to x$, $y_n \to y$, then show that

$$d(x_n, y_n) \to d(x, y).$$

roof :

Suppose $\langle x_n \rangle$ *and* $\langle y_n \rangle$ are sequences in a metric space (X, d).

Let $x_n \to x$, $y_n \to y$.

To prove that

$$d(x_n, y_n) \to d(x, y).$$

Let x, y, z ∈ X be arbitrary.

Firstly we shall prove that

$$| d(x, z) - (y, z)| \le d(x, y) \qquad ...(3)$$

Now write the proof of previous theorem .

$\because \quad x_n \to x,\ y_n \to y.$

$\therefore \quad \varepsilon > 0,\ \exists\ n_0, m_0, \in N$ s.t.

$\forall\ n \ge n_0 \Rightarrow d(x_n, x) < \varepsilon/2$

$\forall\ n \ge m_0 \Rightarrow d(y_n, y) < \varepsilon/2$

Take $\quad k_0 = \max. \{m_0, n_0\}$

Then $\quad d(x_n, x) < \varepsilon/2,\ d(y_n, y) < \varepsilon/2\ \forall\ n \ge k_0$

Now $\quad |d(x_n, y_n) - d(x, y)|$

$= |d(x_n, y_n) - d(x_n, y) + d(x_n, y) - d(x, y)|$

$\le |d(x_n, y_n) - d(x_n, y)| + |d(x_n, y) - d(x, y)|$

$= |d(y_n, x_n) - d(y, x_n)| + |d(x_n, y) - d(x, y)|$

$\le d(y_n, y) + d(x_n, x)$, according to (3)

$$< \varepsilon/2 + < \varepsilon/2 = \varepsilon$$

$\therefore$ $|d(x_n, y_n) - d(x, y)| < \varepsilon/2 \ \forall \ n \geq k_0$

This $\Rightarrow d(x_n, y_n) \to d(x, y)$.

Theorem 23(c):

Let A be a subset of a metric space (X, d). Then $x_0 \in X$ is a limit point of A iff every open sphere centred at x_0 contains an infinitely many; points of A, other than x_0.

Proof:

Let a be a subset of a metric space (X, d) and let $x_0 \in X$.

Step I : Given x_0 is a limit point of A.

Aim. $S_{r(x0)}$ contains an infinite number of points of A other than x_0 ...(*)

By def. of limit point,

$$(S_{r(x0)}) - \{x_0\}) \cap A \neq \phi \qquad ...(1)$$

for every open sphere centred at x_0.

Suppose (*) is not true.

Then $S_{r(x0)}$ contains a finite number of points

$$x_1, x_2, \ldots, x_n, \in A \qquad ...(2)$$

Let $r = \min \{d(x_0, x_i) : = 1, 2. \ldots, n\}$

and $S_{r(x0)} = \{x \in X : (x, x_0) < r_0\}$

Evidently $(S_{r(x0)}) - \{x_0\}) \cap A = \phi$

This $\Rightarrow x_0$ is not a limit point of A.

This is contrary to what is given . This contradiction arises by assuming that $S_{r(x0)}$ contains a finite number of points of A other than x_0. Hence the required result follows.

Step II: Given $S_{r(x0)}$ contains an infinite number of points of A other than x_0.

Aim. x_0 is a limit point of A.

By assumption, $(S_{r(x0)}) - \{x_0\}) \cap A \neq \phi$

This $\Rightarrow x_0$ is a limit point of A.

Theorem 23(d):

Let (X, ρ) be a metric space.

Define $d(x, y) = \dfrac{\rho(x, y)}{1 + \rho(x, y)} \ \forall \ x, y, \in X.$

Then (X, y) is metric space.

Proof:

Let (X, ρ) be a metric space

Let $d(x, y) = \dfrac{\rho(x, y)}{1+\rho(x, y)} \; \forall \; x, y, \in X.$

To prove that (X, d) is a metric space.

Let x, y, z, ∈ X be arbitrary.

∵ (X, ρ) is a metric space.

∴ (i) $\rho(x, y) \geq 0$

(ii) $\rho(x, y) = \rho(y, x)$

(iii) $\rho(x, y) = 0$ iff $x = y$

(iv) $\rho(x, y) + \rho(y, z) \geq \rho(x, z)$.

Now (i)' $d(x, y) \geq 0$ in accordance with (i).

(ii)' $d(x, y) = d(y, x)$.

For $d(y, x) = \dfrac{\rho(y, x)}{1+\rho(y, x)}$

$$= \frac{\rho(x, y)}{1+\rho(x, y)} = d(x, y) \text{ in accordance with (ii)}$$

(iii)' $d(x, y) = 0$, iff $x = y$

$$d(x, y) = 0 \Leftrightarrow \frac{\rho(x, y)}{1+\rho(x, y)} = 0$$

$$\Leftrightarrow \rho(x, y) = 0 \Leftrightarrow x = y, \text{ according to (iii).}$$

(iv)' $d(x, y) + d(y, z) \geq d(x, z)$.

For $d(x, y) + d(y, z) = \dfrac{\rho(x, y)}{1+\rho(x, y)} = \dfrac{\rho(y, z)}{1+\rho(y, z)}$

$$\geq \frac{\rho(x, y)}{1+\rho(x, y)+\rho(y, z)} + \frac{\rho(y, z)}{1+\rho(y, z)+\rho(x, y)}$$

$$\geq \frac{\rho(x, y)+\rho(y, z)}{1+\rho(x, y)+\rho(y, z)}$$

$$\geq \frac{\rho(x, z)}{1+\rho(x, z)}$$

$$= d(x, z)$$

i.e., d (x, y) + d (y, z) ≥ d (x, z).

The results (i)', (ii)', (iii)', (iv)' taken together prove the required result.

Theorem 24:

Let (X, y) be a metric space and let m a positive real number. Show that there exists a metric d' for X such that the metric space (X, y) is bounded with d(X) ≤ m.

Proof:

Let (X, y) be a metric space and m > 0.

we define d (x, y) = $\dfrac{md\,(x,y)}{1+d\,(x,y)}$ $\forall\ x, y \in X$.

To prove that (X) (X, d) is a metric space and m > 0.

Let x, y, x ∈ X be arbitrary.

d is metric ⇒ d (x, y) ∈ R

Also m ∈ R

⇒ d' (x, y) $\dfrac{md\,(x,y)}{1+d\,(x,y)} \in R.$

(X, d) is metric space

(i) d (x, y) ≥ 0

(ii) d (x, y) = d (y, x)

(iii) d (x, y) = 0 iff x = y

(iv) d (x, y) + d (y, z) ≥ d (x, y).

(1) Evidently d' (x, y) ≥ 0, according to (i) as m > 0.

(2) d' (x, y) = d' (y, x)

For d' (y, x) $\dfrac{md\,(y,x)}{1+d\,(y,x)}$

$$= \frac{md\,(x\,y)}{1+d\,(x,y)}$$, according to (ii)

= d' (x, y)

(3) d' (x, y) = 0 iff x = y.

For d' (x, y) = 0 ⇔ $\dfrac{md\,(x\,y)}{1+d\,(x,y)} = 0$

⇔ md (x ,y) = 0

$\Leftrightarrow$ d (x, y) = 0

$\Leftrightarrow$ x = y, according to (iii).

(4) d' (x, y) + d' (y, z) $\geq$ d' (x, z).

For $d'(x, y) + d'(y, z) = \frac{md(x,y)}{1+d(x,y)} + \frac{md(y,z)}{1+d(y,z)}$

$$= m\left[\frac{d(x,y)}{1+d(x,y)} + \frac{d(y,z)}{1+d(y,z)}\right]$$

$$\geq m\left[\frac{d(x,y)}{1+d(x,y)+d(y,z)} + \frac{d(y,z)}{1+d(x,y)+d(y,z)}\right]$$

$$= m\left[\frac{d(x,y)+d(y,z)}{1+d(x,y)+d(y,z)}\right] \geq \frac{md(x,z)}{1+d(x,z)}$$

= d' (x, z)

The results (1), (2), (3) and (4) taken together prove that (X, d') is metric space. Remains to prove that d' is bounded.

$$d'(x, y) = \frac{md(x,y)}{1+d(x,y)} = \frac{m}{1+\frac{1}{d(x,y)}}$$

d' (x, y) $\leq$ m $\forall$ x, y, $\in$ X.

This $\Rightarrow$ (x, d') is a bounded metric space.

If $d'(x, y) = \frac{d(x,y)}{1+d(x,y)}$

Then d' (x, y) $\leq$ 1 so that (X, d') is bounded.

Theorem 25:

If p > 1 and a_i, b_i are non-negative real numbers, then

$$\left[\sum_{i=1}^{n}(a_i + b_i)^p\right]^{1/p} \leq \left(\sum_{i=1}^{n} a_i^p\right)^{1/p} + \left(\sum_{i=1}^{n} b_i^p\right)^{1/p}$$

Proof :

By Holder's inequality,

$$\sum_{i=1}^{n} a_i + b_i \leq \left(\sum_{i=1}^{n} a_i^p\right)^{1/p} \left(\sum_{i=1}^{n} b_i^q\right)^{1/q} \quad \text{... (1)}$$

with $\frac{1}{p} + \frac{1}{q} = 1$ ... (2)

By (2), $q(p-1) = p$

$$\sum_{i=1}^{n} (a_i + b_i)^p = \sum_{i=1}^{n} (a_i + b_i)(a_i + b_i)^{p-1} \qquad ...(3)$$

where $\sum_{i=1}^{n} = \sum$

$$= \sum a_i (a_i + b_i)^{p-1} + \sum b_i (a_i + b_i)^{p-1}$$

Using (1) for each member of R.H.S.,

$$\sum (a_i + b_i)^p \le \left(\sum a_i^p\right)^{1/p} \left[\sum (a_i, b_i)^{q(p-1)}\right]^{1/q}$$

$$+ \left(\sum b_i^p\right)^{1/p} \left[\sum (a_i + b_i)^{q(p-1)}\right]^{1/q}$$

Using (3), $\sum (a_i + b_i)^p \le \left(\sum a_i^p\right)^{1/p} \left[\sum (a_i + b_i)^p\right]^{1/q}$

$$+ \left(\sum b_i^p\right)^{1/p} \left[\sum (a_i + b_i)^p\right]^{1/q}$$

Dividing by $\left[\sum (a_i + b_i)^p\right]^{1/q}$ we get

$$\left[\sum (a_i + b_i)^p\right]^{1-(1/q)} \le \left(\sum a_i^p\right)^{1/p} + \left(\sum b_i^p\right)^{1/p}$$

or, $\left[\sum_{i=1}^{m} (a_i + b_i)^p\right]^{1/p} \le \left(\sum a_i^p\right)^{1/p} + \left(\sum b_i^p\right)^{1/p}$

Theorem 26:

In any metric space X, the empty set ϕ and the full space X are closed sets.

Proof:

The empty set has no limit points, so it contains them all and is therefore closed. Since the full space X contains all points, it automatically contains its own limit points and thus is closed.

The following theorem characterizes closed sets in terms of open sets. We already know a good deal about open sets, so that characterization provides us with a useful tool for establishing properties of closed sets.

Theorem 27:

If a_i, b_i (i = 1, 2 ... n) are real or complex numbers, then

$$\left(\int_a^b fg\,dx \le \left(\int_a^b f^p\,dx\right)^{1/p} \left(\int_a^b g^q\,dx\right)^{1/q}\right)$$

$$\sum_{i=1}^{n} |a_i b_i| \le \left[\sum_{i=1}^{n} |a_i|^2\right]^{1/2} \left[\sum_{i=1}^{n} |b_i|^2\right]^{1/2}$$

Proof :

If we replace a_i, b_i , p, q by $|a_i|$, $|b_i|$, 2, 2 respectively, then we get this inequality. An independent proof is given below :

Write A = $(a_1, a_2, a_3, \ldots a_n)$, b = $(b_1 , b_2 , \ldots , b_n)$.

The magnitudes of these vectors are given by

$$\|a\|^2 = \sum_{i=1}^{n} |a_i|^2 \ \|b\|^2 \sum_{i=1}^{n} |b_i|^2$$

If a = 0 or b = 0, then the inequality reduces to equality.

So we suppose that a ≠ 0. b ≠ 0. We know that

Geometric ≤ mean Arithmetic mean.

This⇒ $\sqrt{(xy)} \le \dfrac{x+y}{2}$

Set $\sqrt{x} = \dfrac{|a_i|}{\|a\|}, \sqrt{y} = \dfrac{|b_i|}{\|b\|}$

Then $$\frac{2|a_i|.|b_i|}{\|a\|.\|b\|} \le \frac{|a_i|^2}{\|a\|^2} + \frac{|b_i|^2}{\|b\|^2}$$

or $$2\sum_{i=1}^{n} \frac{|a_i|.|b_i|}{\|a\|.\|b\|} \le \frac{\sum |a_i|^2}{\|a\|^2} + \frac{\sum |b_i|^2}{\|b\|^2} = 1+1=2$$

or $$\sum_i |a_i|.|b_i| \le \|a\|.\|b\|$$

or $$\sum_{i=1}^{n} |a_i b_i| \le \left(\sum_{i=1}^{n} |a_i|^2\right)^{1/2} \left(\sum_{i=1}^{n} |b_i|^2\right)^{1/2}$$

If a_i , b_i are real numbers, then

$$\sum_i |a_i\, b_i| \le \left(\sum_{i=1}^{n} a_i^2\right)^{1/2} \left(\sum_{i=1}^{n} b_i^2\right)^{1/2}$$

PROJECTIONS

A projection on a Banach space B is an idempotent operator on B, that is, an operator P with the property that P2 = P. It was proved in that section that each projection.

P determines a pair of closed linear subspaces M and N–the range and null space of P–such that $B = M \oplus N$, and also, conversely, that each such pair of closed linear subspaces M and N determines a projection P with range M and null space N.

In this way, there is established a one-to-one correspondence between projections on B and pairs of closed linear subspaces of B which span the whole space and have only the zero vector in common.

The context of our present work, however, is the Hilbert space H, and not a general Banach space, and the structure which H enjoys in addition to being a Banach space enables us to single out for special attention those projections whose range and null space are orthogonal. Our first theorem gives a convenient characterization of these projections.

Theorem 1:

A closed linear subspace M of H is invariant under an operator $T \Leftrightarrow M^{\perp}$ is invariant under T^*.

Proof:

Since $M^{\perp\perp} = M$ and $T^{**} = T$, it suffices by symmetry to prove that if M is invariant under T, then M1- is invariant under T*.

If y is a vector in M1, our conclusion will follow from $(x, T^*y) = 0$ for all x in M. But this is an easy consequence of $(x,T^*y) = (Tx, y)$, for the invariance of M under T implies that (Tx, y) – 0.

Theorem 2:

If P and Q are the projections on closed linear subspaces M and N of H, then $M \perp N \Leftrightarrow PQ = 0 \Leftrightarrow QP = 0$.

Proof:

We first remark that the equivalence of PQ = 0 and QP – 0 is clear by taking adjoints. If $M \perp N$, so that $N \subseteq M^{\perp}$, then the fact that Qx is in N for every x implies that PQx – 0, so PQ = 0. If, conversely, PQ = 0, then for every x in N we have Px = PQx = 0, so $N \subseteq M^{\perp} N$.

Motivated by this result, we say that two projections P and Q are orthogonal if PQ = 0. Our final theorem describes the circumstances under which a sum of projections is also a projection.

Theorem 3:

If $P_1, P_2, \ldots, P_n$ are the projections on closed linear sub-spaces $M_1, M_2, \ldots, M_n$ of H, then $P = P_1 + P_2 + \ldots +$ P» is a projection $\Leftrightarrow$ the P_i's are *pairwise orthogonal (in the sense that $P_iP_j = 0$ whenever $i \neq j$), and in this case, P is the projection on*

$$M = M_1 + M_2 + \ldots + M_n.$$

Proof:

Since P is clearly self-adjoint, it is a projection ⇔ it is idem-potent. If the P_i,s are pairwise orthogonal, then a simple computation shows at once that P is idempotent. To prove the converse, we assume that P is idempotent. Let a; be a vector in the range of P_i, so that $x = P_ix$. Then

$$\| x \|^2 = \| P_i x \|^2 \leq \sum_{j=1}^{n} \|P_j x\|^2 = \sum_{j=1}^{n} (P_j x, x) = (Px, x) = \| Px \|^2 \leq \| x \|^2.$$

We conclude that equality must hold all along the line here, so

$$\sum_{j=1}^{n} \|P_j x\|^2 = \| P_i x \|^2$$

and $\qquad \| P_j x \| = 0$ for $j \neq i$.

Thus the range of P_i is contained in the null space of P_j, that is, $M_i \subseteq M_j{}^{\perp}$L, for every $j \neq i$. This means that $M_i \perp M$, whenever $i \neq j$, and our conclusion that the P_i's are pairwise orthogonal now follows from the preceding theorem. We prove the final statement in two steps. First, we observe that since $\| Px \| = \| x \|$ for every x in M_i, each M_i is contained in the range of P, and therefore M is also contained in the range of P. Second, if x is a vector in the range of P, then

$$x = Px = P_1x + P_2x + \ldots + P_nx$$

is evidently in M.

There are many other ways in which the algebraic structure of the set of all projections on H can be related to the geometry of its closed linear subspaces, and several of these are given in the problems below. The significance of projections in the general theory of operators on H is the theme of the next chapter. As we shall see, the essence of the matter (the spectral theorem) is that every normal operator is made of projections in a way which clearly reveals the geometric nature of its action on the vectors in H.

Theorem 4:

If P is a projection on H with range M and null space N, then $M \perp N$ ⇔ P is self-adjoint; and in this case, $N = M^{\perp}$.

Proof:

Each vector 2 in H can be written uniquely in the form $z = x + y$ with x and y in M and N. If $M \perp N$, so that $x \perp y$, then $P^* = P$ will follow $(P^*z, z) = (Pz, z)$, and this is a consequence of

$(P^*z, z) = (z, Pz) = (z, x) = (x, + y, x) = (x, x) = (y, x) = (x, x)$ and $(Pz, z) = (x, z) = (x, x + y) = (x, x) + (x, y) = (x, x)$. If, conversely, $P^* = P$, then the conclusion that $M \perp N$ follows from the fact that for any x and y in M and N we have

$$(x, y) = (Px, y) = (x, P^*y) = (x, Py) = (x, 0) = 0.$$

All that remains is to see that if $M \perp N$, then $N = M^{\perp}$. It is clear that $N \subseteq M^{\perp}$; and if *N* is a proper subset of $M^{\perp}$, and therefore a proper closed linear subspace of the Hilbert space $M^{\perp}$, then Theorem 2, implies that there exists a non-zero vector z_0 in $M^{\perp}$ such that $z_0 \perp N$. Since $z_0 \perp N$ and $z_0 \perp N$, and since $H = M \oplus N$, it follows that $z_0 \perp H$. This is impossible, so we conclude that $N = M^{\perp}$.

A projection on H whose range and null space are orthogonal is sometimes called a perpendicular projection. The only projections considered in the theory of Hilbert spaces are those which are perpendicular, so it is customary to omit the adjective and to refer to them simply as projections. In the light of this agreement and Theorem 1, a projection on H can be defined as an operator P which satisfies the conditions $P^2 = P$ and $P^* = P$. The operators 0 and I are projections, and they are distinct $\Leftrightarrow H \neq \{0\}$.

The great importance of the projections on H rests mainly which allows us to set up a natural one-to-one correspondence between projections and closed linear subspaces. To each projection P there corresponds its range $M = \{Px : x \varepsilon H\}$, which is a closed linear subspace; and conversely, to each closed linear subspace M there corresponds the projection P with range M defined by $P(x + y) = x$, where x and y are in M and $M^{\perp}$. Either way, we speak of P as the projection on M.

It is clear that P is the projection on $M \Leftrightarrow I - P$ is the projection on $M^{\perp}$. Also, if P is the projection on M, then

$$x \varepsilon M \Leftrightarrow Px = x \Leftrightarrow \| Px \| = \| x \|.$$

The first equivalence here was proved in Problem 44-11; and since for every x in H we have

$$\| x \|^2 = \| Px \mid (I - P) x \|^2 = \| Px \|^2 + \| (I - P) x \|^2. \qquad ...(1)$$

the non-trivial part of the second is given by the following chain of implications:

$$\| Px \| = \| x \| \Rightarrow \| P^* \|^2 = \| * \|^2$$

$$\Rightarrow \qquad \| (I - P) x \|^2 = 0 \Rightarrow Px = x.$$

Relation (20) also shows that $\| Px \| \leq \| x \|$ for every x, so $\| P \| \leq 1$. If x is an arbitrary vector in H, it is easy to see that

$$(Px, x) = (PPx, x) = (Px, P(x) = (Px, Px) = \| px \|^2 \geq 0. \qquad ...(2)$$

so P is a positive operator ($0 \leq P$) in the sense. Since $I - P$ is also a projection, we also have $0 \leq I - P$ or $P \leq I$, so $0 \leq P \leq I$. Let T be an operator on H. A closed linear subspace M of H is said to be invariant under T if $T(M) \subseteq M$. When this happens, the restriction of T to M can be regarded as an operator on M alone, and the action of T on vectors outside of M can be ignored. If both M and Mx are invariant under T, we say that M reduces T, or that T is reduced by M. This situation is much more interesting, for it allows us to replace the study of *T* as a whole by the study of its restrictions to *M* and $M^{\perp}$, and it invites the hope that these restrictions will turn out to be operators of some particularly simple type. In the following four theorems, we translate these concepts into relations between T and the projection on *M.*

SOLVED EXAMPLES

Example 1:

Let l_∞ denote the set of all bounded sequences $\langle x_n \rangle$. If $x = \langle x_n \rangle$, $y = \langle y_n \rangle$ in l_{∞}, then we define

$$d(x, y) = \sup \{|x_n - y_n| : n \varepsilon N\}.$$

Show that d is a metric on l_∞.

Solution:

(1) $d(x, y) = \sup |x_n - y_n|$

$$\leq \sup |x_n| + \sup |y_n| < \infty$$

or d (x, y) is finite. For $<x_n>$ and $<y_n>$ both are bounded.

This $\Rightarrow$ d (x, y) is well defined map.

(2) Evidently $d(x, y) \geq 0$.

(3) $d(x, y) = 0$ iff $x = y$

For $x = y \Leftrightarrow x_n = y_n \ \forall n \Leftrightarrow |x_n - y_n| = 0 \ \forall n$

$$\Leftrightarrow \sup |x_n - y_n| = 0 \Leftrightarrow d(x, y) = 0$$

(4) $d(x, y) = d(y, x)$.

For $|x_n - y_n| = |y_n - x_n|$

(5) $d(x, y) \leq d(x, z) + d(z, y)$

For $d(x, y) = \sup |x_n - y_n| = \sup |x_n - z_n + z_n - y_n|$

$$\leq \sup [|x_n - z_n| + |z_n - y_n|]$$

$$\leq \sup |x_n - z_n| + \sup |z_n - y_n|$$

$$= d(x, z) + d(z, y)$$

$\therefore$ (l_∞, d) is a metric space.

Example 2:

Every derived set in a metric space is a closed set.

Solution:

Let (X, d) be a metric space and $A \subset X$. Let x be a limit point of D (A) . If we show that $x \in D(A)$, the result will be proved.

x is limit point of D (A) $\Rightarrow (S_r(x) - \{x\} \cap D(A) \neq \varnothing$ ...(1)

for every open sphere $S_r(x)$.

Then $\exists\, y_r \neq x \in S_r(x) \cap D(A)$

i.e. $y_r \in D(A)$, $y_r \in S_r(x)$

$\therefore$ $d(x, y) < r$.

Take $r - d(x, y) = r_1 > 0$.

Evidently $S_{r1}(y_r) \subset S_r(x)$ ...(2)

$$\left[\begin{aligned} For\ z \in S_{r1}(y_r) &\Rightarrow d(z, y_r) < r_1 = r - d(x, y_r) \\ &\Rightarrow d(x, y_r) + d(y_r, z) < r \\ &\Rightarrow d(x, z) < r \Rightarrow z \in S_r(x) \end{aligned}\right]$$

$y_r \in D(A) \Rightarrow S_{r1}(y_r) \cap A \neq \varnothing \ \forall\ S_{r1}(y_r)$

$\Rightarrow$ $S_r(x) \cap A \neq \varnothing$, by (2)

$\Rightarrow$ $\in D(A)$.

Example 3(a):

Prove that the mapping defined by $d : R \times R \to R$ *such that*

$$d(x,y) = \begin{cases} o \text{ if } x = y \\ 1 \text{ if } x \neq y \end{cases}$$

is a metric on R.

Solution:

Let x, y, z $\in$ R be arbitrary and let

$$d(x,y) = \begin{cases} o \text{ if } x = y \\ 1 \text{ if } x \neq y \end{cases}$$

Now (i) $d(x, y) \geq 0$

For $d(x, y) = 0$ or 1.

(ii) $d(x, y) = 0$ or iff $x = y$

(iii) $d(x, y) = d(y, x)$

For if d (x, y) = 0. Then d (y, x) = 0 so that d (x, y) = d (y, x); and if d (x, y) = 1, then d (y, x) = 1 so that d (x, y) = d (y, x)

∴ In either case d (y, x) = d (x, y)

(iv) To prove d (x, y) + d (y, z) ≥ d (x, z).

If x = z, then d (x, z) = 0, so that

$$d(x, y) + d(y, z) \geq 0 = d(x, z).$$

If x ≠ z, then d (x, z) = 1 and if y is any real number, then y is at least different from one of x and z ad hence at least one of d (y, x) and d(y, z) be equal to one. Consequently,

$$d(x, z) \leq d(x, y) + d(y, z)$$

The above facts prove that d is a metric on R.

Example 3(b):

Prove that the function $d : R^2 \times R^2 \to R$ *s.t.*

$$d(x, y) = [(x_1 - y_1)^2 + (x_2 - y_2)^2]^{1/2},$$

where $x = (x_1, x_1) = (y_2 - y_2) \in R^2$

Is a metric on R.

[This is called unusual metric on R^2*]*

Solution:

Let $x = (x_1, x_2)$, $y = (y_1, y_2)$, $z = (z_1, z_2)$ be arbitrary elements of R^2

(i) d (x, y) ≥ 0.

For $(x_1 - y_1)^2 \geq 0$, $(x_2 - y_2)^2 \geq 0$ and hence

$$[(x_1 - y_1)^2 + (x_2 - y_2)^2]^{1/2} \geq 0$$

(ii) d (x, y) = 0 iff x = y

$$d(x, y) = 0$$

$$\Leftrightarrow (x_1 - y_1)^2 + (x_2 - y_2)^2 = 0$$

$$\Leftrightarrow x_1 - y_1 = 0, \ x_2 - y_2 = 0$$

$$\Leftrightarrow x_1 = y_1, \ x_2 = y_2$$

$$\Leftrightarrow (x_1, x_2) = (y_1, y_2)$$

$$\Leftrightarrow x = y$$

(iii) d (x, y) = d (y, x).

For $[(x_1 - y)^2 + (x_2 - y_2)^2]^{1/2} = [(y_1 - x_1)^2 + (y_2 - x_2)^2]^{1/2}$

(iv) d (x, y) + d (y, z) ≥ d (x, z)

For $d(x, y) + d(y, z) \geq [(x_1 - y_1)^2 + (x_2 - y_2)^2]^{1/2}$
$+ [(y_1 - z_1)^2 + (y_2 - z_2)^2]^{1/2} \geq [(x_1 - y_1 + y_1 - z_1)^2$
$+ (x_2 - y_{2\,+} y_2 - z_2)^2]^{1/2}$ be Minkowski inequality.
$= [(x_1 - z_1)^2 + (x_2 - z_2)^2]^{1/2} = d(x, z)$

$\therefore d(x, y) + d(y, z) \geq d(x, z)$.

Example 4:

Given an example of two closed subset A and B of the real line r such that d (A, B) = 0, but $A/B\ \varnothing$.

Solution:

Let d be the usual metric on R. Write

$$N_1 = \{2, 3, 4, 5 \ldots\} = \{n + 1 : n \in N\}$$

$$A = \{2, 3, 4, 5 \ldots\},$$

$$B = \left\{2\frac{1}{2}, 3\frac{1}{3}, 4\frac{1}{4}, 5\frac{1}{5}\right\}$$

Then A and B are expressible as

$$A = \{n : n \in N_1\},$$

$$B = \left\{n + \frac{1}{n} : n \in N_1\right\}$$

If $n \in A$, then $n + 1/n \in B$ and the distance between $n + 1/n$ and n is 1/n, which $\to 0$ as $n \to \infty$.

$$d(A, B) = \inf\{d(x, y) : x \in A, y \in B\}$$
$$= \inf\{|x - y| : x \in A, y \in B\}$$
$$= \inf\{1/n : n \in A, n + 1/n \in B\} = 0.$$

Thus, d (A, B) = 0. Also $A\ çB = \varnothing$.

Therefore A and B are the required sets.

Example 5:

Contor's set contains exactly as many points as the interval [0, 1].

Hint : To prove $|E| = |[0,1]|$.

Also $|[0,1]| = C$

E is a closed set.

Since an infinite disjoint union of open sets is open, Hence the family of removed open intervals form an open set.

E = $[0,1]$ – family of removed open intervals.

= $[0,1]$ – open set = closed set.

For complement of open set is a closed set.

Example 6(a):

Every set in a discrete metric space is open.

Solution:

Let (X,d) be a discrete metric space. Let x, y ∈ X be arbitrary.

By def. of discrete metric, $d(x, y) = \begin{cases} 1 \text{ if } x \neq y \\ 0 \text{ if } x = y \end{cases}$.

Let r be any positive real number s.t. r ≤ 1.

Then $Sr(x) = \{y \in X : d(y, x) < r \leq 1\}$

$= \{y \in X : d(y, x) < 1\}$

$= \{y \in X : d(y, x) = 0$, by def. of d.

$= \{y \in X : y = x\} = \{0\}$.

or, $S_{r(x)} = \{0\}$.

But every open sphere is an open set.

∴ {x} is an open set in X ∀ x ∈ X.

If A = $\{x_1, x_2, ..., x_n\}$ = finite set ⊂ X, then

$A = \bigcup_{r=1}^{n} \{x_r\}$ = finite union of open sets.

= open set.

Hence every finite sub set of X is open set. ...(1)

If $B = \{x_1, x_2, x_3, ...\} \subset X$, then

B is an infinite subset of X.

Now $B = \bigcup_{r=1}^{\infty} \{x_r\}$

= Arbitrary union of open sets

= open set, by Theorem 8.

∴ B is an open set, ...(2)

From (1) and (2), it follows that every subset (finite or infinite) is an open set in X.

Example 6(b):

In a discrete metric space, every subset of X is closed.

Solution:

We know that G is open

$\Leftrightarrow$ X – G = G' is closed set.

Let A be a subset of a discrete metric space (X, d), then A is open.

Also A' = X – A is again sub set of X and so A' is open. Consequently A is closed.

Thus every subset of X is open as well as closed.

Example 6(c):

A finite set in any metric space has no limit point.

Solution:

Let A be a finite subset of a metric space (X, d). We know that "x $\in$ X is a limit point of any set B if every open sphere $S_{r(x)}$ contains an infinite number of points of B other than x."

This condition can not be satisfied here as A is finite set.

Hence A has no limit point.

Example 7:

Given an example of Pseudo metric which is not a metric.

Solution:

Suppose C is the family of all Riemann integral functions f; [0, 1] $\rightarrow$ R.

Let f, g, h, $\in$ C be arbitrary. Define a map

d : C × C $\rightarrow$ R s.t. $d(f, g) = \int_0^1 |f - g|(x)\, dx.$

To prove d is a Pseudo metric but; not metric on C.

d (f, g) $\geq$ For $|f - g|$ (x) $\geq$ 0 and so.

$$\int_0^1 |f - g|(x)\, dx \geq 0$$

(1) d (f, g) = d (g, f).

This follows form the fact that $|f - g| = |g - f|$

(2) d (f, g) + d (g, f).

$$= \int_0^1 |f - g|(x)\, dx + \int_0^1 |g - h|(x)\, dx$$

$= \int_0^1 [|f - g| + |g - h|](x)\, dx$

$\geq \int_0^1 |f - g + g - h|(x)\, dx$

$= \int_0^1 |f - h|(x)\, dx = d(f, h)$

(3) d (F, g) = 0 if f = g

If f = g, then

d (f, g) = $\int_0^1 |f - g|(x)\, dx = \int_0^1 |g - g|(x)\, dx = \int_0^1 0\, dx = 0$

(4) d (f, g) = 0 does not necessarily imply f = g.

That is to say, d (f, g) = 0

$\Rightarrow$ f $\neq$ g, in general.

Solution:

Suppose $f(x) = \begin{cases} 4 \text{ where } x = 0 \\ 0 \text{ when } 0 < x \leq 1 \end{cases}$

and $g(x) = \begin{cases} 9 \text{ when } x = 0 \\ 0 \text{ when } 0 < x \leq 1 \end{cases}$

The f(x) $\neq$ g (x), so that f $\neq$ g.

Also $|f - g|(x) = \begin{cases} 9 - 4 = \text{ when } x = 0 \\ 0 \quad \text{ when } 0 < x \leq 1 \end{cases}$

Evidently f – g is Riemann integrable,

x = o is the only point of discontinuity.

$$d(f, g) = \int_0^1 |f - g|(x)\, dx = \int_0^1 0\, dx = 0$$

Thus, d (f, g) = 0 but f $\neq$ g.

This proves that d is a Pseudo metric but metric on C.

Ex : *Let C be the class of bounded continuous functions f : [0, 1 → R. Define a map d : C × C → R such that*

$$d(f, g) = \int_0^1 |f - g|(x)\, dx \;\forall f, g \in C = \|f - g\|$$

Then d is a metric on C.

Example 8:

In any metric space, show that

$x - \overline{A} = (X - A)^{\circ}$

or, $(\overline{A})' = (A')^{\circ}$

Solution:

$(\overline{A})' = X - \overline{A}$

$= X -$ Intersection of all closed super sets of A.

$= X - \bigcap_i F_i$ where F_i is closed and $F_i \supset A$.

$= \bigcup_i (X, F_i)$ where $X - F_i$ is open

and $X - F_i \subset X - A$

= Union of open subsets of $X - A = A'$

$= (A')^{o}$

Example 9:

Describe open spheres for the usual metric on R^2.

Solution:

Let d be the usual metric on R^2 so that d is given by

$$d(x, y) = [(x_1 - x_1)^2 + (y_1 - y_2)^2]^{1/2}$$

where $x = (x_1, y_1) \in R^2$, $y = (x_2, y_2) \in R^2$.

Let $z_0 = (x_0, y_0) \in R^2$ and $r > 0$ be any real number

$$S_r(z_0) = \{(x, y) \in R^2 : z = (x, y), d(z, z_0) < r\}$$

$$= \{(x, y) \in R^2 : z = \sqrt{[(x - x_0)^2 + (y - y_0)^2]} < r\}$$

$$= \{(x, y) \in R^2 : (x - x_0)^2 + (y - y_0)^2 < r^2\}$$

This proves that open sphere $S_r(z_0)$ is the interior of the circle $(x - x_0)^2 + (y - y_0)^2 = r^2$, where $z_0 = (x_0, y_0)$.

Example 10:

Describe open sphere of unit radius having the centre at (0, 0) for each of the following metric on R.

(i) $d_1(z_1, z_2) = |x - x_2| + |y_1 - y_2|$

(ii) $d_2(z_1, z_2) = max.\{|x_1 - x_2|, |y_1 - y_2|\}$

where $z_1 = (x_1, y_1)$, $z_2 = (x_2, y_2) \in R^2$.

Solution:

Let $z_0 = (0, 0)$, $z = (x, y) \in R^2$.

To determine $S_1(z_0)$.

Step (i) $S_1(z_0) = \{(x, y) \in R^2 : d_1(z, z_0) < 1\}$

$$= \{(x, y) \in R^2 : |x - 0| + |y - 0| < 1\}$$

$$= \{(x, y) \in R^2 : |x| + |y| < 1\}$$

This shows that the open sphere $S_1(z_0)$ is the interior of the square bounded by the lines

$$x + y = 1, -x - y = 1, -x + y = 1, x - y = 1.$$

Step (ii) $S_1(z_0) = \{(x, y) \in R^2 : d_1 (z, z_0) < 1\}$

$$= \{(x, y) \in R^2 : \max. \{|x - 0|, |y - 0|\} < 1\}$$

$$= \{(x, y) \in R^2 : \max. \{|x|, |y|\} < 1\}$$

This proves that the open sphere $S_1(z_0)$ is the interior of the square bounded by the lines

$x = 1, -x = 1, y = 1. -y = 1.$

Example 11(a):

Show that the map $d : R^2 \times R^2 \to R$

define by $d (x, y) = \max. \{|x_1 - y_1|, |x_2 - y_2|\}$.

where $x = (x_1, x_2), y = (y_1, y_2) \in R_2$

is a metric on R^2.

Solution:

Let $x = (x_1, x_2)$, $y = (y_1, y_2)$, $z = (z_1, z_2)$ be arbitrary elements or R^2. Also let

$d (x, y) = \max \{|x_1 - y_1|, |x_2 - y_2|\}$

(i) $d (x, y) \geq 0$.

Since modulus of any quantity ;is non-negative.

(ii) $d (x, y) = 0$ iff $x = y$.

For $d (x, y) = 0 \Leftrightarrow \max.\{|x_1 - y_1|, |x_2 - y_2|\}= 0$

$$\Leftrightarrow |x_1 - y_1| = 0 = |x_2 - y_2|$$

$$\Leftrightarrow x_1 = y_1, |x_2 = y_2$$

$$\Leftrightarrow (x_{1,} x_2) =y_{2,} y_2)$$

$$\Leftrightarrow x = y$$

(iii) $d (x, y) = d (y, x)$.

For $|a - b| = |b - a|$ and so

$\max.\{|x_1 - y_1|, |x_2 - y_2|\}= \max \{|y_1 - x_1|, |y_2 - x_2|\}$

(iv) $d (x, y) + d (y, z) \geq d (x, z)$

For $d (x, y) = \max \{|x_1 - z_1|, |x_2 - z_2|\}$

$$= \max. \{|x_1 - y_1 + y_1 - z_1|, |x_2 - y_2 + y_2 - z_2|\}$$
$$\leq \max. \{|x_1 - y_1| + |y_1 - z_1|, |x_2 - y_2| + |y_2 - z_2|\}$$
$$\leq \max. \{|x_1 - y_1|, |x_2 - y_2|\} + \max. \{|y_1 - z_1|, |y_2 - z_2|\}.$$
$$= d(x, y) + d(y, z).$$

Example 11(b):

Let ρ_1 and ρ_2 be metrics on the same set X, show that there exists a metric ρ on X such that

$$\rho(x, y) \geq \rho_i(x, y) \text{ for } i = 1, 2$$

and $\forall x, y \in X$.

Solution:

Let ρ_1 and ρ_2 be metrics on the same set X, Let $x, y, z \in X$ be arbitrary.

To prove that $\exists$ a metric ρ on X s.t.

$$\rho(x, y) \geq \rho_1(x, y) \text{ for } i = 1, 2$$

Write $\rho(x, y) = \rho_1(x, y) + \rho_2(x, y)$. ...(1)

We claim ρ is metric on X.

(i) $\rho(x, y) \geq 0$

$\because$ $\rho_1(x, y) \geq 0$, $\rho_2(x, y) \geq 0$.

(ii) $\rho(x, y) = 0$ iff $x = y$

For $\rho(x, y) = 0 \Leftrightarrow \rho_1(x, y) + \rho_2(x, y) = 0$.

$$\Leftrightarrow \rho_1(x, y) = 0 = \rho_2(x, y).$$

Since ρ_1 and ρ_2 both are non-negative.

$$\Leftrightarrow x = y$$

(iii) $\rho(y, x) = \rho(x, y)$

For $\rho(y, x) = \rho_1(y, x) + \rho_2(y, x) = \rho_1(x, y) + \rho_2(x, y)$

$$\rho(x, y)$$

$\because$ $\rho_1(x, y) = \rho_1(y, x)$, $\rho_2(x, y) = \rho_2(y, x)$

(iv) $\rho(x, y) + \rho(y, z) \geq \rho(x, z)$

For $\rho(x, y) + \rho(y, z) = \rho_1(x, y) + \rho_2(x, y) + \rho_1(y, z) + \rho_2(y, z)$

$$[\rho_1(x, y) + \rho_1(y, z)] + [\rho_2(x, y) + \rho_1(y, z)]$$
$$\geq \rho_1(x, z) + \rho_2(x, z) = \rho(x, z).$$

From what has been done it follows that ρ is a metric on X.

Moreover (1) $\Rightarrow \rho(x, y) \geq \rho_i(x, y)$ for i = 1, 2.

Hence ρ is the required metric.

Example 11(c):

Consider the Hilbert space L_2 associated with the measure space $[0, 2\pi]$, where measure is Lebesgue measure and integrals are Lebesgue integrals. This space essentially consists of all complex functions f defined on $[0,2\pi]$ which are Lebesgue measurable and square integrable, in the sense that

$$\int_0^{2\pi} |f(x)|^2\, dx < \infty$$

Solution:

Its norm and inner product are defined by

$$\| f \| = \left(\int_0^{2\pi} |f(x)|^2\, dx\right)^{1/2}$$

and
$$(f, g) = \int_0^{2\pi} f(x)\overline{g(x)}\, dx .$$

A simple computation shows that the functions e^{inx}, for

$$n = 0, \pm 1, \pm 2, \ldots,$$

are mutually orthogonal in L_2:

$$\int_0^{2\pi} e^{imx} e^{-inx}\, dx = \begin{cases} 0 & m \neq n \\ 2\pi & m = n \end{cases}$$

It follows from this that the functions e_n $(n = 0, \pm 1, \pm 2, \ldots)$ defined by $e_n(x) = e^{inx} / \sqrt{2\pi}$ form an orthonormal set in L_2. For any function f in L_2, the numbers

$$c_n = (f, e_n) = \frac{1}{\sqrt{2\pi}} \int_0^{2\pi} f(x)\, e^{-inx}\, dx \qquad \ldots(9)$$

are its classical *Fourier coefficients*, and *Bessel's inequality* takes the form

$$\sum_{n=-\infty}^{\infty} |c_n|^2 \leq \int_0^{2\pi} |f(x)|^2\, dx .$$

It is a fact of very great importance in the theory of Fourier series that the orthonormal set $\{e_n\}$ is complete in L_2. As we have seen in Theorem F, the completeness of $\{e_n\}$ is equivalent to the assertion that for every f in L_2, Bessel's inequality can be strengthened to *Parseval's equation:*

$$\sum_{n=-\infty}^{\infty} |c_n|^2 = \int_0^{2\pi} |f(x)|^2\, dx$$

Theorem F also tells us that the completeness of $\{e_n\}$ is equivalent to the statement that each / in L_2 has a

Fourier expansion:

$$f(x) = \frac{1}{\sqrt{2\pi}} \sum_{n=-\infty}^{\infty} c_n e^{inx} \quad ...(10)$$

It must be emphasized that this expansion is not to be interpreted as saying that the series converges pointwise to the function. The meaning of (9) is that the partial sums of the series, that is, the vectors f_n in L_2 denned by

$$f_n(x) = \frac{1}{\sqrt{2\pi}} \sum_{k=-n}^{\infty} c_k e^{ikx}, \quad ...(11)$$

Converge to the vector f in the sense of L_2.

$$\| f_n - f \| \to 0.$$

This situation is often expressed by saying that f is the *limit in the mean* of the fn's. We add one final remark to our description of this portion of the theory of Fourier series. If f is an arbitrary function in L_2 with Fourier coefficients c_n defined by (6), then Bessel's inequality tells us that the series $\Sigma_{n=-\infty}^{\infty} |c_n|^2$ converges. The celebrated *Riesz-Fischer theorem* asserts the converse: if c, (n = 0, ± 1, ± 2, . . .) are given complex numbers for which $\Sigma_{n=-\infty}^{\infty} |c_n|^2$ converges, then there exists a function f in L_2whose Fourier coefficients are the c_n's. If we grant the completeness of L_2 as a metric space, this is very easy to prove. All that is necessary is to use the c_n's to define a sequence of f_n's in accordance with (10) . The functions $e^{inx}/\sqrt{2\pi}$ form an orthonormal set, so for m > n we have

$$\|f_m - f_n\|^2 = \sum_{|k|=n+1}^{m} |c_k|^2$$

By the convergence of $\Sigma_{n=-\infty}^{\infty} |c_n|^2$, the sum on the right of (9) can be made as small as we please for all sufficiently large n and all m > n. This tells us that the f_n's form a Cauchy sequence in L_2; and since L_2 is complete, there exists a function f in L_2 such that $f_n \to f$. This function f is given by (7), and the c_n's are clearly its Fourier coefficients. It is apparent from these remarks that the essence of the Riesz-Fischer theorem lies in the completeness of L_2 as a metric space.

We shall have use for one further item in the general theory of orthonormal sets, namely, the Gram-Schmidt orthogonalization process. Suppose that $\{x_1, x_2, \ldots, x_n, \ldots\}$ is a linearly independent set in a Hilbert space H. The problem is to exhibit a constructive procedure for converting this set into a corresponding orthonormal set $\{x_1, x_2, \ldots, x_n, \ldots\}$ with the property that for each n the linear subspace of H spanned by {ei, e2, . . . , en) is the same as that spanned by $\{x_1,$

$x_2, \ldots, x_n, \ldots\}$. Our first step is to normalize x_1 — which is necessarily nonzero — by putting

$$e_1 = \frac{x_1}{\|x_1\|}$$

The next step is to subtract from x_2 its component in the direction of e_1 to obtain the vector $x_2 — (x_2,e_1)\, e_1$ orthogonal to e_1, and then to normalize this by putting

$$e_2 = \frac{x_2 - (x_2, e_1)e_1}{\|x_2 - (x_2, e_1)e_1\|}$$

We observe that since x_2 is not a scalar multiple of x_1, the vector $x_2 - (x_2,e_1)\, e_1$ is not zero, so the definition of e_2 is valid. Also, it is clear that e_2 is a linear combination of x_1 and x_2, and that x_2 is a linear combination of e_1 and e_2. The next step is to subtract from x_3 its components in the directions of e_1 and e_2 to obtain a vector orthogonal to e_1 and e_2, and then to normalize this by putting

$$e_3 = \frac{x_3 - (x_3, e_1)e_1 - (x_3, e_2)e_2}{\|x_3 - (x_3, e_1)\, e_1 - (x_3, e_2)e_2\|}$$

If this process is continued in the same way, it clearly produces an orthonormal set $\{e_1, e_2, \ldots, e_n, \ldots)$ with the required property.

Example 12:

If A is any non-empty subset of a metric space (X, d), prove that

$$|d(x, A) - d(y, A)| \leq d(x, y)$$

for any points $x, y \in X$.

Solution:

By def. of metric,

$$d(x, a) \leq d(x, y) + d(y, a)$$

and $d(y, a) \leq d(y, x) + d(x, a)$

Taking infimum over $a \in A$, we obtain

$$d(x, A) \leq d(x, y) + d(y, A)$$

and $$d(y, A) \leq d(y, x) + d(x, A)$$

i.e., $$d(x, A) - d(y, A) \leq d(x, y) \qquad \ldots(1)$$

$$d(y, A) - d(x, A) \leq d(x, y)$$

The last gives

$$d(x, A) - d(y, A) \geq -d(x, y)$$

From (1) and (2),

$$-d(x, y) \leq d(x, A) - d(y, A) \leq d(x, y)$$

$\Rightarrow |d(x, A) - d(y, A)| \leq d(x, y)$.

Example 13:

Let A, B be non-empty subsets of a metric space (X, d). Then show that

$$d(A \cap B) \leq d(A) + d(B) + d(A, B)$$

Solution:

Let $x, y \in A \cap B$ be arbitrary. Then there arise four cases.

(i) $x, y \in A$

This $\Rightarrow d(x, y) \leq d(A)$

For $d(A) = \sup \{d(x, y) : x, y \in A\}$

(ii) $x, y \in B$.

This $\Rightarrow d(x, y) \leq d(B)$

(iii) $x \in A$ and $y \in B$

(iv) $x \in B$ and $y \in A$

In case (iii), we have

$$d(x, y) \leq d(x, a) + d(a, y), \quad a \in A$$
$$\leq d(x, a) + d(a, b) + d(b, y), \; b \in B$$

$\therefore \quad d(x, y) \leq d(x, a) + d(a, b) + d(b, y)$.

Similarly in case (iv), we have

$$d(x, y) \leq d(x, b) + d(b, a) + d(a, y)$$

$\Rightarrow \quad d(A \cup B) \leq d(B) + d(B, A) + d(A)$

or $\quad d(A \cup B) \leq d(A) + d(B) + d(A, B)$.

Example 14:

Describe open spheres for the discrete metric on a space X.

Solution:

Let (X,d) be a discrete metric space so that d is given by

$$d(x,y) = \begin{cases} 0 \text{ if } x = y \\ 1 \text{ if } x \neq y \end{cases}$$

Let $x_0 \in X$ and $r > 0$ be any real number.

$$S_r(x_0) = \{x \in X : (x, x_0) < r\}.$$

Case (i) : When $r > 1$.

By definition metric, any $x \in X$, $x \neq x_0$

$$\Rightarrow d(x, x_0) = 1 < r$$

$$\Rightarrow d(x, x_0) = < r$$

$$\Rightarrow x \in S_r(x_0)$$

Also $x_0 \in S_r(x_0)$

$\because$ $d(x_0, x_0) = 0 < r$.

Combining these two results, we find that

any $x \in X \Rightarrow x \in S_r(x_0)$

This $\Rightarrow X = S_r(x)$

Case (ii) : when $r \leq 1$.

any $x \in X$, $x \neq x_0 \Rightarrow d(x, x_0) = 1 < r$

$$\Rightarrow d(x, x_0) < r.$$

$$\Rightarrow x \notin S_r(x_0)$$

$$x_0 \in S_r(x_0)$$

Thus, any $x \in X$, $x \neq x_0 \Rightarrow x \notin S_r(x_0)$

and $x_0 \in S_r(x_0)$.

From this it follows that $S_r(x_0) = \{x_0\}$.

Finally, $S_r(x_o) = \begin{cases} X & \text{if } r . 1. \\ \{x_0 & \text{if } r \leq 1.\} \end{cases}$.

Example 15(a):

Let d be a Pseudo metric on X. Define a relation on X by setting X~y iff d (x, y) = 0. Show that this relation is an equivalence relation whose corresponding equivalence class can be made into a metric space in a natural way.

Solution:

Let d be a Pseudo metric on X so that

$\forall$ x, y, z, $\in$ X we have

(1) $d(x, y) ^3 0$

(2) $d(x, y) = d(y, x)$

(3) $d(x, x) = 0$

(4) $d(x, y) + d(y, z) ^3 d(x, z)$

Define a relation ~ on X s.t.

$$x \sim y \text{ iff } d(x, y) = 0$$

To prove that ~ is an equivalence relation

(i) ~ is reflexive

For $\quad d(x, x) = 0$

$\Rightarrow \quad x \sim x$, by (3)

(ii) ~ is symmetric.

For $\quad x \sim y$

$\Rightarrow \quad d(x, y) = 0$

$\Rightarrow \quad d(y, x) = 0$, by (2)

$\Rightarrow y \sim x$

(iii) ~ is transitive.

For $\quad x \sim y, y \sim z$

$\Rightarrow \quad d(x, y) = 0 = d(y, z)$

$\Rightarrow \quad d(x, z) \leq (x, y) + d(y, z) = 0 + 0 = 0$

$\Rightarrow \quad d(x, z) \leq 0$

But $\quad d(x, z) \geq 0$, by (1)

$\Rightarrow \quad d(x, z) = 0$

$\Rightarrow \quad x \sim z$

Thus we have proved that ~ is an equivalence relation of X.

Hence the relation ~ decomposes X into family of disjoint subsets of X, known as equivalence classes.

Let [X] denote the equivalence class determined by $x \in X$ and let C be the set of all equivalence classes,

i.e., $\quad C = \{[x] : x \in X\}$

Define a map $\quad f : C \times C \to R$ s.t.

$$f([x_1], [x_2]) = d(x_1, x_2)$$

where $\quad [x_1], [x_2] \in C$

Then f is a metric for C on account of the fact that d is a pseudometric on X. The metric f defined as above is called natural metric for C.

Example 15(b):

Let C be the set of all real valued bounded continuous functions defined on [0, 1].

If d is defined by

d (f, g0 = sup {|f (x) – g (x) | : x ∈ [0, 1]}

then d is a metric on C,

where f, g∈ C.

Solution:

Let f, g, h ∈ C and x ∈ [0, 1] be arbitrary. Then

(1) d (f, g) ≥ 0. For mod of any thing ≥ 0.

(2) d (f, g) = d (g, f). For |f (x) – g (x)| = |g (x) – f (x) |.

(3) d (f, g) = 0 ⇔ f = g

For sup {|f (x) – g (x) | : x ∈ [0, 1]} = 0

⇔ |f (x) – g (x) | = 0

⇔ f (x) – g (x) ∀ x

⇔ f = g

(4) To prove d (f, h) ≤ d (f, g) + d (g, h)

Since |f – h| = |f – g + g – h|

≤ |f – g| + |g – h|

This ⇒ sup |f (x) – h (x)| ≤ sup |f (x) – g (x)| + sup |g (x) – h(x)|

⇒ d (f, h) ≤ d (f, g) + d (g, h)

From what has been it follows that d is a metric on C.

Example 16:

Describe open spheres for the usual metric on R^2.

Solution:

Let d be the usual metric on R^2 so that d is given by

$$d (x, y) = [(x_1 - x_1)^2 + (y_1 - y_2)^2]^{1/2}$$

where $x = (x_1, y_1) \in R^2$, $y = (x_2, y_2) \in R^2$.

Let $z_0 = (x_0, y_0) \in R^2$ and $r > 0$ be any real number

$$S_r(z_0) = \{(x, y) \in R^2 : z = (x, y),\ d (z, z_0) < r\}$$

$$= \{(x, y) \in R^2 : z = \sqrt{[(x - x_0)^2 + (y - y_0)^2]} < r\}$$

$= \{(x, y) \in R^2 : (x - x_0)^2 + (y - y_0)^2 < r^2\}$

This proves that open sphere $S_r(z_0)$ is the interior of the circle $(x - x_0)^2 + (y - y_0)^2 = r^2$, where $z_0 = (x_0, y_0)$.

Example 17:

Show that C the set of complex numbers is metric space under the metric d defined as follows:

$$d(z_1, z_2) = \frac{|z_1 - z_2|}{\left[1 + |z_1|^2\right]^{1/2}\left[1 + |z_2|^2\right]^{1/2}}$$

Solution:

Let z_1, z_2, z_3 be arbitrary complex numbers . It is given that

$$d(z_1, z_2) = \frac{|z_1 - z_2|}{\left[1 + |z_1|^2\right]^{1/2}\left[1 + |z_2|^2\right]^{1/2}}$$

To prove that (C, d) is a metric space.

We know that $|z| \geq 0$ for every complex number z.

If view of this, (1) says that $d(z_1, z_2) \geq 0$. ...(1)

Since $|z_1 - z_2| = |z_2 - z_1|$. In view of this, (1) $\Rightarrow$

$$d(z_1, z_2) = d(z_2, z_1) \quad ...(2)$$

$$d(z_1, z_2) = 0 \Leftrightarrow \frac{|z_1 - z_2|}{\left[1 + |z_1|^2\right]^{1/2}\left[1 + |z_2|^2\right]^{1/2}} = 0$$

$$\Leftrightarrow |z_1 - z_2| = 0 \Leftrightarrow z_1 = z_2$$

$\therefore$ $\quad d(z_1, z_2) = 0 \Leftrightarrow z_1 = z_2$ ...(3)

$d(z_1, z_2) + d(z_2, z_3)$

$$= \frac{|z_1 - z_2|}{\left[1 + |z_1|^2\right]^{1/2}\left[1 + |z_2|^2\right]^{1/2}} + \frac{|z_2 - z_3|}{\left[1 + |z_2|^2\right]^{1/2}\left[1 + |z_3|^2\right]^{1/2}}$$

$$\geq \frac{|z_1 - z_2|}{\left[1 + |z_1|^2\right]^{1/2}\left[1 + |z_2|^2\right]^{1/2}\left[1 + |z_3|^2\right]^{1/2}}$$

$$+ \frac{|z_2 - z_3|}{\left[1 + |z_1|^2\right]^{1/2}\left[1 + |z_2|^2\right]^{1/2}\left[1 + |z_3|^2\right]^{1/2}}$$

$$= \frac{|z_1 - z_2| + |z_2 - z_3|}{\left[1+|z_1|^2\right]^{1/2}\left[1+|z_2|^2\right]^{1/2}\left[1+|z_3|^2\right]^{1/2}}$$

$$\geq \frac{|z_1 - z_3|}{\left[1+|z_1|^2\right]^{1/2}\left[1+|z_2|^2\right]^{1/2}\left[1+|z_3|^2\right]^{1/2}}$$

$$\geq \frac{|z_1 - z_3|}{\left[1+|z_1|^2\right]^{1/2}\left[1+|z_3|^2\right]^{1/2}} = d\left(z_{1,}z_3\right)$$

or $d(z_1, z_2) + d(z_2, z_3) \geq d(z_1, z_3)$ ...(4)

From (1), (2), (3) and (4), it follows that (C, d) is a metric space.

Example 18(a):

Show that the function d defined by d (p, d) = $\|p - q\|$, where p, q are vectors in a normal space V, is a metric on V.

Solution:

If p, q, r ∈ V and K is any scalar, then by def. of normed vector space,

(N_1) $\|p - q\| \geq 0$, (N_2) $\|p - q\| = 0 \Leftrightarrow p = 0$

(N_3) $\|kp\| = |k|\ \|p\|$, (N_4) $\|p + q\| \leq \|p\| + \|q\|$

We have $d(p, q) = \|p - q\|$ ...(1)

Using (N_1) in (1), we find that $d(p, q) \geq 0$ ...(M_1)

Using (N_2) in (1), we see that

$$d(p, q) = 0 \Leftrightarrow \|p - q\| = 0 \Leftrightarrow p = q = 0$$

$$\Leftrightarrow p = q \quad ...(M_2)$$

or $d(p, q) = 0 \Leftrightarrow p = q$ ...(M_3)

Using (N_3) in (1), we see that

$$d(q,p) = \|q - p\| = \|(-1)(p - q)\| = \|p - q\| = d(p,q)$$

or $d(q, p) = d(p, q)$ ...(M_3)

Using (N_4) in (1), we see that

$$d(p,r) = \|p - r\| = \|(p - q) + (q, r)\| \leq \|p - q\| + \|q - r\|$$

$$d(p, q) + d(q, r)$$

or $d(p, r) \leq d(p, q) + d(q, r)$

From (M_1), (M_2), (M_3) and (M_4), it follows that (C, d) is a metric space.

Example 18(b):

Let X be a non-empty set and let d be a real valued function of ordered pairs of elements of X which satisfies the following conditions:

(i) $d(x, y) = 0 \Leftrightarrow x = y$

(ii) $d(x, y) \leq d(x, z) + d(y, z)$

Then prove that d is a metric on X.

Theorem 19:

If A and B are subsets of a metric space (X, d), then

(i) $(A \cap B)^o = A^o \cap B^o$.

(ii) $A^o \cup B^o \subset (A \subset B)^o$.

Also show that $A^o \cup B^o \neq (A \cup B)^o$ *by means of an example.*

Proof:

(i) Any $x \in (A \cap B)^o$

$\Rightarrow$ x is an interior point of $A \cap B$

$\Rightarrow S_r(x) \subset A \cap B$

$\Rightarrow S_r(x) \subset A \cap B \subset A,\ S_r(x) \subset A \cap B \subset B$

$\Rightarrow S_r(x) \subset A,\ S_r(x) \subset B$

$\Rightarrow x \in A^o,\ x \in B^o$

$\Rightarrow x \in A^o \cap B^o$

$\therefore$ $(A \cap B^o) \subset A^o \cap B^o$.

Conversely, any $x \Rightarrow A^o \cap B^o$. ...(1)

$\Rightarrow x \in A^o$ and $x \in B^o$

$\Rightarrow \exists$ open spheres $S_{r1}(x)$, $S_{r2}(x)$, s.t.

$S_{r1}(x) \subset A,\ S_{r2}(x) \subset B$

Choose $r = \min.(r_1, r_2)$

$\Rightarrow S_r(x) \subset S_{r1}(x) \subset A,\ S_r(x) \subset S_{r2}(x) \subset B$

$\Rightarrow S_r(x) \subset A,\ S_r(x) \subset B$

$\Rightarrow S_r(x) \subset A \cap B$

$\Rightarrow x \in (A \cap B)^o$

$\therefore$ $A^o \cap B^o \subset (A \cap B)^o$ ... (2)

Combining (1) and (2), we get the result (i)

(ii) Any $x \in A^o \cup B^o$

$$\Rightarrow x \in A^o \text{ or } x \in B^o.$$

$$\Rightarrow S_{r1}(x) \subset A \text{ or } S_{r2}(x) \subset B$$

Choose the two spheres s.t. one is contained in the other so that $r_2 \geq r_1$ and $S_{r1}(x) \cup S_{r2}(x) = S_{r2}(x)$.

$$\Rightarrow S_{r2}(x) = S_{r1}(x) \cup S_{r2}(x) \subset A \cup B$$

$$\Rightarrow S_{r2}(x) \subset A \cup B$$

$$\Rightarrow x \in (A \cup B)^o$$

$\therefore \quad A^o \cup B^o \subset (A \cup B)^o$.

Solution:

(1) To prove $d(x, y) \geq 0$.

By (ii), $d(x, x) \leq d(x, z) + d(x, z)$, on putting $x = y$.

or, $0 \leq 2d(x, z)$ by (i)

or, $d(x, z) \geq 0$

or, $d(x, y) \geq 0$

(2) To prove $d(x, y) = d(y, x)$.

putting $z = x$ in (ii),

$$d(x, y) \leq 0 + d(y, x)$$

or, $d(x, y) \leq d(y, x)$

Interchanging x and y,

$d(y, x) \leq d(x, y)$

Combining the last two, $d(x, y) = d(y, x)$.

(3) $d(x, y) = 0 \Leftrightarrow x = y$ by (i)

(4) To prove $d(x, z) + d(z, y) \geq d(x, y)$

Using (2) in (ii), $d(x, y) \leq d(x, z) + d(z, y)$

or, $d(x, z) + d(z, y) \geq d(x, y)$

Example 20:

In any metric space (X, d), show that $A \subset X$ is closed iff $A = \overline{A}$

Solution:

Let (X, d) be metric space and $A \subset X$. Firstly we shall establish the following lemmas :

(i) $\overline{A}$ is closed and $A \subset \overline{A}$.

(ii) $F \supset A$, F is closed $\Rightarrow F \supset C$

By definition, $\overline{A} = \cap\{F \subset X : F \text{ is closed}, F \supset A\}$, ...(1)

(i) Since an arbitrary; intersection of closed sets is closed and hence $\overline{A}$ is closed.

By (1), $\overline{A} \supset A$

(ii) Suppose $F \subset X$ is closed and $F \supset A$.

$\therefore$ F is one of those members whose intersection is $\overline{A}$ and hence $F \supset \overline{A}$. .E.D. lemma (ii)

Now we come to the proof of actual theorem.

Let $A = \overline{A}$. We claim A is closed. By (i). $\overline{A}$ is closed.

$\overline{A}$ is closed, $A = \overline{A} \Rightarrow$ is closed.

Conversely let A be closed. To prove that $A = \overline{A}$.

By the lemma (i), $A \subset \overline{A}$. ...(2)

By the lemma (ii), f is closed, $F \supset A \Rightarrow F \supset \overline{A}$. ...(3)

Since $A \supset A$ for every set A.

In particular (3) gives A is closed, $A \supset A \Rightarrow A \supset \overline{A}$ ...(4)

From (2) and (4), $A = \overline{A}$.

This concludes the problem.

Example 21:

In a metric space (X, d), $S = S(x_0, r)$ is an open sphere and p is any point of it. Prove that $\exists$ an open sphere $T = S(p, \varepsilon)$ s.t. $T \subset S$.

Solution:

$p \in S(x_0, r) \Rightarrow d(p, x_0) < r \Rightarrow r - d(p, x_0) = r_1 > 0$.

Take $\varepsilon < r_1$, then $x \in S\, p\, \varepsilon \Rightarrow d(x, p) < \varepsilon < r_1 \Rightarrow d(x, p) < r_1$

or $\quad d(x, p) < r - d(p, x_0)$

or $\quad d(x, p) + d(p, x_0) < r$

or $\quad d(x, x_0) \le d(x, p) + d(p, x_0) < r$

or $\quad d(p, x_0) < r$

or $\quad x \in S(x_0, r)$ ...(M_4)

Thus $\quad x \in S(p, \varepsilon)$

$\Rightarrow \quad x \in S(x_0, r)$.

Hence $\quad S(p, \varepsilon) \subset S(x_0, r)$ or $T \subset S$.

Example 22(a):

In any metric space (X, d), prove that A is open

$$\Leftrightarrow A^\circ = A.$$

Solution:

Let A be a subset of a metric space (X, d). By def. of interior,

$$A^\circ = \cup \{S_{r(x)} : S_{r(x)} \subset A\} \quad ...(1)$$

Since every open sphere is an open set and arbitrary union of open sets is open. Consequently,

A° is an open set. ...(2)

By (1), it is clear that $A^\circ \subset A$ ...(3)

and A° is largest open subset of A. ...(4)

(i) Given $A = A^\circ$ ...(5)

Aim. A is an open set

(2) and (5) $\Rightarrow$ A is an open set.

(ii) Given A is an open set ...(6)

Aim, $A = A^\circ$

(4) and (6) $\Rightarrow = A\ A^\circ$

Example 22(b):

Let $A = [0, 1], B = [1, 2]$

Then $A^o = (0, 1), B^o = (1, 2).$

$A^\circ \cup B^\circ = (0, 1), \cup (1, 2)$

$= \{x \in R : 0 < x < 1 \text{ or } 1 < x < 2\}$

$\because$ $(A \cup B) = [0, 2]$

$\therefore$ $(A \cup B)^o = (0, 2).$

Evidently $(A \cup B)^o \neq A^o \cup B^o.$

Solution:

Do yourself.

Example 22(c):

If G_1, G_2 *are open sets in metric space* (X_1, d_1) *and* (X_2, d_2) *then* $G_1 \times G_2$ *is open in the product metric space (X, d).*

Solution:

Let $x = (x_1, x_2)$, $y\ (y_1, y_2) \in X = X_1 \times X_2$, then $x_1, y_1 \in X_1$ and x_2 $y_2 \in X_2$. Let $x \in G_1 \times G_2$, then $x_1 \in G_1$, $x_2 \in G_2$.

By def. of open set,

$x \in G_1 \Rightarrow \exists\, r_1 > 0$ s.t. $S(x_1, r_1) \subset G_1$ ...(1)

Similarly, $S(x_2, r_2) \subset G_2$ ...(2)

By (1) and (2),

$S(x_1, r_1) \times S(x_2, r_2) \subset G_1 \times G_1$ ...(3)

Take $r = \min\{r_1, r_2\}$, then $r \le r_1$, $r \le r_2$

$\therefore \quad S(x_1, r) \times S(x_2, r) \subset S(x_1, r_1) \times S(x_2, r_2) \subset G_1 \times G_2$

or $\quad S(x, r) \subset G_1 \times G_2$,

for any $x \in G_1 \times G_2$.

By def., this proves that $G_1 \times G_2$ is open.

Example 23:

In a metric space, prove that $d(A) = d(\overline{A})$.

Solution:

Evidently $A \subset B \Rightarrow d(A) \le d(B)$.

But $A \subset \overline{A}$ and so that $d(A) \le d(\overline{A})$. ...(1)

Let $x, y \in \overline{A}$, then \$ $u, v \in A$ s.t.

$$d(x, u) < r,\ d(u, v) < r \text{ where } r > 0.$$

But $d(x, y) \le d(x, u) + d(u, v) + d(v, y)$

$< d(u, v) + 2r \le d(A) + 2r$

or $\quad d(x, y) \le d(A) + 2r$.

For $d(A) = \sup d(x, y)$ where $x, y \in A$.

This$\Rightarrow d(\overline{A}) \le d(A)$. ...(2)

Combining (1) and (2), $d(A) = d(\overline{A})$.

Example 24:

Let X be a non-empty set and let d be a real valued function of ordered pairs of elements of X which satisfies the following conditions:

(i) $d(x, y) = 0 \Leftrightarrow x = y$

(ii) $d(x, y) \le d(x, z) + d(y, z)$

Then prove that d is a metric on X.

Solution:

(1) To prove $d(x, y) \ge 0$.

By (ii), d (x, x) $\leq$ d (x, z) + d (x, z), on putting x = y.

or, 0 $\leq$ 2d (x, z) by (i)

or, d (x, z) ≥ 0

or, d (x, y) ≥ 0

(2) To prove d (x, y) = d (y, x).

putting z = x in (ii),

$$d(x, y) \leq 0 + d(y, x)$$

or, d (x, y) $\leq$ d (y, x)

Interchanging x and y,

d (y, x) $\leq$ d (x, y)

Combining the last two, d (x, y) = d (y, x).

(3) d (x, y) = 0 $\Leftrightarrow$ x = y by (i)

(4) To prove d (x, z) + d (z, y) $\geq$ d (x, y)

Using (2) in (ii), d (x, y) $\leq$ d (x, z) + d (z, y)

or, d (x, z) + d (z, y) $\geq$ d (x, y).

Example 25(a):

Every set in a discrete metric space is open.

Solution:

Let (X,d) be a discrete metric space. Let x, y $\in$ X be arbitrary.

By def. of discrete metric, d (x, y) = $\begin{cases} 1 \text{ if } x \neq y \\ 0 \text{ if } x = y \end{cases}$.

Let r be any positive real number s.t. r $\leq$ 1.

Then Sr(x) = {y $\in$ X : d (y, x) < r $\leq$ 1}

= {y $\in$ X : d (y, x) < 1}

= {y $\in$ X : d (y, x) = 0, by def. of d.

= {y $\in$ X : y= x} = {0}.

or, $S_{r(x)}$ = {0}.

But every open sphere is an open set.

$\therefore$ {x} is an open set in X $\forall$ x $\in$ X.

If A = {x_1, x_2, ..., x_n} = finite set $\subset$ X, then

$A = \bigcup_{r=1}^{n} \{x_r\}$ = finite union of open sets.

= open set.

Hence every finite sub set of X is open set. ...(1)

If $B = \{x_1, x_2, x_3, ...\} \subset X$, then

B is an infinite subset of X.

Now $$B = \overset{\infty}{\underset{r=1}{\cup}} \{x_r\}$$

= Arbitrary union of open sets

= open set, by Theorem 8.

$\therefore$ B is an open set, ...(2)

From (1) and (2), it follows that every subset (finite or infinite) is an open set in X.

Example 25(b):

Show that the map $d : R^2 \times R^2 \to R$

define by $d(x, y) = max. \{|x_1 - y_1|, |x_2 - y_2|\}$.

where $x = (x_1, x_2), y = (y_1, y_2) \in R_2$

is a metric on R^2.

Solution:

Let $x = (x_1, x_2)$, $y = (y_1, y_2)$, $z = (z_1, z_2)$ be arbitrary elements or R^2.

Also let

$$d(x, y) = \max \{|x_1 - y_1|, |x_2 - y_2|\}$$

(i) $d(x, y) \geq 0$.

Since modulus of any quantity ;is non-negative.

(ii) $d(x, y) = 0$ iff $x = y$.

For $d(x, y) = 0 \Leftrightarrow \max.\{|x_1 - y_1|, |x_2 - y_2|\} = 0$

$\Leftrightarrow |x_1 - y_1| = 0 = |x_2 - y_2|$

$\Leftrightarrow x_1 = y_1, |x_2 = y_2$

$\Leftrightarrow (x_1, x_2) = y_2, y_2)$

$\Leftrightarrow x = y$

(iii) $d(x, y) = d(y, x)$.

For $|a - b| = |b - a|$ and so

$\max.\{|x_1 - y_1|, |x_2 - y_2|\} = \max \{|y_1 - x_1|, |y_2 - x_2|\}$

(iv) $d(x, y) + d(y, z) \geq d(x, z)$

For $d(x, y) = \max \{|x_1 - z_1|, |x_2 - z_2|\}$

$$= \max.\ \{|x_1 - y_1 + y_1 - z_1|, |x_2 - y_2 + y_2 - z_2|\}$$
$$\le \max.\ \{|x_1 - y_1| + |y_1 - z_1|, |x_2 - y_2| + |y_2 - z_2|\}$$
$$\le \max.\ \{|x_1 - y_1|, |x_2 - y_2|\} + \max.\ \{|y_1 - z_1|, |y_2 - z_2|\}.$$
$$= d(x, y) + d(y, z).$$

The above facts lead to the conclusion that d is a metric on R^2.

Example 26(a):

Let X be a non-empty set and define d by

$$d(x,y) = \begin{cases} 0\, if\ x = y \\ 1\, if\ x \neq y \end{cases}$$

Prove that d is metric on X.

Solution:

Let x, y, z ∈ X be arbitrary.

Then

(1) $d(x, y) \ge 0$

This follows from the definition of d.

(2) $d(x, y) = 0$ iff $x = y$, by def. of d.

(3) $d(x, y) = d(y, x)$ as value of d is free from order or x and y.

(4) $d(x, y) + d(y, z) \ge d(x, z)$

For if $x \neq y \neq z$, then

$$d(x, y) + d(y, z) = 1 + 1 = 2 > 1 = d(x, z)$$

or, $$d(x, y) + d(y, z) > d(x, z) \quad ...(a)$$

Again if $x = y \neq z$, then

$$d(x, y) + d(y, z) = 0 + 1$$
$$= d = (x, z)$$

or, $$d(x, y) + d(y, z) = d(x, z) \quad ...(b)$$

Combining (a) and (b), we get result (4).

∴ d is a metric on X.

Example 26(b):

Describe open spheres for the usual metric on R, *i.e.*, open spheres on the real line are precisely the bounded open intervals.

Solution:

Let d be the usual metric on R; then

$$d(x, y) = |x - y| \ \forall \ x, y \in R$$

Let $x_0 \in R$ and $r > 0$ be any real number.

$$\begin{aligned} S_r(x_0) &= \{x \in R : d(x, x_0) < r\} \\ &= \{x \in R : |x - x_0| < r\} \\ &= \{x \in R : -r < x - x_0 < r\} \\ &= \{x \in R : x_0 - r < x < x_0 + r\} \\ &= (x_0 - r, x_0 + r). \end{aligned}$$

This proves that the open spheres on the real line R are open intervals.

Example 26(c):

Is the subset [0, 3] open in X = [0, 3] under the metric d given by d (x, y) = | x – y?

Solution:

Since Φ and X both are open sets in any metric space (X, d).

$\therefore$ [0, 3] is open in x = [0, 3] .

Example 27:

Given two different metrics for the set R of all real numbers.

Solution:

Consider the map

$$d : R \times R \to R \text{ s. t.}$$

$$d(x,y) = \begin{cases} 0 \text{ if } x = y \\ 1 \text{ if } x \neq y \end{cases}$$

where $(x, y) \in R^2$.

Then d is metric on R.

Next consider the map

$$d' : R \times R \to R$$

$$d'(x, y) |x - y|.$$

where $\quad x, y \in R$

Then d' is metric on R.

Thus d and d' are two different metrics on R.

Show that a subset F of a metric space X is closed iff

$\{x \in X : d(x, F) = 0\} \subset F$.

Solution:

Let (X, d) be a metric space.

Let $F \subset X$ be closed.

To prove that

$$\{x \in X : d(x, F) = 0\} \subset F.$$

We know that $\overline{F} = \{x \in X : d(x, F) = 0\}$.

F is closed $\Rightarrow \overline{F} = F \Rightarrow \overline{F} \subset F$

$\Rightarrow \{x \in X : d(x, F) = 0\} \subset F.$ on using (1)

Conversely suppose that F is a subset of a metric space (X, d) with the property that

$$\{x \in X : d(x, F) = 0\} \subset F \quad ...(2)$$

To prove that F is closed, it suffices to show that $F = \overline{F}$.

But $\overline{F} = \{x \in X : d(x, F) = 0\}$ for any set $F \subset X$. ...(3)

Now (2) and (3) $\Rightarrow \overline{F} \subset F$ ...(4)

But $F \subset \overline{F}\ \forall F$ ...(5)

Combining (4) and (5), $\overline{F} = F$.

Example 28:

In every metric space, all finite sets are closed

Solution:

Let (X, d) be a metric space.

To prove that every finite subset of X is closed.

For any subset A of X, we know that

$\overline{A} = \{x \in X : d(x, A) = 0\}$

Let $p \in X$ be arbitrary.

Then $\overline{\{p\}} = \{x \in X : d(x, \{p\}) = 0)$

$= \{p\}$

$\therefore\quad d(x, \{p\}) = 0 \Rightarrow x \in \{p\} \Rightarrow x = p$

$\{p\} = \overline{\{p\}}$,

Showing thereby, {p} is closed in X.

Hence every singleton set in X is closed. Moreover a finite union of closed sets is closed. A finite set is a finite union of singleton sets in which every singleton set is closed. Accordingly every finite set in a metric space is closed.

Example 29:

Let R be the set of all real numbers and let

$$d(x,y) = \frac{|x-y|}{1+|x-y|} \text{ for all } x, y, \in R.$$

Prove that d is a metric for R.

Solution:

Let x, y, z ∈ R be arbitrary.

Let $\rho(x,y) = |x-y|$.

Then (1) ⇒ ρ (x, y) ≥ 0.

For modulus of any quantity is always non-negative.

(2) ρ (x, y) = 0 iff x = y.

For ρ (x, y) = 0

⇒ $|x-y| = 0$ ⇒ x = y.

(3) ρ (x, y) = ρ (y, x).

For |x − y| = |y − x|

(4) ρ (x, y) + ρ (y, z) ≥ ρ (x, z).

For ρ (x, y) + ρ (y, z) = |x − y| + |y − z|.

≥ |x − y + y − z| = |x − z| = ρ (x, z).

i.e., ρ (x, y) + ρ (y, z) ≥ ρ (x, z).

The above arguments prove that ρ is a metric on R.

By what is given $d(x, y) = \frac{|x-y|}{1+|x-y|}$.

This is expressible as

$$d(x, y) = \frac{\rho(x,y)}{1+\rho(x,y)}$$

where ρ is a metric on R. We have to prove that d is a metric on R.

Example 30:

Let ρ_1 and ρ_2 be metrics on the same set X, show that there exists a metric ρ on X such that

$\rho(x, y) \geq \rho_i(x, y)$ for i = 1, 2

and ∀ x, y ∈ X.

Solution:

Let ρ_1 and ρ_2 be metrics on the same set X, Let x, y, z $\in$ X be arbitrary. To prove that $\exists$ a metric ρ on X s.t.

$$\rho (x, y) \geq \rho_1 (x, y) \text{ for } i = 1, 2$$

Write $\rho (x, y) = \rho_1 (x, y) + \rho_2 (x, y)$. ...(1)

We claim ρ is metric on X.

(i) $\rho (x, y) \geq 0$

$\because$ $\rho_1 (x, y) \geq 0$, $\rho_2 (x, y) \geq 0$.

(ii) $\rho (x, y) = 0$ iff $x = y$

For $\rho (x, y) = 0 \Leftrightarrow \rho_1 (x, y) + \rho_2 (x, y) = 0$.

$$\Leftrightarrow \rho_1 (x, y) = 0 = \rho_2 (x, y).$$

Since ρ_1 and ρ_2 both are non-negative.

$$\Leftrightarrow x = y$$

(iii) $\rho (y, x) = \rho (x, y)$

For $\rho (y, x) = \rho_1 (y, x) + \rho_2 (y, x) = \rho_1 (x, y) + \rho_2 (x, y)$

$$\rho (x, y)$$

$\because$ $\rho_1 (x, y) = \rho_1 (y, x)$, $\rho_2 (x, y) = \rho_2 (y, x)$

(iv) $\rho (x, y) + \rho (y, z) \geq \rho (x, z)$

For $\rho (x, y) + \rho (y, z) = \rho_1 (x, y) + \rho_2 (x, y) + \rho_1 (y, z) + \rho_2 (y, z)$

$$[\rho_1 (x, y) + \rho_1 (y, z)] + [\rho_2 (x, y) + \rho_1 (y, z)]$$

$$\geq \rho_1 (x, z) + \rho_2 (x, z) = \rho (x, z).$$

From what has been done it follows that ρ is a metric on X.

Moreover (1) $\Rightarrow \rho (x, y) \geq \rho_1 (x, y)$ for i = 1, 2.

Hence ρ is the required metric.

EXERCISES

1. Show that the unitary operators on H form a group.
2. Define a new operation of "multiplication" for self-adjoint operators by $A_1 \circ A_2 = (A_1A_2 + A_2A_1)/2$, and note that $A_1 \circ A_2$ is always self-adjoint and that it equals A_1A_2 whenever A_1 and A_2 commute. Show that this operation has the following properties:

$$A_1 \circ A_2 = A_2 \circ A_1,$$

$$A_1 o (A_2 + A_3) = A_1 o A_2 + A_1 o A_3,$$

$$\alpha (A_1 \circ A_2) = (\alpha A_1) \text{ o } A_2 = A_1 \text{ o } (\alpha A_2),$$

and A I = I o A = A. Show also that $A_1 \circ (A_2 \circ A_3) = (A_1 \text{ o } A_2)\ 0\ A_3$ whenever A_1 and A_3 commute.

3. If H is finite-dimensional, show that every isometric isomorphism of H into itself is unitary.1. If T is an arbitrary operator on.H, and if α and β are scalars such that $| a | = | \beta |$, show that $\alpha T + \beta T^*$ is normal.
4. If H is finite-dimensional, show that every isometric isomorphism of H into itself is unitary.
5. Show that an operator T on H is unitary ⇔ T({e_i}) is a complete orthonormal set whenever {e_i} is.
6. Show that an operator T on H is unitary ⇔ T({e_i}) is a complete orthonormal set whenever {e_i} is.
7. Show that the unitary operators on H form a group.
8. Prove parts (1), (2), and (4) of Theorem 1.
9. Show that the adjoint operation is one-to-one onto as a mapping of $\mathcal{B}$(H) into itself.
10. Show that 0* = 0 and 7* = I. Use the latter to show that if T is non-singular, then T* is also non-singular, and that in this case $(T^*)^{-1} = (T^{-1})^*$.
11. Show that $\| TT^* \| = \| T |^2$.
12. Show that every orthonormal set in a Hilbert space is contained in some complete orthonormal set, and use this fact to give an alternative proof.
13. Prove that a Hilbert space *H* is separable *of* every orthonormal set in *H* is countable.
14. Show that an orthonormal set in a Hilbert space is linearly independent, and use this to prove that a Hilbert space is finite-dimensional ⇔ every complete orthonormal set is a basis.
15. Prove that any two complete orthonormal sets in a Hilbert space *H* have the same cardinal number. This cardinal number is called the *orthogonal dimension* of *H* (if *H* has no complete orthonormal sets, into orthogonal dimension is said to be 0).
16. If *H* and *H'* are Hilbert spaces, prove that *H* is isometrically isomorphic to *H'* «=> they have the same orthogonal dimension. *(Hint:* by Eq. (2), an isometric isomorphism *T* preserves inner products,] in the sense that (T(x),T(y)) = (x, y).)

17. Let $S = \{e_i\}$ be a complete orthonormal set in a Hilbert space H. Each vector x in H determines a function f defined on S by
$$f(e_i) = (x, e_i),$$
and Theorems 2 and 3 tell us that f is in $l_2(S)$. Show that the mapping $x \to f$ is an isometric isomorphism of H onto $l_2(S)$.
18. If S is a non-empty subset of a Hilbert space, show that $S^{\perp} = S^{\perp\perp\perp}$.
19. If M is a linear subspace of a Hilbert space, show that M is closed $\Leftrightarrow M = M^{\perp\perp}$.
20. If S is a non-empty subset of a Hilbert space H, show that the set of all linear combinations of vectors in S is dense in $H \Leftrightarrow S^{\perp} = \{0\}$.
21. If S is a non-empty subset of a Hilbert space H, show that $S^{\perp\perp}$ is the closure of the set of all linear combinations of vectors in S. This is usually expressed by saying that $S^{\perp\perp}$ is the smallest closed linear subspace of H which contains S.
22. Show that the series which defines the inner product in Example 2 is convergent.
23. The *Hilbert cube* is the subset of l_2 consisting of all sequences $x = \{x_1, x_2, \ldots, x_n, \ldots\}$ such that $|x_n| \leq 1/n$ for all n. Show that this set is compact as a subspace of l_2.
24. For the special Hilbert space l_2^n, use Cauchy's inequality to prove Schwarz's inequality.
25. Show that the parallelogram law is not true in l_1^n $(n > 1)$.
26. In a Hilbert space, show that if $\|x\| = \|y\| - 1$, and if $e > 0$ is given, then there exists $\delta > 0$ such that $\|(x + y/)/2\| > 1 - \delta \Rightarrow \|x - y\| < \epsilon$. A Banach space with this property is said to be *uniformly convex*.
27. Verify relations (3).
28. Let H be a Hilbert space, and show that H* is also a Hilbert space with respect to the inner product defined by $(f_x, f_y) = (y, x)$. In just the same way, the fact that H* is a Hilbert space implies that H** is a Hilbert space whose inner product is-given by $(F_f, F_0) = (g, f)$.
29. Let H be a Hilbert space. We have two natural mappings of H into H**, the second of which is onto: the Banach space natural embedding $x \to F_x$, where $F_x(f) = f(x)$, and the product mapping $x \to fx \to F_{f_x}$, where $f_x(y) = (y, x)$ and $F_{y_x}(f) = (f, f_x)$. Show that these mappings are equal, and conclude that H is reflexive, Show also that $(F_x, F_y) = (x, y)$.